INTERNATIONAL UNION OF PURE AND APPLIED CHEMISTRY

ANALYTICAL CHEMISTRY DIVISION
COMMISSION ON SOLUBILITY DATA

SOLUBILITY DATA SERIES

Volume 31

ALKALI METAL
ORTHOPHOSPHATES

SOLUBILITY DATA SERIES

Editor-in-Chief

A. S. KERTES

The Hebrew University
Jerusalem, Israel

EDITORIAL BOARD

Managing Editor
P. D. GUJRAL
IUPAC Secretariat, Oxford, UK

INTERNATIONAL UNION OF PURE AND APPLIED CHEMISTRY
IUPAC Secretariat: Bank Court Chambers, 2–3 Pound Way,
Cowley Centre, Oxford OX4 3YF, UK

NOTICE TO READERS

Dear Reader

If your library is not already a standing-order customer or subscriber to the Solubility Data Series, may we recommend that you place a standing order or subscription order to receive immediately upon publication all new volumes published in this valuable series. Should you find that these volumes no longer serve your needs, your order can be cancelled at any time without notice.

Robert Maxwell
Publisher at Pergamon Press

A complete list of volumes published in the Solubility Data Series will be found on p. 348.

SOLUBILITY DATA SERIES

Editor-in-Chief
A. S. KERTES

Volume 31

⌊ALKALI METAL ORTHOPHOSPHATES

Volume Editors

JITKA EYSSELTOVÁ
Charles University, Prague
Czechoslovakia

THEDFORD P. DIRKSE
Calvin College
Grand Rapids, MI, USA

Contributors

JIŘÍ MAKOVIČKA
Charles University, Prague
Czechoslovakia

MARK SALOMON
US Army ET & DL (LABCOM)
Fort Monmouth, NJ, USA

PERGAMON PRESS

OXFORD · NEW YORK · BEIJING · FRANKFURT
SÃO PAULO · SYDNEY · TOKYO · TORONTO

U.K.	Pergamon Press plc, Headington Hill Hall, Oxford OX3 0BW, England
U.S.A.	Pergamon Press, Inc., Maxwell House, Fairview Park, Elmsford, New York 10523, U.S.A.
PEOPLE'S REPUBLIC OF CHINA	Pergamon Press, Room 4037, Qianmen Hotel, Beijing, People's Republic of China
FEDERAL REPUBLIC OF GERMANY	Pergamon Press, GmbH Hammerweg 6, D-6242 Kronberg, Federal Republic of Germany
BRAZIL	Pergamon Editora Ltda, Rua Eça de Queiros, 346, CEP 04011, Paraiso, São Paulo, Brazil
AUSTRALIA	Pergamon Press Pty Ltd. Australia, P.O. Box 544, Potts Point, N.S.W. 2011, Australia
JAPAN	Pergamon Press, 8th Floor, Matsuoka Central Building, 1-7-1 Nishishinjuku, Shinjuku-ku, Tokyo 160, Japan
CANADA	Pergamon Press Canada Ltd., Suite No. 271, 253 College Street, Toronto, Ontario, Canada M5T 1R5

First edition 1988

The Library of Congress has catalogued this serial title as follows:

Solubility data series.—Vol. 1 —Oxford; New York: Pergamon, c 1979-
v.; 28 cm.
Separately cataloged and classified in LC before no. 18.
ISSN 0191-5622 = Solubility data series.
1. Solubility—Tables—Collected works.
QD543.S6629   541.3'42'05-dc19   85-641351
AACR 2   MARC-S

British Library Cataloguing in Publication Data
Alkali metal orthophosphates.
1. Alkali metal orthophposphates. Solubility
I. Eysseltová, Jitka   II. Dirkse, T.P.
III. Makovicka, Jeorge   IV. Saloman, Mark
V. Series
546'.38
ISBN 0–08–035937–X

Printed in Great Britain by A. Wheaton & Co. Ltd., Exeter

CONTENTS

(continued next page)

FOREWORD

If the knowledge is
undigested or simply wrong,
more is not better

How to communicate and disseminate numerical data effectively in chemical science and technology has been a problem of serious and growing concern to IUPAC, the International Union of Pure and Applied Chemistry, for the last two decades. The steadily expanding volume of numerical information, the formulation of new interdisciplinary areas in which chemistry is a partner, and the links between these and existing traditional subdisciplines in chemistry, along with an increasing number of users, have been considered as urgent aspects of the information problem in general, and of the numerical data problem in particular.

Among the several numerical data projects initiated and operated by various IUPAC commissions, the *Solubility Data Project* is probably one of the most ambitious ones. It is concerned with preparing a comprehensive critical compilation of data on solubilities in all physical systems, of gases, liquids and solids. Both the basic and applied branches of almost all scientific disciplines require a knowledge of solubilities as a function of solvent, temperature and pressure. Solubility data are basic to the fundamental understanding of processes relevant to agronomy, biology, chemistry, geology and oceanography, medicine and pharmacology, and metallurgy and materials science. Knowledge of solubility is very frequently of great importance to such diverse practical applications as drug dosage and drug solubility in biological fluids, anesthesiology, corrosion by dissolution of metals, properties of glasses, ceramics, concretes and coatings, phase relations in the formation of minerals and alloys, the deposits of minerals and radioactive fission products from ocean waters, the composition of ground waters, and the requirements of oxygen and other gases in life support systems.

The widespread relevance of solubility data to many branches and disciplines of science, medicine, technology and engineering, and the difficulty of recovering solubility data from the literature, lead to the proliferation of published data in an ever increasing number of scientific and technical primary sources. The sheer volume of data has overcome the capacity of the classical secondary and tertiary services to respond effectively.

While the proportion of secondary services of the review article type is generally increasing due to the rapid growth of all forms of primary literature, the review articles become more limited in scope, more specialized. The disturbing phenomenon is that in some disciplines, certainly in chemistry, authors are reluctant to treat even those limited-in-scope reviews exhaustively. There is a trend to preselect the literature, sometimes under the pretext of reducing it to manageable size. The crucial problem with such preselection - as far as numerical data are concerned - is that there is no indication as to whether the material was excluded by design or by a less than thorough literature search. We are equally concerned that most current secondary sources, critical in character as they may be, give scant attention to numerical data.

On the other hand, tertiary sources - handbooks, reference books and other tabulated and graphical compilations - as they exist today are comprehensive but, as a rule, uncritical. They usually attempt to cover whole disciplines, and thus obviously are superficial in treatment. Since they command a wide market, we believe that their service to the advancement of science is at least questionable. Additionally, the change which is taking place in the generation of new and diversified numerical data, and the rate at which this is done, is not reflected in an increased third-level service. The emergence of new tertiary literature sources does not parallel the shift that has occurred in the primary literature.

With the status of current secondary and tertiary services being as briefly stated above, the innovative approach of the *Solubility Data Project* is that its compilation and critical evaluation work involve consolidation and reprocessing services when both activities are based on intellectual and scholarly reworking of information from primary sources. It comprises compact compilation, rationalization and simplification, and the fitting of isolated numerical data into a critically evaluated general framework.

The *Solubility Data Project* has developed a mechanism which involves a number of innovations in exploiting the literature fully, and which contains new elements of a more imaginative approach for transfer of reliable information from primary to secondary/tertiary sources. *The fundamental trend of the Solubility Data Project is toward integration of secondary and tertiary services with the objective of producing in-depth critical analysis and evaluation which are characteristic to secondary services, in a scope as broad as conventional tertiary services.*

Fundamental to the philosophy of the project is the recognition that the basic element of strength is the active participation of career scientists in it. Consolidating primary data, producing a truly critically-evaluated set of numerical data, and synthesizing data in a meaningful relationship are demands considered worthy of the efforts of top scientists. Career scientists, who themselves contribute to science by their involvement in active scientific research, are the backbone of the project. The scholarly work is commissioned to recognized authorities, involving a process of careful selection in the best tradition of IUPAC. This selection in turn is the key to the quality of the output. These top experts are expected to view their specific topics dispassionately, paying equal attention to their own contributions and to those of their peers. They digest literature data into a coherent story by weeding out what is wrong from what is believed to be right. To fulfill this task, the evaluator must cover *all* relevant open literature. No reference is excluded by design and every effort is made to detect every bit of relevant primary source. Poor quality or wrong data are mentioned and explicitly disqualified as such. In fact, it is only when the reliable data are presented alongside the unreliable data that proper justice can be done. The user is bound to have incomparably more confidence in a succinct evaluative commentary and a comprehensive review with a complete bibliography to both good and poor data.

It is the standard practice that the treatment of any given solute-solvent system consists of two essential parts: I. Critical Evaluation and Recommended Values, and II. Compiled Data Sheets.

The Critical Evaluation part gives the following information:

(i)   a verbal text of evaluation which discusses the numerical solubility information appearing in the primary sources located in the literature. The evaluation text concerns primarily the quality of data after consideration of the purity of the materials and their characterization, the experimental method employed and the uncertainties in control of physical parameters, the reproducibility of the data, the agreement of the worker's results on accepted test systems with standard values, and finally, the fitting of data, with suitable statistical tests, to mathematical functions;

(ii) a set of recommended numerical data. Whenever possible, the set of recommended data includes weighted average and standard deviations, and a set of smoothing equations derived from the experimental data endorsed by the evaluator;

(iii) a graphical plot of recommended data.

The Compilation part consists of data sheets of the best experimental data in the primary literature. Generally speaking, such independent data sheets are given only to the best and endorsed data covering the known range of experimental parameters. Data sheets based on primary sources where the data are of a lower precision are given only when no better data are available. Experimental data with a precision poorer than considered acceptable are reproduced in the form of data sheets when they are the only known data for a particular system. Such data are considered to be still suitable for some applications, and their presence in the compilation should alert researchers to areas that need more work.

The typical data sheet carries the following information:

    (i)   components - definition of the system - their names, formulas and Chemical Abstracts registry numbers;

    (ii)   reference to the primary source where the numerical information is reported. In cases when the primary source is a less common periodical or a report document, published though of limited availability, abstract references are also given;

    (iii) experimental variables;

    (iv)   identification of the compiler;

    (v)   experimental values as they appear in the primary source. Whenever available, the data may be given both in tabular and graphical form. If auxiliary information is available, the experimental data are converted also to SI units by the compiler.

Under the general heading of Auxiliary Information, the essential experimental details are summarized:

    (vi)   experimental method used for the generation of data;

    (vii) type of apparatus and procedure employed;

    (viii) source and purity of materials;

    (ix)   estimated error;

    (x)   references relevant to the generation of experimental data as cited in the primary source.

This new approach to numerical data presentation, formulated at the initiation of the project and perfected as experience has accumulated, has been strongly influenced by the diversity of background of those whom we are supposed to serve. We thus deemed it right to preface the evaluation/compilation sheets in each volume with a detailed discussion of the principles of the accurate determination of relevant solubility data and related thermodynamic information.

Finally, the role of education is more than corollary to the efforts we are seeking. The scientific standards advocated here are necessary to strengthen science and technology, and should be regarded as a major effort in the training and formation of the next generation of scientists and engineers. Specifically, we believe that there is going to be an impact of our project on scientific-communication practices. The quality of consolidation adopted by this program offers down-to-earth guidelines, concrete examples which are bound to make primary publication services more responsive than ever before to the needs of users. The self-regulatory message to scientists of the early 1970s to refrain from unnecessary publication has not achieved much. A good fraction of the literature is still cluttered with poor-quality articles. The Weinberg report (in 'Reader in Science Information', ed. J. Sherrod and A. Hodina, Microcard Editions Books, Indian Head, Inc., 1973, p. 292) states that 'admonition to authors to restrain themselves from premature, unnecessary publication can have little effect unless the climate of the entire technical and scholarly community encourages restraint...' We think that projects of this kind translate the climate into operational terms by exerting pressure on authors to avoid submitting low-grade material. The type of our output, we hope, will encourage attention to quality as authors will increasingly realize that their work will not be suited for permanent retrievability unless it meets the standards adopted in this project. It should help to dispel confusion in the minds of many authors of what represents a permanently useful bit of information of an archival value, and what does not.

If we succeed in that aim, even partially, we have then done our share in protecting the scientific community from unwanted and irrelevant, wrong numerical information.

                        A. S. Kertes

PREFACE

This volume presents and evaluates solubility data for the orthophosphates of lithium, sodium, potassium, rubidium and cesium. There are two exceptions to this: (a) data are presented for the solubility of sodium metaphosphate in water (1) on page 46 in chapter 3; and (b) solubility values for the $NH_3-K_2O-H_3PO_4-H_4P_2O_7-H_2O$ system (2) are given on pp. 269-270 in chapter 8. Neither of these systems is evaluated because no other comparable data are given in this volume, nor, especially in the latter case, have any similar data been reported.

The orthophosphates have been known and used for many years, but interest in these substances has varied according to their use as, e.g., for fertilizers, corrosion inhibitors and piezoelectricity.

So far as we are aware, all the relevant articles dealing with the alkali metal orthosphosphates as a solid phase and published up to 1984 have been reviewed.

Chemical Abstracts was used to search for relevant articles published in the years 1920-84. The following three sources were used to locate articles published prior to 1920.

1. The 1928 edition of Gmelin's *Handbuch der Anorganischen Chemie*.
2. References cited in the articles that have been reviewed.
3. The review article of Wendrow and Kobe (3).

The various systems are treated in the order in which the alkali metals are listed in Group I of the Periodic Table. Most of the available solubility data are for the ortho-phosphates of sodium and potassium, and for these two systems an introductory chapter on the $MOH-H_3PO_4-H_2O$ (M = Na or K) system is given. Each of these chapters (chapters 2 and 7) also refers to compounds to be considered in later chapters. Following each of these introductory chapters there are chapters dealing with the solubility data for individual orthophosphates having different M/P ratios, and the ternary and multicomponent systems in which these orthophosphates are components. Only one chapter is devoted to each of the orthophosphates of lithium, rubidium and cesium.

A considerable amount of help was given to us in the preparation of this volume, and we wish to acknowledge this help and express our thanks to those who provided it. Dr. Mark Salomon kindly coordinated the work of the editors. Dr. Kurt Loening of Chemical Abstracts Service gave indispensable help by providing copies of articles that were difficult for us to locate, and supplying also the CAS Registry Numbers for many of the substances mentioned in this volume. Drs. G. Bohnsack, J.W. Lorimer, and H. Miyamoto provided us with copies of some of the articles reviewed in this volume. Sue Sweetman, in her patient and efficient way, typed the entire manuscript.

We also wish to thank the institutions with which we are affiliated for assistance in many ways during the work of this project. And one of us (J.E.) wishes to express special thanks to IUPAC Commission V.8 and to USSR Minister of Education Academician Prof. G.A. Yagodin for their help in making arrangements for her to spend some time in Moscow in 1984 to search the literature there for many of the articles that have been compiled in this volume. She also wishes to thank Prof. Dr. M. Ebert, Head of the Department of Inorganic Chemistry at Charles University of Prague for making her participation in this project possible and for providing good conditions in which to carry out this work.

References

1. Morey, G.W. *J. Am. Chem. Soc.* <u>1953</u>, *75*, 5794.
2. Frazier, A.W.; Dillard, E.F.; Thrasher, R.D.; Waerstad, K.R. *J. Agr. Food Chem.* <u>1973</u>, *21*, 700.
3. Wendrow, B.; Kobe, K.A. *Chem. Rev.* <u>1954</u>, *54*, 891.

Jitka Eysseltová
Charles University
Prague, Czechoslovakia

Thedford P. Dirkse
Calvin College
Grand Rapids, Michigan   USA

October 1986

INTRODUCTION TO THE SOLUBILITY OF SOLIDS IN LIQUIDS

Nature of the Project

   The Solubility Data Project (SDP) has as its aim a comprehensive search of the literature for solubilities of gases, liquids, and solids in liquids or solids. Data of suitable precision are compiled on data sheets in a uniform format. The data for each system are evaluated, and where data from different sources agree sufficiently, recommended values are proposed. The evaluation sheets, recommended values, and compiled data sheets are published on consecutive pages.

Definitions

   A *mixture* (1, 2) describes a gaseous, liquid, or solid phase containing more than one substance, when the substances are all treated in the same way.

   A *solution* (1, 2) describes a liquid or solid phase containing more than one substance, when for convenience one of the substances, which is called the *solvent*, and may itself be a mixture, is treated differently than the other substances, which are called *solutes*. If the sum of the mole fractions of the solutes is small compared to unity, the solution is called a *dilute solution*.

   The *solubility* of a substance B is the relative proportion of B (or a substance related chemically to B) in a mixture which is saturated with respect to solid B at a specified temperature and pressure. *Saturated* implies the existence of equilibrium with respect to the processes of dissolution and precipitation; the equilibrium may be stable or metastable. The solubility of a substance in metastable equilibrium is usually greater than that of the corresponding substance in stable equliibrium. (Strictly speaking, it is the activity of the substance in metastable equilibrium that is greater.) Care must be taken to distinguish true metastability from supersaturation, where equilibrium does not exist.

   Either point of view, mixture or solution, may be taken in describing solubility. The two points of view find their expression in the quantities used as measures of solubility and in the reference states used for definition of activities, activity coefficients and osmotic coefficients.

   The qualifying phrase "substance related chemically to B" requires comment. The composition of the saturated mixture (or solution) can be described in terms of any suitable set of thermodynamic components. Thus, the solubility of a salt hydrate in water is usually given as the relative proportion of anhydrous salt in solution, rather than the relative proportions of hydrated salt and water.

Quantities Used as Measures of Solubility

   1. *Mole fraction of substance B*, x_B:

$$x_B = n_B / \sum_{s=1}^{c} n_s \qquad [1]$$

where n_s is the amount of substance of s, and c is the number of distinct substances present (often the number of thermodynamic components in the system). *Mole per cent of B is* $100 \, x_B$.

   2. *Mass fraction of substance B*, w_B:

$$w_B = m_B' / \sum_{s=1}^{c} m_s' \qquad [2]$$

where m_s' is the mass of substance s. *Mass per cent is* $100 \, w_B$. The equivalent terms *weight fraction* and *weight per cent* are not used.

   3. *Solute mole (mass) fraction of solute B* (3, 4):

$$x_{s,B} = m_B / \sum_{s=1}^{c'} m_s = x_B / \sum_{s=1}^{c'} x_s \qquad [3]$$

$$w_{s,B} = m_B' / \sum_{s=1}^{c} m_s' = w_B / \sum_{s=1}^{c'} w_s \qquad [3a]$$

where the summation is over the solutes only. For the solvent A, $x_{S,A} = x_A/(1 - x_A)$, $w_{S,A} = w_A/(1 - w_A)$. These quantities are called *Jänecke mole (mass) fractions* in many papers.

4. *Molality of solute B* (1, 2) *in a solvent A:*

$$m_B = n_B/n_A M_A \qquad \text{SI base units: mol kg}^{-1} \qquad [4]$$

where M_A is the molar mass of the solvent.

5. *Concentration of solute B* (1, 2) *in a solution of volume V:*

$$c_B = [B] = n_B/V \qquad \text{SI base units: mol m}^{-3} \qquad [5]$$

The symbol c_B is preferred to $[B]$, but both are used. The terms *molarity* and *molar* are not used.

Mole and mass fractions are appropriate to either the mixture or the solution point of view. The other quantities are appropriate to the solution point of view only. Conversions among these quantities can be carried out using the equations given in Table 1-1 following this Introduction. Other useful quantities will be defined in the prefaces to individual volumes or on specific data sheets.

In addition to the quantities defined above, the following are useful in conversions between concentrations and other quantities.

6. *Density:* $\rho = m/V$ \qquad SI base units: kg m^{-3} \qquad [6]

7. *Relative density: d;* the ratio of the density of a mixture to the density of a reference substance under conditions which must be specified for both (1). The symbol d_t will be used for the density of a mixture at $t°C$, 1 bar divided by the density of water at $t'°C$, 1 bar. (In some cases 1 atm = 101.325 kPa is used instead of 1 bar = 100 kPa.)

8. *A note on nomenclature.* The above definitions use the nomenclature of the IUPAC *Green Book* (1), in which a solute is called B and a solvent A In compilations and evaluations, the first-named component (component 1) is the solute, and the second (component 2 for a two-component system) is the solvent. The reader should bear these distinctions in nomenclature in mind when comparing nomenclature and theoretical equations given in this Introduction with equations and nomenclature used on the evaluation and compilation sheets.

Thermodynamics of Solubility

The principal aims of the Solubility Data Project are the tabulation and evaluation of: (a) solubilities as defined above; (b) the nature of the saturating phase. Thermodynamic analysis of solubility phenomena has two aims: (a) to provide a rational basis for the construction of functions to represent solubility data; (b) to enable thermodynamic quantities to be extracted from solubility data. Both these are difficult to achieve in many cases because of a lack of experimental or theoretical information concerning activity coefficients. Where thermodynamic quantities can be found, they are not evaluated critically, since this task would involve critical evaluation of a large body of data that is not directly relevant to solubility. The following is an outline of the principal thermodynamic relations encountered in discussions of solubility. For more extensive discussions and references, see books on thermodynamics, e.g., (5-12).

Activity Coefficients (1)

(a) *Mixtures.* The activity coefficient f_B of a substance B is given by

$$RT \ln (f_B x_B) = \mu_B - \mu_B^* \qquad [7]$$

where μ_B^* is the chemical potential of pure B at the same temperature and pressure. For any substance B in the mixture,

$$\lim_{x_B \to 1} f_B = 1 \qquad [8]$$

(b) *Solutions.*

(i) *Solute B.* The molal activity coefficient γ_B is given by

$$RT \ln(\gamma_B m_B) = \mu_B - (\mu_B - RT \ln m_B)^\infty \qquad [9]$$

where the superscript ∞ indicates an infinitely dilute solution. For any solute B,

$$\gamma_B^\infty = 1 \qquad [10]$$

Activity coefficients y_B connected with concentrations c_B, and $f_{x,B}$ (called the *rational activity coefficient*) connected with mole fractions x_B are defined in analogous ways. The relations among them are (1, 9), where ρ^* is the density of the pure solvent:

$$f_B = (1 + M_A \sum_s m_s)\gamma_B = [\rho + \sum_s (M_A - M_s)c_s]y_B/\rho^* \qquad [11]$$

$$\gamma_B = (1 - \sum_s x_s)f_{x,B} = (\rho - \sum_s M_s c_s)y_B/\rho^* \qquad [12]$$

$$y_B = \rho^* f_{x,B}[1 + \sum_s (M_s/M_A - 1)x_B]/\rho = \rho^*(1 + \sum_s M_s m_s)\gamma_B/\rho \qquad [13]$$

For an electrolyte solute $B \equiv C_{\nu_+}A_{\nu_-}$, the activity on the molality scale is replaced by (9)

$$\gamma_B m_B = \gamma_\pm^\nu m_B^\nu Q^\nu \qquad [14]$$

where $\nu = \nu_+ + \nu_-$, $Q = (\nu_+^{\nu_+}\nu_-^{\nu_-})^{1/\nu}$, and $\gamma_\pm$ is the mean ionic activity coefficient on the molality scale. A similar relation holds for the concentration activity, $y_B c_B$. For the mole fractional activity,

$$f_{x,B} x_B = Q^\nu f_\pm^\nu x_\pm^\nu \qquad [15]$$

where $x_\pm = (x_+ x_-)^{1/\nu}$. The quantities x_+ and x_- are the ionic mole fractions (9), which are

$$x_+ = \nu_+ x_B/[1 + \sum_s (\nu_s - 1)x_s]; \quad x_- = \nu_- x_B[1 + \sum_s (\nu_s - 1)x_s] \qquad [16]$$

where ν_s is the sum of the stoichiometric coefficients for the ions in a salt with mole fraction x_s. Note that the mole fraction of solvent is now

$$x_A' = (1 - \sum_s \nu_s x_s)/[1 + \sum_s (\nu_s - 1)x_s] \qquad [17]$$

so that

$$x_A' + \sum_s \nu_s x_s = 1 \qquad [18]$$

The relations among the various mean ionic activity coefficients are:

$$f_\pm = (1 + M_A \sum_s \nu_s ms)\gamma_\pm = [\rho + \sum_s (\nu_s M_A - M_s)c_s]y_\pm/\rho^* \qquad [19]$$

$$\gamma_\pm = \frac{(1 - \sum_s x_s)f_\pm}{1 + \sum_s (\nu_s - 1)x_s} = (\rho - \sum_s M_s c_s)y_\pm/\rho^* \qquad [20]$$

$$y_\pm = \frac{\rho^*[1 + \sum_s (M_s/M_A - 1)xs]f_\pm}{\rho[1 + \sum_s (\nu_s - 1)x_s]} = \rho^*(1 + \sum_s M_s m_s)\gamma_\pm/\rho \qquad [21]$$

(ii) *Solvent*, A:

The *osmotic coefficient*, ϕ, of a solvent A is defined as (1):

$$\phi = (\mu_A^* - \mu_A)/RT\, M_A \sum_s m_s \qquad [22]$$

where μ_A^* is the chemical potential of the pure solvent.

The *rational osmotic coefficient*, ϕ_x, is defined as (1):

$$\phi_x = (\mu_A - \mu_A^*)/RT\ln x_A = \phi M_A \sum_s m_s/\ln(1 + M_A \sum_s m_s) \qquad [23]$$

The activity, a_A, or the activity coefficient, f_A, is sometimes used for the solvent rather than the osmotic coefficient. The activity coefficient is defined relative to pure A, just as for a mixture.

For a mixed solvent, the molar mass in the above equations is replaced by the average molar mass; i.e., for a two-component solvent with components J, K, M_A becomes

$$M_A = M_J + (M_K - M_J)x_{v,K} \qquad [24]$$

where $x_{v,K}$ is the solvent mole fraction of component K.

The osmotic coefficient is related directly to the vapor pressure, p, of a solution in equilibrium with vapor containing A only by (12, p.306):

$$\phi M_A \sum_s \nu_s m_s = -\ln(p/p_A^*) + (V_{m,A}^* - B_{AA})(p - p_A^*)/RT \qquad [25]$$

where p_A^*, $V_{m,A}^*$ are the vapor pressure and molar volume of pure solvent A, and B_{AA} is the second virial coefficient of the vapor.

The Liquid Phase

A general thermodynamic differential equation which gives solubility as a function of temperature, pressure and composition can be derived. The approach is similar to that of Kirkwood and Oppenheim (7); see also (11, 12). Consider a solid mixture containing c thermodynamic components i. The Gibbs-Duhem equation for this mixture is:

$$\sum_{i=1}^{c} x_i'(S_i'dT - V_i'dp + d\mu_i') = 0 \qquad [26]$$

A liquid mixture in equilibrium with this solid phase contains c' thermodynamic components i, where $c' \geqslant c$. The Gibbs-Duhem equation for the liquid mixture is:

$$\sum_{i=1}^{c} x_i(S_idT - V_idp + d\mu_i') + \sum_{i=c+1}^{c'} x_i(S_idT - V_idp + d\mu_i) = 0 \qquad [27]$$

Subtract [26] from [27] and use the equation

$$d\mu_i = (d\mu_i)_{T,p} - S_idT + V_idp \qquad [28]$$

and the Gibbs-Duhem equation at constant temperature and pressure:

$$\sum_{i=1}^{c} x_i(d\mu_i')_{T,p} + \sum_{i=c+1}^{c'} x_i(d\mu_i)_{T,p} = 0 \qquad [29]$$

The resulting equation is:

$$RT\sum_{i=1}^{c} x_i'(d\ln a_i)_{T,p} = \sum_{i=1}^{c} x_i'(H_i - H_i')dT/T - \sum_{i=1}^{c} x_i'(V_i - V_i')dp \qquad [30]$$

where

$$H_i - H_i' = T(S_i - S_i') \qquad [31]$$

is the enthalpy of transfer of component i from the solid to the liquid phase at a given temperature, pressure and composition, with H_i and S_i the partial molar enthalpy and entropy of component i.

Use of the equations

$$H_i - H_i^0 = -RT^2(\partial \ln a_i/\partial T)_{x,p} \qquad [32]$$

and

$$V_i - V_i^0 = RT(\partial \ln a_i/\partial p)_{x,T} \qquad [33]$$

where superscript o indicates an arbitrary reference state gives:

$$RT\sum_{i=1}^{c} x_i'd\ln a_i = \sum_{i=1}^{c} x_i'(H_i^0 - H_i')dT/T - \sum_{i=1}^{c} x_i'(V_i^0 - V_i')dp \qquad [34]$$

where

$$d\ln a_i = (d\ln a_i)_{T,p} + (\partial \ln a_i/\partial T)_{x,p} + (\partial \ln a_i/\partial p)_{x,T} \qquad [35]$$

The terms involving enthalpies and volumes in the solid phase can be written as:

$$\sum_{i=1}^{c} x_i'H_i' = H_s^* \qquad \sum_{i=1}^{c} x_i'V_i' = V_s^* \qquad [36]$$

With eqn [36], the final general solubility equation may then be written:

$$R\sum_{i=1}^{c} x_i'd\ln a_i = (H_s^* - \sum_{i=1}^{c} x_i'H_i^0)d(1/T) - (V_s^* - \sum_{i=1}^{c} x_i'V_i^0)dp/T \qquad [37]$$

Note that those components which are not present in both phases do not appear in the solubility equation. However, they do affect the solubility through their effect on the activities of the solutes.

Several applications of eqn [37] (all with pressure held constant) will be discussed below. Other cases will be discussed in individual evaluations.

(a) *Solubility as a function of temperature.*

Consider a binary solid compound A_nB in a single solvent A. There is

no fundamental thermodynamic distinction between a binary compound of A
and B which dissociates completely or partially on melting and a solid
mixture of A and B; the binary compound can be regarded as a solid mixture
of constant composition. Thus, with $c = 2$, $x_A' = n/(n + 1)$,
$x_B' = 1/(n + 1)$, eqn [37] becomes:

$$d\ln(a_A^n a_B) = -\Delta H_{AB}^0 d(1/RT) \qquad [38]$$

where

$$\Delta H_{AB}^0 = nH_A + H_B - (n + 1)H_s^* \qquad [39]$$

is the molar enthalpy of melting and dissociation of pure solid A_nB to
form A and B in their reference states. Integration between T and T_0,
the melting point of the pure binary compound A_nB, gives:

$$\ln(a_A^n a_B) = \ln(a_A^n a_B)_{T=T_0} - \int_{T_0}^{T} \Delta H_{AB}^0 d(1/RT) \qquad [40]$$

(i) Non-electrolytes

In eqn [32], introduce the pure liquids as reference states. Then,
using a simple first-order dependence of ΔH_{AB}^* on temperature, and
assuming that the activitity coefficients conform to those for a simple
mixture (6):

$$RT \ln f_A = wx_B^2 \qquad RT \ln f_B = wx_A^2 \qquad [41]$$

then, if w is independent of temperature, eqn [32] and [33] give:

$$\ln\{x_B(1-x_B)^n\} + \ln\left\{\frac{n^n}{(1 + n)^{n+1}}\right\} = G(T) \qquad [42]$$

where

$$G(T) = -\left\{\frac{\Delta H_{AB}^* - T^* \Delta C_p^*}{R}\right\}\left\{\frac{1}{T} - \frac{1}{T^*}\right\}$$
$$+ \frac{\Delta C_p^*}{R}\ln(T/T^*) - \frac{w}{R}\left\{\frac{x_A^2 + nx_B^2}{T} - \frac{n}{(n + 1)T^*}\right\} \qquad [43]$$

where ΔC_p^* is the change in molar heat capacity accompanying fusion plus
decomposition of the pure compound to pure liquid A and B at temperature
T^*, (assumed here to be independent of temperature and composition), and
ΔH_{AB}^* is the corresponding change in enthalpy at $T = T^*$. Equation [42]
has the general form:

$$\ln\{x_B(1-x_B)^n\} = A_1 + A_2/(T/K) + A_3\ln(T/K) + A_4(x_A^2 + nx_B^2)/(T/K) \qquad [44]$$

If the solid contains only component B, then $n = 0$ in eqn [42] to [44].

If the infinite dilution reference state is used, then:

$$RT \ln f_{x,B} = w(x_A^2 - 1) \qquad [45]$$

and [39] becomes

$$\Delta H_{AB}^\infty = nH_A^* + H_B^\infty - (n + 1)H_s^* \qquad [46]$$

where ΔH_{AB}^∞ is the enthalpy of melting and dissociation of solid compound
A_nB to the infinitely dilute reference state of solute B in solvent A; H_A^*
and H_B^∞ are the partial molar enthalpies of the solute and solvent at
infinite dilution. Clearly, the integral of eqn [32] will have the same
form as eqn [35], with ΔH_{AB}^∞ replacing ΔH_{AB}^*, ΔC_p^∞ replacing ΔC_p^*, and
$x_A^2 - 1$ replacing x_A^2 in the last term.

See (5) and (11) for applications of these equations to experimental
data.

(ii) Electrolytes

(a) Mole fraction scale

If the liquid phase is an aqueous electrolyte solution, and the
solid is a salt hydrate, the above treatment needs slight modification.
Using rational mean activity coefficients, eqn [34] becomes:

$$\ln\left\{\frac{x_B{}^{\nu}(1-x_B)^n}{[1+(\nu-1)x_B]^{n+\nu}}\right\} - \ln\left\{\frac{n^n}{(n+\nu)^{n+\nu}}\right\} + \ln\left\{\left[\frac{f_B}{f_B{}^*}\right]^{\nu}\left[\frac{f_A}{f_A{}^*}\right]^n\right\}$$

$$= -\left\{\frac{\Delta H_{AB}{}^* - T^*\Delta C_p{}^*}{R}\right\}\left\{\frac{1}{T} - \frac{1}{T^*}\right\} + \frac{\Delta C_p{}^*}{R}\ln(T/T^*)$$

[47]

where superscript * indicates the pure salt hydrate. If it is assumed that the activity coefficients follow the same temperature dependence as the right-hand side of eqn [47] (13-16), the thermochemical quantities on the right-hand side of eqn [47] are not rigorous thermodynamic enthalpies and heat capacities, but are apparent quantities only. Data on activity coefficients (9) in concentrated solutions indicate that the terms involving these quantities are not negligible, and their dependence on temperature and composition along the solubility-temperature curve is a subject of current research.

A similar equation (with $\nu = 2$ and without the heat capacity terms or activity coefficients) has been used to fit solubility data for some MOH-H_2O systems, where M is an alkali metal (13); enthalpy values obtained agreed well with known values. The full equation has been deduced by another method in (14) and applied to MCl$_2$-H_2O systems in (14) and (15). For a summary of the use of equation [47] and similar equations, see (14).

(2) Molality scale

Substitution of the mean activities on the molality scale in eqn [40] gives:

$$\nu\ln\left\{\frac{\gamma_{\pm}m_B}{\gamma_{\pm}{}^*m_B{}^*}\right\} - \nu(m_B/m_B{}^* - 1) - \nu\{m_B(\phi-1)/m_B{}^* - \phi^* + 1\}$$

$$= G(T)$$

[48]

where $G(T)$ is the same as in eqn [47], $m_B{}^* = 1/nM_A$ is the molality of the anhydrous salt in the pure salt hydrate and $\gamma_{\pm}$ and ϕ are the mean activity coefficient and the osmotic coefficient, respectively. Use of the osmotic coefficient for the activity of the solvent leads, therefore, to an equation that has a different appearance to [47]; the content is identical. However, while eqn [47] can be used over the whole range of composition ($0 \leqslant x_B \leqslant 1$), the molality in eqn [48] becomes infinite at $x_B = 1$; use of eqn [48] is therefore confined to solutions sufficiently dilute that the molality is a useful measure of composition. The essentials of eqn [48] were deduced by Williamson (17); however, the form used here appears first in the Solubility Data Series. For typical applications (where activity and osmotic coefficients are not considered explicitly, so that the enthalpies and heat capacities are apparent values, as explained above), see (18).

The above analysis shows clearly that a rational thermodynamic basis exists for functional representation of solubility-temperature curves in two-component systems, but may be difficult to apply because of lack of experimental or theoretical knowledge of activity coefficients and partial molar enthalpies. Other phenomena which are related ultimately to the stoichiometric activity coefficients and which complicate interpretation include ion pairing, formation of complex ions, and hydrolysis. Similar considerations hold for the variation of solubility with pressure, except that the effects are relatively smaller at the pressures used in many investigations of solubility (5).

(b) Solubility as a function of composition.

At constant temperature and pressure, the chemical potential of a saturating solid phase is constant:

$$\mu_{A_nB}{}^* = \mu_{A_nB}(sln) = n\mu_A + \mu_B$$

[49]

$$= (n\mu_A{}^* + \nu_+\mu_+{}^{\infty} + \nu_-\mu_-{}^{\infty}) + nRT\ln f_A x_A$$

$$+ \nu RT\ln(\gamma_{\pm}m_{\pm}Q)$$

for a salt hydrate A_nB which dissociates to water (A), and a salt (B), one mole of which ionizes to give ν_+ cations and ν_- anions in a solution in which other substances (ionized or not) may be present. If the saturated solution is sufficiently dilute, $f_A = x_A = 1$, and the quantity K_s in

$$\Delta G^{\infty} = (\nu_+\mu_+{}^{\infty} + \nu_-\mu_-{}^{\infty} + n\mu_A{}^* - \mu_{AB}{}^*)$$

$$= -RT\ln K_s$$

$$= -\nu RT \ln(Q\gamma_{\pm}m_B) \qquad [50]$$

is called the *solubility product* of the salt. (It should be noted that it
is not customary to extend this definition to hydrated salts, but there is
no reason why they should be excluded.) Values of the solubility product
are often given on mole fraction or concentration scales. In dilute
solutions, the theoretical behaviour of the activity coefficients as a
function of ionic strength is often sufficiently well known that reliable
extrapolations to infinite dilution can be made, and values of K_S can be
determined. In more concentrated solutions, the same problems with
activity coefficients that were outlined in the section on variation of
solubility with temperature still occur. If these complications do not
arise, the solubility of a hydrate salt $C_\nu A_\nu \cdot nH_2O$ in the presence of
other solutes is given by eqn [50] as

$$\nu \ln\{m_B/m_B(0)\} = -\nu\ln\{\gamma_{\pm}/\gamma_{\pm}(0)\} - n \ln\{a_A/a_A(0)\} \qquad [51]$$

where a_A is the activity of water in the saturated solution, m_B is the
molality of the salt in the saturated solution, and (0) indicates
absence of other solutes. Similar considerations hold for non-
electrolytes.

Consideration of *complex mixed ligand equilibria* in the solution phase
are also frequently of importance in the interpretation of solubility
equilibria. For nomenclature connected with these equilibria (and
solubility equilibria as well), see (19, 20).

The Solid Phase

The definition of solubility permits the occurrence of a single solid
phase which may be a pure anhydrous compound, a salt hydrate, a non-
stoichiometric compound, or a solid mixture (or solid solution, or
"mixed crystals"), and may be stable or metastable. As well, any
number of solid phases consistent with the requirements of the phase
rule may be present. Metastable solid phases are of widespread
occurrence, and may appear as polymorphic (or allotropic) forms or
crystal solvates whose rate of transition to more stable forms is very
slow. Surface heterogeneity may also give rise to metastability, either
when one solid precipitates on the surface of another, or if the size of
the solid particles is sufficiently small that surface effects become
important. In either case, the solid is not in stable equilibrium
with the solution. See (21) for the modern formulation of the effect of
particle size on solubility. The stability of a solid may also be
affected by the atmosphere in which the system is equilibrated.

Many of these phenomena require very careful, and often prolonged,
equilibration for their investigation and elimination. A very general
analytical method, the "wet residues" method of Schreinemakers (22),
is often used to investigate the composition of solid phases in
equilibrium with salt solutions. This method has been reviewed in (23),
where [see also (24)] least-squares methods for evaluating the composition
of the solid phase from wet residue data (or initial composition data)
and solubilities are described. In principle, the same method can be used
with systems of other types. Many other techniques for examination of
solids, in particular X-ray, optical, and thermal analysis methods, are
used in conjunction with chemical analyses (including the wet residues
method).

COMPILATIONS AND EVALUATIONS

The formats for the compilations and critical evaluations have been
standardized for all volumes. A brief description of the data sheets
has been given in the FOREWORD; additional explanation is given below.

Guide to the Compilations

The format used for the compilations is, for the most part, self-
explanatory. The details presented below are those which are not found
in the FOREWORD or which are not self-evident.

Components. Each component is listed according to IUPAC name, formula,
and Chemical Abstracts (CA) Registry Number. The formula is given either
in terms of the IUPAC or Hill (25) system and the choice of formula is
governed by what is usual for most current users: i.e., IUPAC for
inorganic compounds, and Hill system for organic compounds. Components
are ordered according to:
   (a) saturating components;
   (b) non-saturating components in alphanumerical order;
   (c) solvents in alphanumerical order.

The saturating components are arranged in order according to a
18-column periodic table with two additional rows:
    Columns 1 and 2: H, alkali elements, ammonium, alkaline earth elements
          3 to 12: transition elements
        13 to 17: boron, carbon, nitrogen groups; chalcogenides, halogens
            18: noble gases
         Row 1: Ce to Lu
         Row 2: Th to the end of the known elements, in order of
             atomic number.

Salt hydrates are generally not considered to be saturating components
since most solubilities are expressed in terms of the anhydrous salt. The
existence of hydrates or solvates is carefully noted in the text, and CA
Registry Numbers are given where available, usually in the critical
evaluation. Mineralogical names are also quoted, along with their CA
Registry Numbers, again usually in the critical evaluation.

Original Measurements. References are abbreviated in the forms given
by *Chemical Abstracts Service Source Index* (CASSI). Names originally in
other than Roman alphabets are given as transliterated by *Chemical
Abstracts.*

Experimental Values. Data are reported in the units used in the
original publication, with the exception that modern *names* for units
and quantities are used; e.g., mass per cent for weight per cent;
mol dm^{-3} for molar; etc. Both mass and molar values are given. Usually,
only one type of value (e.g., mass per cent) is found in the original
paper, and the compiler has added the other type of value (e.g., mole
per cent) from computer calculations based on 1983 atomic weights (26).

Errors in calculations and fitting equations in original papers have
been noted and corrected, by computer calculations where necessary.

Method. Source and Purity of Materials. Abbreviations used in
Chemical Abstracts are often used here to save space.

Estimated Error. If these data were omitted by the original authors,
and if relevant information is available, the compilers have attempted
to estimate errors from the internal consistency of data and type of
apparatus used. Methods used by the compilers for estimating and
and reporting errors are based on the papers by Ku and Eisenhart (27).

Comments and/or Additional Data. Many compilations include this
section which provides short comments relevant to the general nature of
the work or additional experimental and thermodynamic data which are
judged by the compiler to be of value to the reader.

References. See the above description for Original Measurements.

Guide to the Evaluations

The evaluator's task is to check whether the compiled data are correct,
to assess the reliability and quality of the data, to estimate errors
where necessary, and to recommend "best" values. The evaluation takes
the form of a summary in which all the data supplied by the compiler
have been critically reviewed. A brief description of the evaluation
sheets is given below.

Components. See the description for the Compilations.

Evaluator. Name and date up to which the literature was checked.

Critical Evaluation
(a) Critical text. The evaluator produces text evaluating all the
published data for each given system. Thus, in this section the
evaluator reviews the merits or shortcomings of the various data. Only
published data are considered; even published data can be considered only
if the experimental data permit an assessment of reliability.
(b) Fitting equations. If the use of a smoothing equation is
justifiable the evaluator may provide an equation representing the
solubility as a function of the variables reported on all the
compilation sheets.
(c) Graphical summary. In addition to (b) above, graphical summaries
are often given.
(d) Recommended values. Data are *recommended* if the results of at
least two independent groups are available and they are in good
agreement, and if the evaluator has no doubt as to the adequacy and
reliability of the applied experimental and computational procedures.
Data are considered as *tentative* if only one set of measurements is

available, or if the evaluator considers some aspect of the computational
or experimental method as mildly undesirable but estimates that it should
cause only minor errors. Data are considered as *doubtful* if the
evaluator considers some aspect of the computational or experimental
method as undesirable but still considers the data to have some value
in those instances where the order of magnitude of the solubility is
needed. Data determined by an inadequate method or under ill-defined
conditions are *rejected*. However references to these data are included
in the evaluation together with a comment by the evaluator as to the
reason for their rejection.

   (e) References. All pertinent references are given here. References
to those data which, by virtue of their poor precision, have been
rejected and not compiled are also listed in this section.
   (f) Units. While the original data may be reported in the units
used by the investigators, the final recommended values are reported
in S.I. units (1, 28) when the data can be accurately converted.

References

1. Whiffen, D.H., ed., *Manual of Symbols and Terminology for Physico-chemical Quantities and Units. Pure Applied Chem.* 1979, 51, No. 1.
2. McGlashan, M.L. *Physicochemical Quantities and Units.* 2nd ed. Royal Institute of Chemistry. London. 1971.
3. Jänecke, E. *Z. Anorg. Chem.* 1906, 51, 132.
4. Friedman, H.L. *J. Chem. Phys.* 1960, 32, 1351.
5. Prigogine, I.; Defay, R. *Chemical Thermodynamics.* D.H. Everett, transl. Longmans, Green. London, New York, Toronto. 1954.
6. Guggenheim, E.A. *Thermodynamics.* North-Holland. Amsterdam. 1959. 4th ed.
7. Kirkwood, J.G.; Oppenheim, I. *Chemical Thermodynamics.* McGraw-Hill. New York, Toronto, London. 1961.
8. Lewis, G.N.; Randall, M. (rev. Pitzer, K.S.; Brewer, L.). *Thermodynamics.* McGraw Hill. New York, Toronto, London. 1961. 2nd. ed.
9. Robinson, R.A.; Stokes, R.H. *Electrolyte Solutions.* Butterworths. London. 1959. 2nd ed.
10. Harned, H.S.; Owen, B.B. *The Physical Chemistry of Electrolytic Solutions.* Reinhold. New York. 1958. 3rd ed.
11. Haase, R.; Schönert, H. *Solid-Liquid Equilibrium.* E.S. Halberstadt, trans. Pergamon Press, London, 1969.
12. McGlashan, M.L. *Chemical Thermodynamics.* Academic Press. London. 1979.
13. Cohen-Adad, R.; Saugier, M.T.; Said, J. *Rev. Chim. Miner.* 1973, 10, 631.
14. Counioux, J.-J.; Tenu, R. *J. Chim. Phys.* 1981, 78, 815.
15. Tenu, R.; Counioux, J.-J. *J. Chim. Phys.* 1981, 78, 823.
16. Cohen-Adad, R. *Pure Appl. Chem.* 1985, 57, 255.
17. Williamson, A.T. *Faraday Soc. Trans.* 1944, 40, 421.
18. Siekierski, S.; Mioduski, T.; Salomon, M. *Solubility Data Series.* Vol. 13. *Scandium, Yttrium, Lanthanum and Lanthanide Nitrates.* Pergamon Press. 1983.
19. Marcus, Y., ed. *Pure Appl. Chem.* 1969, 18, 459.
20. IUPAC Analytical Division. *Proposed Symbols for Metal Complex Mixed Ligand Equilibria (Provisional). IUPAC Inf. Bull.* 1978, No. 3, 229.
21. Enüstün, B.V.; Turkevich, J. *J. Am. Chem. Soc.* 1960, 82, 4502.
22. Schreinemakers. F.A.H. *Z. Phys. Chem., Stoechiom. Verwandschaftsl.* 1893, 11, 75.
23. Lorimer, J.W. *Can. J. Chem.* 1981, 59, 3076.
24. Lorimer, J.W. *Can. J. Chem.* 1982, 60, 1978.
25. Hill, E.A. *J. Am. Chem. Soc.* 1900, 22, 478.
26. IUPAC Commission on Atomic Weights. *Pure Appl. Chem.* 1984, 56, 653.
27. Ku, H.H., p. 73; Eisenhart, C., p. 69; in Ku, H.H., ed. *Precision Measurement and Calibration.* NBS Special Publication 300. Vol. 1. Washington. 1969.
28. *The International System of Units.* Engl. transl. approved by the BIPM of *Le Système International d'Unités.* H.M.S.O. London. 1970.

September, 1986   R. Cohen-Adad,
                                                   Villeurbanne, France

                                                   J. W. Lorimer,
                                                   London, Ontario, Canada

                                                   M. Salomon,
                                                   Fair Haven, New Jersey, U.S.A.

Table I-1

Quantities Used as Measures of Solubility of Solute B
Conversion Table for Multicomponent Systems
Containing Solvent A and Solutes s

	mole fraction $x_B =$	mass fraction $w_B =$	molality $m_B =$	concentration $c_B =$
x_B	x_B	$\dfrac{M_B x_B}{M_A + \sum\limits_{s}(M_s - M_A)x_s}$	$\dfrac{x_B}{M_A(1 - \sum\limits_{s} x_s)}$	$\dfrac{\rho x_B}{M_A + \sum\limits_{s}(M_s - M_A)x_s}$
w_B	$\dfrac{w_B/M_B}{1/M_A + \sum\limits_{s}(1/M_s - 1/M_A)w_s}$	w_B	$\dfrac{w_B}{M_B(1 - \sum\limits_{s} w_s)}$	$\rho w_B/M_B$
m_B	$\dfrac{M_A m_B}{1 + M_A\sum\limits_{s} m_s}$	$\dfrac{M_B m_B}{1 + \sum\limits_{s} m_s M_s}$	m_B	$\dfrac{\rho m_B}{1 + \sum\limits_{s} M_s m_s}$
c_B	$\dfrac{M_A c_B}{\rho + \sum\limits_{s}(M_A - M_s)c_s}$	$M_B c_B/\rho$	$\dfrac{c_B}{\rho - \sum\limits_{s} M_s c_s}$	c_B

ρ = density of solution
M_A, M_B, M_s = molar masses of solvent, solute B, other solutes s
Formulas are given in forms suitable for rapid computation; all
calculations should be made using SI base units.

COMPONENTS:	EVALUATOR: J. Eysseltová
(1) Lithium phosphate; Li_3PO_4; [10377-52-3]	Charles University
(2) Ammonia; NH_3; [7664-41-7]	Prague, Czechoslovakia
(3) Lithium hydroxide; LiOH; [1310-65-2]	and
(4) Phosphoric acid; H_3PO_4; [7664-38-2]	M. Salomon
(5) Water; H_2O; [7732-18-5]	U.S. Army Research Laboratories Ft. Monmouth, NJ 07703 U.S.A. December 1981

CRITICAL EVALUATION:

There have been relatively few solubility studies on Li_3PO_4-aqueous systems (1-5). Four of the studies (1-4) report solubilities based on direct analysis of the binary system, and three studies (2, 3, 5) report the solubility of Li_3PO_4 in ternary systems.

THE BINARY SYSTEM

1. Solubilities at 273 K to 298 K. The direct determination of the solubility of Li_3PO_4 in pure water has proved difficult due to the formation of a fine colloid which cannot be removed by filtration (1,4). Thus Rammelsberg's value (1) of 0.0104 mol kg^{-1} at 288 K is undoubtedly much too high, and this work has therefore not been compiled. Although Mayer's average value of 0.00340 mol kg^{-1} (0.394 g kg^{-1}) appears slightly high, he certainly was aware of this problem (2,3): in a footnote on page 201 of reference (2), Mayer states that turbidity in saturated solutions does not occur if the solution is heated. Although not stated, the evaluators assume that Mayer prepared his saturated solutions by first heating the solutions (i.e. from supersaturation). Since Mayer's value is slightly high, the evaluators were considering rejecting these data but decided against this as his data constitute one of two direct measurements on the binary system [the other being from ref (4)], and his results are the most widely quoted ones in various important handbooks (e.g. see references 6-8). Mayer's value of 0.0394 g per 100 g H_2O is usually quoted for 291 K when his solutions were equilibrated over the temperature range of 288 - 291 K.

Rollet and Lauffenburger (5) obtained solubility values for the binary system at 273 K and 293 K by extrapolation of the isotherms in the Li_2O-P_2O_5-H_2O ternary systems. Their results for the solubility of Li_3PO_4 in pure water are: m_{satd} = 0.0019 mol kg^{-1} at 273 K, and m_{satd} = 0.0026 mol kg^{-1} at 293 K. These data suggest that the solubility of Li_3PO_4 in pure water increases with temperature and therefore casts some doubt on the accuracy of Mayer's value of 0.00340 mol kg^{-1} at 288-291 K. Adding to the uncertainty in the solubility of this salt is the value of 0.00227 mol dm^{-3} at 298 K calculated by the evaluators (see below) using the conductivity data of Rosenheim and Reglin (4). If we assume the solubility follows a log m_{satd} vs 1/(T/K) relation, then using Rollet and Laufenburger's extrapolated solubility values for 273 K and 293 K, the evaluators calculate m_{satd} = 0.0028 ± 0.0002 mol kg^{-1} at 298 K (this corresponds to c_{satd} = 0.00279 mol dm^{-3} (evaluators)).

Rosenheim and Reglin (4) measured the electrolytic conductivity of saturated Li_3PO_4 slns. Their calculation of κ_{salt} (corrected for the electrolytic conductivity of water) is in error, and the correct average determined by the compilers is 9.30 x 10^{-4} S cm^{-1}. In computing the solubility from κ_{salt}, one cannot neglect, as did Rosenheim and Reglin, the hydrolysis of the phosphate ion according to

$$PO_4^{3-} + H_2O = HPO_4^{2-} + OH^- \qquad [1]$$

The thermodynamic equilibrium constant for this hydrolysis reaction is obtained from

$$K_h^\circ = K_w^\circ / K_{a3}^\circ \qquad [2]$$

(continued next page)

COMPONENTS:	EVALUATOR: J. Eysseltová
(1) Lithium phosphate; Li_3PO_4; [10377-52-3]	Charles University
(2) Ammonia; NH_3; [7664-41-7]	Prague, Czechoslovakia
(3) Lithium hydroxide; LiOH; [1310-65-2]	and
(4) Phosphoric acid; H_3PO_4 [7664-38-2]	M. Salomon
(5) Water; H_2O; [7732-18-5]	U.S. Army Research Laboratories
	Ft. Monmouth, NJ 07703
	U.S.A.
	December 1981

CRITICAL EVALUATION:

In eq [2] $K_w^\circ = 1.005 \times 10^{-14}$ (9), and the third acid dissociation constant of H_3PO_4 is $K_{a_3}^\circ = 4.217 \times 10^{-13}$ (10): thus $K_h^\circ = 0.02383$ at 298 K, and the major ions in solution are Li^+, HPO_4^{2-}, OH^- and a small but significant amount of PO_4^{3-}. It is reasonable to assume that further hydrolysis to $H_2PO_4^-$ is negligible (11), that ion association of Li^+ and OH^- is negligible (12-14), and that ion association of Li^+ and HPO_4^{2-} is negligible (15). For saturated solutions, we have

$$K_h^\circ\{y_\pm(PO_4^{3-})/[y_\pm(OH^-)y_\pm(HPO_4^{2-})]\} = [OH^-][HPO_4^{2-}]/[PO_4^{3-}] = \alpha^2 c_{satd}/(1-\alpha) \qquad [3]$$

where α is the degree of hydrolysis and $y_\pm$ is the mean molar activity coefficient of the indicated species. The solubility of Li_3PO_4 in pure water can be calculated from Rosenheim and Reglin's average κ_{salt} value from

$$10^3\kappa_{salt}/c_{satd} = 2\alpha\Lambda(Li_2HPO_4) + \alpha\Lambda(LiOH) + 3(1-\alpha)\Lambda(Li_3PO_4) \qquad [4]$$

where the molar conductivities, Λ, can be calculated for a given concentration, c_i, of the various species using Robinson and Stokes' equation (9)

$$\Lambda = \Lambda^\infty - \{0.77816|z_1 z_2|q\Lambda^\infty/(1+q^2) + 30.16(|z_1| + |z_2|)\}c_i \qquad [5]$$

The reader is referred to reference (9) for definition of q and the origins of the numerical terms. Limiting molar conductivities are separated into individual ionic contributions, e.g. for Li_2HPO_4, we have $\Lambda^\infty = \lambda^\infty(Li^+) + \lambda^\infty(\frac{1}{2}HPO_4^{2-})$ and $c_i = \alpha c_{satd}$. We first assume a value for c_{satd} and compute α iteratively from eq [3]. In solving for α from eq [3], the Davies eq

$$\log y_\pm = -0.5115z^2\{I^{\frac{1}{2}}/(1 + I^{\frac{1}{2}}) - 0.3I\} \qquad [6]$$

was used (16). The ionic strength I was calculated from

$$I = 2c_{satd}(3 - \alpha) \qquad [7]$$

For the assumed c_{satd} and the corresponding α, eq [4] is solved for κ_{salt} and the calcs repeated until the experimental value $\kappa_{salt} = 9.30 \times 10^{-4}$ S cm^{-1} is obtained. Although the solution of eq [4] employing eqs [3], [5]-[7] is fairly straight forward, there is at least a 5% uncertainty in the refined c_{satd} due to the uncertainties in the molar conductivities at infinite dilution for HPO_4^{2-} and PO_4^{3-}. $\lambda^\infty(\frac{1}{2}HPO_4^{2-})$ values of 53.4 S $cm^2 mol^{-1}$ (17) and 57 S $cm^2 mol^{-1}$ (18) have been reported, and the uncertainty is probably higher than indicated by the closeness of these two values. For the present calculations, the evaluators have taken $\lambda^\infty(\frac{1}{2}HPO_4^{2-}) = 53.4$ S $cm^2 mol^{-1}$. Greater uncertainty is associated with the value for $\lambda^\infty(1/3PO_4^{3-})$ as reported values range from 69.0 S $cm^2 mol^{-1}$ (17) to 82.3 S $cm^2 mol^{-1}$ (19) and 92.8 S $cm^2 mol^{-1}$ (20). Milazzo's value (20) appears much too high, and for consistency in the calculations, we have used Prideaux's value (17) of $\lambda^\infty(1/3PO_4^{3-})$ and 67 S $cm^2 mol^{-1}$. For the remaining molar conductivities, we have used $\lambda^\infty(Li^+) = 38.71$ S $cm^2 mol^{-1}$ (13) and $\lambda^\infty(OH^-) = 199.18$ S $cm^2 mol^{-1}$ (21). Our final results are: $c_{satd} = 0.00227$ mol dm^{-3} and $\alpha = 0.8875$. Thus the extent of hydrolysis of PO_4^{3-} in the saturated solution is about 89%, and the fact that Rosenheim and Reglin obtain a c_{satd} close to 0.00227 mol dm^{-3} neglecting hydrolysis is attributed to their use of inaccurate λ^∞ values.

(continued next page)

COMPONENTS:	EVALUATOR: J. Eysseltová
(1) Lithium phosphate; Li_3PO_4; [10377-52-3]	Charles University
(2) Ammonia; NH_3; [7664-41-7]	Prague, Czechoslovakia
(3) Lithium hydroxide; LiOH; [1310-65-2]	and
(4) Phosphoric acid; H_3PO_4; [7664-38-2]	M. Salomon
(5) Water; H_2O; [7732-18-5]	U.S. Army Research Laboratories
	Ft. Monmouth, NJ 07703
	U.S.A.
	December 1981

CRITICAL EVALUATION:

We now review the status of the solubility of Li_3PO_4 in pure water. Mayer's oft quoted value of 0.0034 mol kg^{-1} at 288-293 K is probably too high by at least 15%. Using the conductivity data of Rosenheim and Reglin, the evaluators have calculated a solubility value of 0.0023 ± 0.0001 mol dm^{-3} at 298 K. This value is not in very good agreement with the extrapolated value of 0.0028 ± 0.0002 mol kg^{-1} estimated by the evaluators from Rollet and Lauffenburger's extrapolated values of 0.0019 mol kg^{-1} at 273 K and 0.0026 mol kg^{-1} at 293K.

2. The Solubility Product Constant. For the reaction

$$3Li^+ + PO_4^{3-} = Li_3PO_4(s) \tag{8}$$

the thermodynamic solubility product constant is defined by

$$K_{s0}^° = \{Li^+\}^3\{PO_4^{3-}\} = 27(1 - \alpha)c_{satd}^4 y_\pm^3(Li^+)y_\pm(PO_4^{3-}) \tag{9}$$

Based on the solubility of 0.00227 mol dm^{-3} calculated by the evaluators from the conductivity data, it is found that $K_{s0}^° = 2.370 \times 10^{-11}$ mol^2 dm^{-6}, or $pK_{s0}^° = 10.625$. In the latest revision of Lange's Handbook, Dean (8) reports $pK_{s0} = 8.5$. It is not stated whether this is a pK or a pK° value, what the temperature is, and the origin of this value. It appears (see below) that this pK_{s0} was incorrectly calculated from Mayer's m_{satd}, and since earlier versions of Lange's Handbook (e.g. see ref. 7) do not report a K_{s0} for Li_3PO_4, the value $pK_{s0} = 8.5$ is probably one of Dean's contributions to the revised Handbook. If we incorrectly neglect hydrolysis and activity coefficients, K_{s0} would be given by

$$K_{s0} = [Li^+]^3[PO_4^{3-}] = 27 c_{satd}^4 \tag{10}$$

Using Mayer's value of $m_{satd} = 0.00340$ mol kg^{-1} at 288-291 K, eq [10] give $K_{s0} = 3.61 \times 10^{-9}$ mol^2 kg^{-2}, or $pK_{s0} = 8.44$ which is practically identical to Dean's value of 8.5.

Two additional sources (22,23) quote a value of $pK_{s0} = 12.5$ but fail to state whether this is a pK or pK° value, and fail to cite the original publication. The evaluators could not find the source of this pK_{s0} in spite of an exhaustive literature search. Fitting eq [3] to this pK_{s0} requires $c_{satd} = 0.000875$ mol dm^{-3}, and we therefore conclude that the pK_{s0} value of 12.5 is in serious error and must be rejected.

MULTICOMPONENT SYSTEMS

Of the few solubility studies on multicomponent systems (1-3, 5, 24) only ternary systems have been investigated and only references 2,3 and 5 report quantitative data. Qualitative studies state that Li_3PO_4 is soluble in strong acids (1-3, 24), is difficult to dissolve in acetic acid (2,3) and that addition of NH_4Cl tends to increase the solubility (2,3). The quantitative studies are discussed below.

1. The Li_3PO_4-NH_3-H_2O system. Mayer (2,3) reported only one data point for this system at 288-291 K. In approximately 1.6 mol kg^{-1} NH_3 solution, the average value of the soly as calculated by the compilers is 0.0015 mol kg^{-1} ($\sigma = 0.0001$).

2. The Li_2O-P_2O_5-H_2O System. This is the most complete phase study available for Li_3PO_4 systems. Rollet and Laueffenburger (5) reported the compositions of saturated solutions at 273 K and 293 K in mass% of Li_2O and P_2O_5. The compiler separated appropriate data

(continued next page)

COMPONENTS:	EVALUATOR: J. Eysseltová
(1) Lithium phosphate; Li_3PO_4; [10377-52-3]	Charles University
(2) Ammonia; NH_3; [7664-41-7]	Prague, Czechoslovakia
(3) Lithium hydroxide; LiOH; [1310-65-2]	and
(4) Phosphoric acid; H_3PO_4; [7664-38-2]	M. Salomon
(5) Water; H_2O; [7732-18-5]	U.S. Army Research Laboratories Ft. Monmouth, NJ 07703 U.S.A. December 1981

CRITICAL EVALUATION:

into two compilations corresponding to the ternary systems Li_3PO_4-LiOH-H_2O and Li_3PO_4-H_3PO_4-H_2O. The original phase diagrams are reproduced in Figures 1 and 2. Note that Figure 2 is an expanded detail of the initial portion of Figure 1. Numerical data corresponding to the points A-E are given in the compilations as well as the compiler's conversions from mass% to mol% and mol kg^{-1}. At 273 K, LiOH·H_2O is the initial solid phase up to invariant point A where Li_3PO_4 is also in equilibrium with the solution. The solubility then decreases rapidly and then increases slowly to invariant point B at which point both Li_3PO_4 and LiH_2PO_4 solid phases are in equilibrium with the solution. Between B and C, LiH_2PO_4 precipitates. The 293 K isotherm is similar to the 273 K isotherm. The invariant points are D and E. Note that point E is on the acid side of the line for P_2O_5/Li_2O = 1.00 which means that LiH_2PO_4 will dissolve incongruently to form Li_3PO_4 and H_3PO_4 until the composition of point E is reached.

SOLID PHASES

A number of solid phases have been reported or suggested to be in equilibrium with saturated Li_3PO_4 solutions. They are

lithium hydroxide hydrate; LiOH·H_2O; [1310-66-3]
lithium phosphate; Li_3PO_4; [10377-52-3]
lithium phosphate dihydrate; Li_3PO_4·$2H_2O$; [74893-09-7]
lithium phosphate hemihydrate; Li_3PO_4·$\frac{1}{2}H_2O$; [10102-26-8]
lithium dihydrogen phosphate; LiH_2PO_4; [13453-80-0]

Rollet and Lauffenburger's detailed phase study (5) reports the absence of any phosphate hydrates as well as the absense of Li_2HPO_4 for their experimental conditions. On the other hand, Rosenheim and Reglin (4) state that their solid phase is the dihydrate which forms by precipitation from aqueous H_3PO_3 with excess LiOH. These conflicting results are difficult to assess since neither study describes sufficient details of the analyses of the solid phases. Presumably Rollet and Lauffenburger used a wet residue method such as Schreinemakers' method, and Rosenheim and Reglin simply air dried their solid at 289 K so that it is quite possible that the water they found in the solid was not water of hydration. Upon drying at 333 K for several days, Rosenheim and Reglin state that they obtain the hemihydrate. It may indeed be possible that the hemihydrate is stable under the conditions reported by Rosenheim and Reglin since Sanfourche (25) reported that the neutralization method of preparation of Li_3PO_4 actually yields the hemihydrate, and that the water of hydration can be removed only at red heat. These results combined with Rollet and Lauffenburger's findings that no hydrate is formed at ambient temperatures casts some doubt on the nature of the solid phases present in all of the reported solubility studies. Because of this situation, and of the uncertainties in the reported solubility data, the evaluators feel that new studies are required before recommended data can be specified.

(continued next page)

COMPONENTS:	EVALUATOR:
(1) Lithium phosphate; Li_3PO_4; [10377-52-3]	J. Eysseltová
(2) Ammonia; NH_3; [7664-41-7]	Charles University
(3) Lithium hydroxide; LiOH; [1310-65-2]	Prague, Czechoslovakia
(4) Phosphoric acid; H_3PO_4; [7664-38-2]	and
(5) Water; H_2O, [7732-18-5]	M. Salomon
	U.S. Army Research Laboratories
	Ft. Monmouth, NJ 07703
	U.S.A.
	December 1981

CRITICAL EVALUATION:

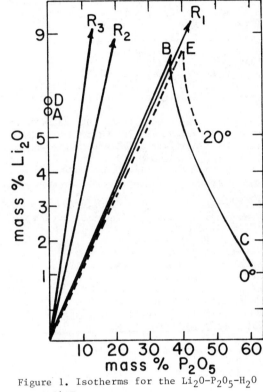

Figure 1. Isotherms for the $Li_2O-P_2O_5-H_2O$
    system.

$$R_1 = Li_2O/P_2O_5 = 1$$
$$R_2 = Li_2O/P_2O_5 = 2$$
$$R_3 = Li_2O/P_2O_5 = 3$$

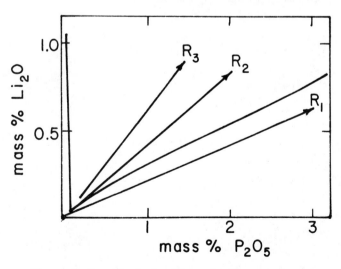

Figure 2. Detail of the 0° isotherm on Figure 1.

(continued next page)

COMPONENTS:	EVALUATOR: J. Eysseltová
(1) Lithium phosphate; Li_3PO_4; [10377-52-3]	Charles University Prague, Czechoslovakia
(2) Ammonia; NH_3; [7664-41-7]	and M. Salomon
(3) Lithium hydroxide; LiOH; [1310-65-2]	U.S. Army Research Laboratories
(4) Phosphoric acid; H_3PO_4; [7664-38-2]	Ft. Monmouth, NJ 07703
(5) Water; H_2O; [7732-18-5]	U.S.A. December 1981

CRITICAL EVALUATION:

REFERENCES

1. Rammelsberg, F.C. *Wied. Ann.* <u>1882</u>, *16*, 707.
2. Mayer, W. *Ann. Chem. u. Pharm.* <u>1856</u>, *98*, 193.
3. Mayer, W. *Ann. Chim.* <u>1856</u>, *47*, 288.
4. Rosenheim, A.; Reglin, W. *Z. Anorg. Chem.* <u>1921</u>, *120*, 103.
5. Rollet, A.P.; Lauffenburger, R. *Bull. Soc. Chim. France* <u>1934</u>, 146.
6. Latimer, W.M.; Hildebrand, J.H. *Reference Book of Inorganic Chemistry*. 3rd Edit. Macmillan, N.Y. <u>1951</u>.
7. Lange, N.A.; Forker, G.M. *Handbook of Chemistry*. 10th Edit. McGraw-Hill. N.Y. <u>1961</u>.
8. Dean, J.A. *Lange's Handbook of Chemistry*. 11th Edit. McGraw-Hill. N.Y. <u>1973</u>.
9. Robinson, R.A.; Stokes, R.H. *Electrolyte Solutions*. Butterworths. London. <u>1955</u>.
10. Vanderzee, C.E.; Quist, A.S. *J. Phys. Chem.* <u>1961</u>, *65*, 118.
11. Butler, J.N. *Ionic Equilibrium: A Mathematical Approach*. Addison Wesley. Reading, Mass. <u>1964</u>.
12. Gimblett, F.G.R.; Monk, C.B. *Trans. Faraday Soc.* <u>1954</u>, *50*, 965.
13. Corti, H.; Crovetto, R.; Fernandez-Prini, R. *J. Solution Chem.* <u>1979</u>, *8*, 897.
14. Note that for the LiOH association constant, ref. 12 gives K_A = 1.504 and ref. 13 gives K_A = 0.97 mol^{-1} dm^3 at 298 K.
15. Smith, R.M.; Alberty, R.A. *J. Phys. Chem.* <u>1956</u>, *60*, 150. These authors report $K_A(LiHPO_4^-)$ = 5.2 mol^{-1} dm^3 at 298 K.
16. Davies, C.W. *Ion Association*. Butterworths. London. <u>1962</u>.
17. Prideaux, E.B.R. *J. Chem. Soc.* <u>1944</u>, 606.
18. Landolt-Bornstein. *Zahlenwerte und Funktionen*. 3rd Erg. Band III. <u>1936</u>.
19. Bottger, W. *Z. Phys. Chem.* <u>1903</u>, *46*, 596.
20. Milazzo, C. *Electrochemistry*. Elsevier. Amsterdam. <u>1963</u>.
21. Marsh, K.N.; Stokes, R.H. *Austr. J. Chem.* <u>1964</u>, *17*, 740.
22. Moeller, T. *Qualitative Analysis*. McGraw-Hill, N.Y. <u>1958</u>.
23. Jaulmes, P.; Brun, S. *Trav. Soc. Pharm. Montpellier* <u>1965</u>, *25*, 98.
24. de Schulten, M.A. *Bull. Soc. Chim.* <u>1889</u>, *1*, 479.
25. Sanfourche, A.A. *Bull. Soc. Chim.* <u>1938</u>, *5*, 1669.

COMPONENTS:	ORIGINAL MEASUREMENTS:
(1) Lithium phosphate; Li_3PO_4; [10377-52-3] (2) Water; H_2O; [7732-18-5]	Rosenheim, A.; Reglin, W. Z. Anorg. Chem. 1921, 120, 103-19.
VARIABLES: One temperature: 25°C	PREPARED BY: J. Eysseltová and M. Salomon

EXPERIMENTAL VALUES:

The electrolytic conductances of satd Li_3PO_4 slns at 25°C were reported

experiment No.	$10^4\kappa_{sln}$ /S cm^{-1}	experiment No.	$10^4\kappa_{sln}$ /S cm^{-1}
1	20.1	5	9.43
2	11.5	6	9.24
3	10.8	7	9.24
4	9.51	8	9.25

The high κ_{sln} values for expts 1-3 were attributed to impurities and neglected. Based on the data from expts 4-8, the authors reported an ave κ_{sln} = 9.40 x 10^{-4}S cm^{-1} and $\kappa_{salt} = \kappa_{sln} - \kappa_{H_2O}$ = 9.37 x 10^{-4}S cm^{-1}. The soly of Li_3PO_4 was calcd from

$$soly = \frac{1000\kappa_{salt}}{3(\lambda^\infty_{Li} + \lambda^\infty_{PO_4})} = (7.688/3) \times 10^{-3} mol\ dm^{-3} = 2.563 \times 10^{-3}\ mol\ dm^{-3}$$

$\lambda^\infty(Li^+)$ = 39.7 S cm^{2}mol^{-1} and was taken from Kohlraush and Holborn (1). $\lambda^\infty(\frac{1}{3}PO_4^{3-})$ = 82.3 S cm^{2}mol^{-1} was estimated by Böttger (2): both values correspond to 25 C. In the original calculation, the authors neglected to multiply κ_{salt} by 1000, and hence report a solubility too low by this factor. The author's calcns are also subject to rounding off errors amounting to an error of around +1% in the final value for the soly. Additional errors involve the uncertainties in the λ^∞ values. Although these errors are significant, they are relatively minor to the error involved in neglecting the hydrolysis of the PO_4^{3-} ion. The effect of hydrolysis on the calcn of the soly from conductivity data is discussed in detail in the critical evaluation.

AUXILIARY INFORMATION

METHOD/APPARATUS/PROCEDURE:	SOURCE AND PURITY OF MATERIALS:
The soly could not be detd by "standard" methods due to the formation of a fine colloid which could not be removed by filtration. The soly was therefore detd by the conductivity method. Equilibration was attained by shaking at 25°C for 14-21 d. Eight slns were prepared using the same solid phase, but with successive renewal of the water. Initial impurities, as implied by the high κ values of slns 1-3, were assumed to have been completely removed by this washing by the fourth experiment. The electrolytic conductivity of the water was reported to be κ_{H_2O} = 3 x 10^{-6}S cm^{-1}. Based on the results for experiments 4-8, the authors reported an average electrolytic conductivity of κ_{sln} = 9.40 x 10^{-4}S cm^{-1}. However the compilers compute an average value of κ_{sln} = 9.33 x 10^{-4}S cm^{-1}, and the electrolytic conductivity of the salt is then $\kappa_{salt} = \kappa_{sln} - \kappa_{H_2O}$ = 9.30 x 10^{-4}S cm^{-1}	$Li_3PO_4 \cdot 2H_2O$ was pptd from aq H_3PO_4 with excess LiOH. The dihydrate was washed, air dried at about 16°C and analysed with the following results: Li 13.50, 13.66 mass% found (16.67% calcd); PO_4 62.53, 62.46 mass% found (62.58% calcd); H_2O 23.90, 23.78% found (23.72% calcd). Drying at 60°C for several days gave the hemihydrate which analysed as $Li_3PO_4 \cdot \frac{1}{2}H_2O$. Presumably conductivity water was used for prep of slns and washing of ppts. The compilers assume that $Li_3PO_4 \cdot 2H_2O$ was used as the starting material for all experiments.
	ESTIMATED ERROR:
	Nothing specified. The compilers assume the experimental precision to be around ± 1 x 10^{-6} S cm^{-1}. The std dev in κ_{salt} is 4.2 x 10^{-6} S cm^{-1}.
	REFERENCES: 1. Kohlrausch, F.; Holborn, O. Das Leitvermögen der Elektrolyte. II Auflage, 1916, Tab. 8a. 2. Böttger, W. Z. Phys. Chem. 1903, 46, 596.

COMPONENTS:	ORIGINAL MEASUREMENTS:
(1) Lithium phosphate; Li_3PO_4; [10377-52-3] (2) Ammonia; NH_3; [7664-41-7] (3) Water; H_2O; [7732-18-5]	Mayer, W. *Ann. Chem. u. Pharm. <u>1856</u>, 98, 192-212; Ann. Chim. <u>1856</u>, 288- .
VARIABLES: Room temperature: 15 - 18°C	PREPARED BY: J. Eysseltová and M. Salomon

EXPERIMENTAL VALUES: Composition of saturated solutions:

solvent composition	sln mass/g	Li_3PO_4 mass/g	g Li_3PO_4/100 g H_2O [b]	m(Li_3PO_4)/mol kg^{-1} [b]
pure H_2O	45	0.0176	0.0391	0.00338
	45	0.0178	0.0396	0.00342
	75	0.0296	0.0395	0.00341
2 vol H_2O + 1 vol NH_4OH [a]	74.12	0.0174	0.023	0.0014
	74.12	0.0190	0.026	0.0015
	44.47	0.0124	0.028	0.0016
	44.47	0.0117	0.026	0.0015

[a] For this NH_3 sln, the specific gravity = 0.965 (author), but temp not specified.
Assuming temp = 20 C, the NH_3 concn in the final sln is about 1.6 mol kg^{-1} (compilers).

[b] Compilers' calculations. Average values and their standard deviations are given below.

<u>In pure water</u>.   solubility = 0.0394 g/100 g H_2O (σ = 0.002)

= 0.00340 mol kg^{-1} (σ = 0.00002)

<u>In ~ 1.6 mol kg^{-1} NH_3 sln</u>.

solubility = 0.025 g/100 g H_2O (σ = 0.002)

= 0.0015 mol kg^{-1} (σ = 0.0001)

Note: in converting to mol kg^{-1} in 1.6 mol kg^{-1} NH_3 slns, the compilers calculated the mass of water from g(H_2O) = g(sln) - g(NH_3) - g(Li_3PO_4).

AUXILIARY INFORMATION

METHOD/APPARATUS/PROCEDURE:	SOURCE AND PURITY OF MATERIALS:
Each determination consisted of equilibrating solid + liquid for 10-14 days at 15-18°C with frequent shaking. Analysis not described, but probably was either evaporation of a satd sln followed by weighing, or by pptn of $Ba_3(PO_4)_2$ by addn of $Ba(OH)_2$ followed by weighing as described elsewhere in the paper for the stoichiometric analysis of the ppt. Although not stated, it is possible that the approach to equilibrium was from supersaturation (see discussion in the critical evaluation).	Li_3PO_4 pptd from a mixture of Na_2HPO_4, Li_2SO_4, and NH_4OH. The ppt was washed with boiled water until the wash water was free of SO_4^{2-} (tested with $BaCl_2$ sln).
	ESTIMATED ERROR: Nothing specified. The reproducibility appears satisfactory, but the overall accuracy of the solubility is probably no better than 15%.
	REFERENCES:

COMPONENTS:	ORIGINAL MEASUREMENTS:
(1) Lithium phosphate; Li_3PO_4; [10377-52-3] (2) Phosphoric acid; H_3PO_4; [7664-38-2] (3) Lithium oxide; Li_2O; [12057-24-8] (4) Water; H_2O; [7732-18-5]	Rollet, A.P.; Lauffenburger, R. *Bull. Soc. Chim. France* <u>1934</u>, 146-52.
VARIABLES: Temperature and Composition	PREPARED BY: J. Eysseltová

EXPERIMENTAL VALUES:

Composition of saturated solutions of Li_3PO_4 at 20°C.

Li_2O mass %	P_2O_5 mass %	H_3PO_4[a] mass %	H_3PO_4[a] mol/kg H_2O	Li_3PO_4[a] mass %	Li_3PO_4[a] mol/kg H_2O	Solid phase [b]
7.05	0.017	----	----	0.0277	0.0026 (D)	A + B
0.077	0.016	----	----	0.0261	0.0023	B
0.0134	0.021	----	----	0.0343	0.0030	"
0.0165	0.0272	0.00152	0.00015	0.0426	0.0037	"
0.0197	0.0345	0.00456	0.00047	0.0509	0.0044	"
0.0203	0.0360	0.00538	0.00054	0.0524	0.0045	"
0.050	0.116	0.0508	0.00521	0.129	0.0111	"
0.096	0.242	0.123	0.0126	0.249	0.0216	"
0.118	0.305	0.163	0.0167	0.305	0.0264	"
0.150	0.409	0.236	0.0243	0.388	0.0337	"
0.205	0.620	0.407	0.0420	0.530	0.0461	"
0.262	0.875	0.635	0.0657	0.677	0.0593	"
5.73	27.5	25.4	4.34	14.8	2.14	"
7.63	37.1	34.5	7.70	19.7	3.72	"
7.95	38.8	36.2	8.53	20.5	4.10	"
8.45	41.6	38.9	10.14	21.8	4.81 (E)	B + C
7.73	41.7	40.7	10.55	20.0	4.38	C
6.62	43.5	45.5	12.4	17.1	3.96	"

[a] All these values were calculated by the compiler.

[b] The solid phases are: A = $LiOH \cdot H_2O$; B = Li_3PO_4; C = LiH_2PO_4.

(continued next page)

AUXILIARY INFORMATION

METHOD/APPARATUS/PROCEDURE:	SOURCE AND PURITY OF MATERIALS:
Phosphoric acid, lithia and water were placed in glass tubes and sealed with a Hg stirrer. The tubes, equipped with Pt electrodes for conductivity measurements, were placed in a thermostat and stirred for at least 8 h. The attainment of equil was ascertained by the constancy in the conductivity. P_2O_5 was detd in a few cases by titrn with NaOH soln; usually P_2O_5 was detd gravimetrically by ppting as $(NH_4)_3PMo_{12}O_{40}$, re-pptd as $NH_4MgPO_4 \cdot 6H_2O$, calcined and weighed as $Mg_2P_2O_7$. Additional gravimetric analyses were performed when the P_2O_5/Li_2O ratio corresponded to the formulas Li_3PO_4 and LiH_2PO_4. In this case the soln was evaporated and the residue calcined. The total weight of the calcined residue is the sum of P_2O_5 and Li_2O. No other details given.	Nothing specified.
	ESTIMATED ERROR: Temp control at best is ± 0.1 K Soly: exptl error not specified. For binary system $Li_3PO_4-H_2O$ extrapolations give errors of ± 4.5% at 0° and ± 6.7% at 20°C (authors)
	REFERENCES:

COMPONENTS:	ORIGINAL MEASUREMENTS:
(1) Lithium phosphate; Li_3PO_4; [10377-52-3]	Rollet, A.P.; Lauffenburger, R.
(2) Phosphoric acid; H_3PO_4; [7664-38-2]	*Bull. Soc. Chim. France* 1934, 146-52.
(3) Lithium oxide; Li_2O, [12057-24-8]	
(4) Water, H_2O; [7732-18-5]	

EXPERIMENTAL VALUES, cont'd:

Composition of saturated solutions of Li_3PO_4 at 0°C.

Li_2O mass %	P_2O_5 mass %	H_3PO_4[a] mass %	H_3PO_4[a] mol/kg H_2O	Li_3PO_4[a] mass %	Li_3PO_4[a] mol/kg H_2O	Solid phase[b]
6.70	0.00	----	----	0.00	0.00	A
6.72	0.015	----	----	0.024	0.0023 (A)	A + B
5.88	0.020	----	----	0.033	0.0030	B
2.40	0.016	----	----	0.026	0.0023	"
0.49	0.020	----	----	0.033	0.0028	"
0.0088	0.0148	0.0012	0.0001	0.0227	0.0020	"
0.0098	0.0163	0.0011	0.0001	0.0253	0.0021	"
0.0185	0.0375	0.0113	0.0012	0.0478	0.0041	"
0.025	0.058	0.0248	0.0026	0.0646	0.0056	"
0.149	0.388	0.210	0.0216	0.385	0.0334	"
0.167	0.414	0.207	0.0212	0.432	0.0375	"
0.174	0.452	0.240	0.0250	0.450	0.0399	"
0.2285	0.635	0.377	0.0389	0.590	0.0515	"
0.266	0.785	0.502	0.0519	0.687	0.0601	"
0.270	0.80	0.513	0.0531	0.698	0.0610	"
0.330	0.96	0.603	0.0626	0.863	0.0747	"
0.403	1.355	0.989	0.1031	1.04	0.0928	"
0.520	1.88	1.46	0.153	1.34	0.119	"
0.740	2.74	2.17	0.230	1.91	0.172	"
2.55	11.03	9.65	1.18	6.59	0.679	"
3.90	17.2	15.3	2.08	10.1	1.165	"
4.58	20.26	18.0	2.61	11.8	1.46	"
4.97	22.16	19.7	2.99	12.8	1.64	"
5.42	24.30	21.7	3.44	14.0	1.88	"
6.88	31.13	28.0	5.26	17.7	2.82	"
7.55	34.43[c]	31.0	6.40	19.5	3.41	"
8.30[c]	38.0[c]	34.2	7.92	21.4	4.19 (B)	B + C
8.19	37.95	34.5	7.94	21.2	4.12	C
8.08	38.10	34.9	8.07	20.9	4.08	"
7.68	38.20	35.9	8.30	19.8	3.88	"
7.54	38.45	36.6	8.50	19.5	3.83	"
7.18	38.85	37.9	8.90	18.6	3.68	"
4.62	47.5	55.5	17.6	11.9	2.86	"
3.21	53.05	66.2	26.5	8.30	2.81	"
2.74	55.65	70.4	32.8	7.60	2.80	"
2.22	58.8	76.3	43.4	5.74	2.76 (C)	"

[a] All these values were calculated by the compiler.

[b] The solid phases are: A = $LiOH \cdot H_2O$; B = Li_3PO_4; C = LiH_2PO_4.

[c] Read from the intersection of branch lines. For discussion of points A, B, C, D, E see the discussion of the phase diagram in the Critical Evaluation.

By interpolation of the isotherms, the authors report the solubilites of Li_3PO_4 at 0° and 20° in the binary system to be 0.022 ± 0.001 mass % and 0.030 ± 0.002 mass %, respectively. The compiler has calculated these solubility values to be 0.0019 mol kg^{-1} at 0°C and 0.0026 mol kg^{-1} at 20°C.

From the data in the region rich in H_3PO_4 the solubility of LiH_2PO_4 [13453-80-0] was determined to be 55.8 ± 0.1 mass % or 12.15 ± 0.02 mol kg^{-1} (compiler's calculation) at 0°C.

COMPONENTS:	EVALUATOR:
(1) Trisodium Phosphate; Na_3PO_4; [7601-54-9]	
(2) Phosphoric acid; H_3PO_4; [7664-38-2]	J. Eysseltová
	Charles University
(3) Sodium hydroxide; NaOH; [1310-73-2]	Prague, Czechoslovakia
(4) Water; H_2O; [7732-18-5]	July, 1986

CRITICAL EVALUATION:

Solubility data for the $Na_2O-P_2O_5-H_2O$ system have been reported in 13 different publications (1-13). Some of these[2] (1-7) report the solubility in systems in which there is a range of Na/P ratios. Others (8-13) are limited to one Na/P ratio, i.e., the solubility of a given sodium phosphate in water is reported.

Many solid phases have been reported or suggested as being in equilibrium with saturated solutions in the $Na_2O-P_2O_5-H_2O$ system. These are:

NaOH; [1310-73-2]	$Na_2HPO_4 \cdot 2H_2O$; [10028-24-7]
$Na_3PO_4 \cdot 1/4NaOH \cdot 12H_2O$; [12362-10-6]	Na_2HPO_4; [7558-79-4]
$Na_3PO_4 \cdot 1/7NaOH \cdot 12H_2O$; [101056-44-4]	$Na_2HPO_4 \cdot NaH_2PO_4$; [65185-91-3]
$Na_3PO_4 \cdot 12H_2O$; [10101-89-0]	$Na_2HPO_4 \cdot 2NaH_2PO_4 \cdot 2H_2O$; [66905-89-3]
$Na_3PO_4 \cdot 10H_2O$; [10361-89-4]	$NaH_2PO_4 \cdot 4H_2O$; [101056-45-5]
$Na_3PO_4 \cdot 8H_2O$; [60593-59-1]	$NaH_2PO_4 \cdot 2H_2O$; [13472-35-0]
$Na_3PO_4 \cdot 6H_2O$; [15819-50-8]	$NaH_2PO_4 \cdot H_2O$; [10049-21-5]
$Na_3PO_4 \cdot 0.5H_2O$; [60593-58-0]	NaH_2PO_4; [7558-80-7]
Na_3PO_4; [7601-54-9]	$Na_3H_3(PO_4)_2 \cdot 7.5H_2O$; [101056-46-6]
$Na_2HPO_4 \cdot 12H_2O$; [10039-32-4]	$Na_3H_3(PO_4)_2 \cdot 1.5H_2O$; [101917-67-3]
$Na_2HPO_4 \cdot 8H_2O$; [67417-37-2]	$NaH_2PO_4 \cdot H_3PO_4$; [14887-48-0]
$Na_2HPO_4 \cdot 7H_2O$; [7782-85-6]	$H_3PO_4 \cdot 0.5H_2O$; [16271-20-8]

The conditions under which these phosphates exist is discussed in the Critical Evaluation of the respective binary systems.

$Na_3PO_4-NaOH-H_2O$ system. Menzel and von Sahr (1) studied this system at 298 K. They found that as the mole ratio Na_2O/P_2O_5 in the saturated solutions varied from 2.69 to 3.68, the same ratio in the solid phase increased from 3.11 to 3.22. Thereafter as the Na/P ratio in the solution increased to about 145, the same ratio in the solid phase changed only from 3.22 to 3.24. They concluded that the equilibrium solid phases were solid solutions although X-ray diffraction diagrams of four such solid phases showed little difference among them.

Kobe and Leipper (2) suggested that the commercial trisodium phosphate has the formula $Na_3PO_4 \cdot 1/7NaOH \cdot 12H_2O$. Later Kobe returned to this problem and studied systems of high Na_2O/P_2O_5 ratios (3). He and his co-worker found the system to be a complex one. The anyhdrous form of Na_3PO_4 as well as the hemihydrate, the hexahydrate and the octa-hydrate were identified as equilibrium solid phases. An alkaline complex salt was also observed. The complex was studied further and, in agreement with Bell (14), they suggested that at 273-333 K two different complexes are present: $Na_3PO_4 \cdot 1/7NaOH \cdot 12H_2O$ and $Na_3PO_4 \cdot 1/4NaOH \cdot 12H_2O$ but they included only the latter in their Tables. At 353-373 K they found only the hydrates of Na_3PO_4 in the highly alkaline solutions. This agrees with the opinion of others (4) who mention no complex formation in this system at 423, 523 and 623 K.

The transition of the different hydrates and the identification of these hydrates at increasing NaOH concentrations cannot be evaluated because of lack of corroborating work by others. However, Ravich and Shcherbakova (23) did present X-ray evidence for the formation of solid solutions m $Na_3PO_4 \cdot n$ Na_2HPO_4. These solid solutions are reported to coexist with saturated solutions having Na/P ratios even greater than 3 at 523, 573 and 638 K. This is in agreement with the observations of Broadbent, et al. (18) who found equilibrium solid phases in which the Na/P ratio varied from 2.64 to 2.82 at 524 and 573 K.

As noted above, this system has been studied at temperatures of 293 K (1), 298 K (2, 3, 5) and at elevated temperatures 423, 523 and 623 K (4) and 523, 573 and 638 K (23). The lower temperature results are shown on Figure 1. The data agree fairly well with each other except for one data point (5) which is obviously incorrect. Except for this one data point these results can be accepted tentatively. A similar comparison cannot

(continued next page)

COMPONENTS:	EVALUATOR:
(1) Trisodium phosphate; Na$_3$PO$_4$; [7601-54-9] (2) Phosphoric acid; H$_3$PO$_4$; [7664-38-2] (3) Sodium hydroxide; NaOH; [1310-73-2] (4) Water; H$_2$O; [7732-18-5]	J. Eysseltová Charles University Prague, Czechoslovakia July, 1986

CRITICAL EVALUATION:

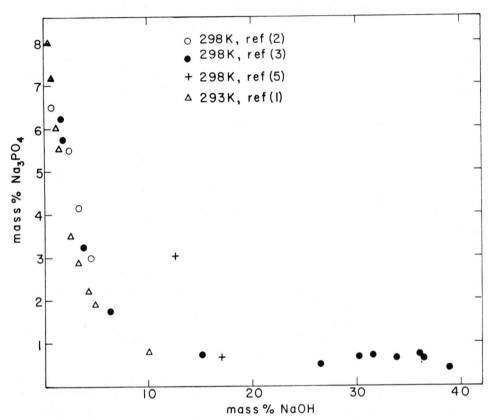

Figure 1. Solubility of Na$_3$PO$_4$ in aqueous NaOH.

COMPONENTS:	EVALUATOR:
(1) Trisodium phosphate; Na_3PO_4; [7601-54-9] (2) Phosphoric acid; H_3PO_4; [7664-38-2] (3) Sodium hydroxide; NaOH; [1310-73-2] (4) Water; H_2O; [7732-18-5]	J. Eysseltová Charles University Prague, Czechoslovakia July, 1986

CRITICAL EVALUATION:

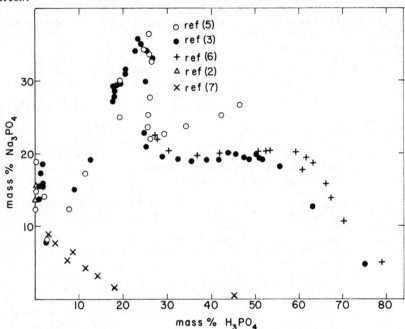

Figure 2. Solubility of Na_3PO_4 in aqueous H_3PO_4 at 298 K.

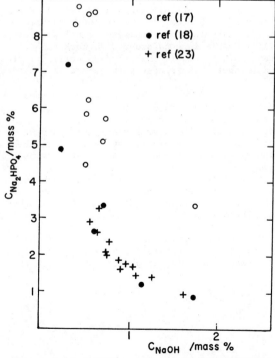

Figure 3. Solubility of Na_2HPO_4 in aqueous NaOH at 573 K.

COMPONENTS:	EVALUATOR:
(1) Trisodium Phosphate; Na$_3$PO$_4$; [7601-54-9] (2) Phosphoric acid; H$_3$PO$_4$; [7664-38-2] (3) Sodium hydroxide; NaOH; [1310-73-2] (4) Water; H$_2$O; [7732-18-5]	J. Eysseltová Charles University Prague, Czechoslovakia July, 1986

CRITICAL EVALUATION: (cont'd)

be made for the results obtained at elevated temperatures. The data points of Ravich and Shcherbakova (23) are concentrated in a narrow range of Na/P ratios and can be compared with only one data point of Broadbent, et al. (18) and of Panson, et al. (17).

Na$_3$PO$_4$–H$_3$PO$_4$–H$_2$O system. Solubility studies have been made at 298 K in systems which have a wide range of Na/P ratios (3, 5). Studies at 298 K in a more limited range have also been made: the Na$_3$PO$_4$–Na$_2$HPO$_4$–H$_2$O system (2); the NaH$_2$PO$_4$–H$_3$PO$_4$–H$_2$O system (6); and the Na$_2$HPO$_4$–H$_3$PO$_4$–H$_2$O system (7). The Na$_3$PO$_4$–Na$_2$HPO$_4$–H$_2$O system has also been studied at 293 K (1).

Figure 2 shows the solubility results obtained at 298 K. It is apparent that the work of Beremzhanov, et al. (7) ignores the existence of NaH$_2$PO$_4$ and the results are clearly incorrect. The data in the region where P/Na>1 (5) probably are for super-saturated solutions. For this region the results of Wendrow and Kobe (3) and Lilich, et al. (6) are tentatively accepted. Between Na$_2$HPO$_4$ and NaH$_2$PO$_4$ the solubility data of Wendrow and Kobe (3) and D'Ans and Schreiner (5) are very close to each other and are tentatively accepted as describing the solubility in this region. With respect to the identity of the solid phases in equilibrium with these saturated solutions, more work is needed before a decision can be made.

Phosphates in which the Na/P ratio is other than 3, 2 or 1, e.g., Na$_2$HPO$_4$·NaH$_2$PO$_4$ and Na$_2$HPO$_4$·2NaH$_2$PO$_4$·2H$_2$O, have been reported as existing in a very narrow concentration range (3). Their existence has not been confirmed by others, and more work is needed before a judgment about the existence of such phosphates can be made.

The hydrates Na$_3$H$_3$(PO$_4$)$_2$·7.5H$_2$O and Na$_3$H$_3$(PO$_4$)$_2$·1.5H$_2$O have been reported to exist in equilibrium with very concentrated solutions having a pH = 7 (15). There is no other report on the existence of these phosphates. They probably are metastable (5).

The existence of NaH$_2$PO$_4$·H$_3$PO$_4$ in strongly acid solutions has been reported by several investigators (3,6,8). The solubility of this substance has been measured over a range of temperatures. As a result of such a study Paravano and Mieli (8) state that the system is glass-forming in the temperature range 235 to 307 K. By extrapolating their values it appears that the composition of the system at 323 K is 45 mass% NaH$_2$PO$_4$ and 37.4 mass% H$_3$PO$_4$. This is in good agreement with the value reported by Lilich, et al. (6), especially if their value of 28.7 mass% H$_3$PO$_4$ at 323 K is a typographical error and the correct value should be 38.7 mass%. Their reported value for NaH$_2$PO$_4$ is 46 mass%. Paravano and Mieli (8) state that at temperatures below about 373 K, NaH$_2$PO$_4$·H$_3$PO$_4$ has an incongruent solubility. This is consistent with work reported by others (3,6). However, there are significant differences in the 298 K solubility isotherms reported for this substance (3,6) and further work is necessary before this matter can be resolved.

Solubility measurements have also been made at elevated temperatures (17-19, 23). Some of the data are shown on Figure 3. At 573 K the solubility results of Ravich, et al. (23) and of Panson, et al. (17) agree fairly well with each other while the values reported by Broadbent, et al. (18) have a significant amount of scatter and generally report a larger NaOH content. Therefore, the data of Braodbent, et al. (18) should probably be rejected because of an apparent systematic error. There is another report of solubility data under these conditions (20) but the data are presented only in graphical form. The author states that tetrasodium diphosphate and sodium triphosphate are equilibrium solid phases, but this seems unlikely in view of the conditions for the preparation of pyro- and tri-phosphates (21). In still another report (22), Na$_4$P$_2$O$_7$ was the only solid phase found at 573 K and its existence was estimated to be limited to the 563-573 K temperature interval.

Marshall (19) reviewed all this work and considered that the reported compositions of the saturated solutions were in fairly good agreement.

Liquid-liquid immiscibility is the phenomenon that characterizes this system at high temperatures.

(continued next page)

COMPONENTS:	EVALUATOR:
(1) Trisodium Phosphate; Na$_3$PO$_4$; [7601-54-9] (2) Phosphoric acid; H$_3$PO$_4$; [7664-38-2] (3) Sodium hydroxide; NaOH; [1310-73-2] (4) Water; H$_2$O; [7732-18-5]	J. Eysseltová Charles University Prague, Czechoslovakia July, 1986

CRITICAL EVALUATION: (cont'd)

References

1. Menzel, H.; von Sahr, E. *Z. Elektrochem.* <u>1937</u>, *43*, 104.
2. Kobe, K.A.; Leipper, A. *Ind. Eng. Chem.* <u>1940</u>, *32*, 198.|
3. Wendrow, B.; Kobe, K.A. *Ind. Eng. Chem.* <u>1952</u>, *44*, 1439.
4. Schroeder, W.C.; Berk, A.A.; Gabriel, A. *J. Am. Chem. Soc.* <u>1937</u>, *59*, 1783.
5. D'Ans, J.; Schreiner, O. *Z. Anorg. Chem.* <u>1911</u>, *75*, 95.
6. Lilich, L.S.; Vanjusheva, L.N.; Chernykh, L.V. *Zh. Neorg. Khim.* <u>1971</u>, *16*, 2782.
7. Beremzhanov, B.A.; Savich, R.F.; Kunanbaeva, G.S. *Prikl. Teor. Khim.* <u>1978</u>, 8.
8. Paravano, N.; Mieli, A. *Gaz. Chim. Ital.* <u>1908</u>, *38*, 535.
9. Imadsu, A. *Mem. Col. Sci. Emp. Kyoto* <u>1911-12</u>, *3*, 257.
10. Shiomi, Ts. *Mem. Col. Sci. Emp. Kyoto* <u>1908</u>, 406.
11. Hammick, D.L.; Goadby, H.K.; Booth, H. *J. Chem. Soc.* <u>1920</u>, *67*, 1589.
12. Menzel, G.; Gabler, C. *Z. Anorg. Chem.* <u>1928</u>, *177*, 187.
13. Mulder, G.J. *Bijdragen tot de geschiedenis van het scheikundig gebonden water,* Rotterdam <u>1894</u>; quoted in Landolt-Bornstein, p. 558.
14. Bell, R.N. *Ind. Eng. Chem.* <u>1949</u>, *41*, 2901.
15. Filhol, E.; Senderens, J.B. *Compt. rend.* <u>1882</u>, *94*, 649.
16. Staudenmayer, L. *Z. Anorg. Chem.* <u>1894</u>, *5*, 395.
17. Panson, A.J.; Economy, J.; Liu, Chin-sun; Bulischeck, T.S.; Lindsay Jr., W.T. *J. Electrochem. Soc.* <u>1975</u>, *122*, 915.
18. Broadbent, D.; Lewis, G.G.; Wetton, E.A.M. *J. Chem. Soc., Dalton trans.* <u>1977</u>, 464.
19. Marshall, W.L. *J. Chem. Eng. Data* <u>1982</u>, *27*, 175.
20. Wetton, E.A.M. *Power Industry Research* <u>1981</u>, *1*, 151.
21. Osterheld, R.; Audrieth, L. *J. Phys. Chem.* <u>1952</u>, *56*, 38.
22. Taylor, P.; Tremaine, P.R.; Bailey, M.G. *Inorg. Chem.* <u>1979</u>, *18*, 2947.
23. Ravich, M.I.; Shcherbakova, L.G. *Izv. Sektora Fiz.-Khim. Analiza, Inst. Obshch. Neorgan., Khim. Akad. Nauk SSSR* <u>1955</u>, *26*, 248.

COMPONENTS:	ORIGINAL MEASUREMENTS:
(1) Sodium dihydrogenphosphate; NaH_2PO_4; [7558-80-7] (2) Phosphoric acid; H_3PO_4; [7664-38-2] (3) Water; H_2O; [7732-18-5]	Paravano, N.; Mieli, A. *Gazz. Chim. Ital.* <u>1908</u>, 38, 535-44.

VARIABLES:	PREPARED BY:
Composition and temperature. One ratio NaH_2PO_4/H_3PO_4 = 1.	J. Eysseltová

EXPERIMENTAL VALUES:

Saturation temperatures of solutions of $NaH_2PO_4 \cdot H_3PO_4$ in water.

	$NaH_2PO_4 \cdot H_3PO_4$		NaH_2PO_4[a]		H_3PO_4[a]		
$t/°C.$	mass%	mol%	mass%	mol/kg	mass%	mol/kg	solid phase[b]
0	0	0	0	0	0	0	ice
-5.7	20.77	1.81	11.43	1.20	9.34	1.20	"
-7.9	26.92	2.95	14.82	1.69	12.10	1.69	"
-11.4	34.15	4.11	18.80	2.38	15.35	2.38	"
-38[c]	56.66	9.75	31.19	6.00	25.47	6.00	"
----	70.52	16.50	38.82	10.97	31.70	10.97	
34.0	80.46	25.39	44.29	18.89	36.17	18.89	NaH_2PO_4
41.0	81.82	27.11	45.04	20.64	36.78	20.64	"
51.7	83.68	29.75	46.06	23.52	37.61	23.52	"
79.7	87.48	36.62	48.16	32.05	39.32	32.05	"
85.0	88.65	39.22	48.80	35.83	39.85	35.83	"
101.7	91.47	46.98	50.35	49.18	41.12	49.18	$NaH_2PO_4 \cdot H_3PO_4$
104.5	92.67	51.09	51.01	57.99	41.66	57.99	"
110.0	95.79	65.28	52.73	87.15	43.06	87.15	"
110.7	95.86	65.68	52.77	106.2	43.09	106.2	"
119.0	97.99	80.12	53.94	223.6	44.05	223.6	"
126.5	100	100	55.04	----	44.96	----	"

[a] These values were calculated by the compiler.

[b] The phases were not given by the authors. The compiler derived them from a graph.

[c] The temperature was about -38°C.

AUXILIARY INFORMATION

METHOD/APPARATUS/PROCEDURE:	SOURCE AND PURITY OF MATERIALS:
Saturation temperatures were determined visually as the temperature at which the last crystal disappeared.	$NaH_2PO_4 \cdot H_3PO_4$ was prepared from an equimolar mixture of concentrated solutions of NaH_2PO_4 and H_3PO_4 by slow crystallization. The analysis was:

	found	calculated
P_2O_5	65.37%	65.12%
Na_2O	14.41	14.23

ESTIMATED ERROR:

Nothing is stated.

REFERENCES:

COMPONENTS:	ORIGINAL MEASUREMENTS:
(1) Trisodium phosphate; Na$_3$PO$_4$; [7601-54-9]	Schroeder, W.C.; Berk, A.A.; Gabriel, A.
(2) Sodium hydroxide; NaOH; [1310-73-2]	J. Am. Chem. Soc. 1937, 59, 1783-90.
(3) Water; H$_2$O; [7732-18-5]	

VARIABLES:	PREPARED BY:
Three temperatures: 150, 250, 350°C	J. Eysseltová
Composition	

EXPERIMENTAL VALUES:

Solubility of sodium phosphate in sodium hydroxide solutions.

concn of NaOH			concn of Na$_3$PO$_4$			C$_{H_2O}$
g/100g H$_2$O	mass%[a]	mol/kg[a]	g/100g H$_2$O	mass%[a]	mol/kg[a]	mass%[a]
			temp = 150°C.			
----	----	----	82	45.05	5.00	54.95
8.2	5.21	2.07	49.0	31.77	3.07	63.01
20.0	14.22	5.04	20.6	15.28	1.32	70.50
			temp = 250°C.			
----	----	----	8.6	7.92	0.52	92.08
8.2	7.12	2.05	7.0	6.13	0.43	86.75
20.6	16.34	5.16	5.5	4.51	0.35	79.15
29.5	21.82	7.40	5.7	4.47	0.37	73.71
			temp = 350°C.			
----	----	----	0.15	0.15	0.01	99.85
8.0	7.38	2.00	0.44	0.41	0.03	92.21
21.9	17.68	5.48	2.0	1.67	0.13	80.65
21.3	17.25	5.33	2.2	1.84	0.14	80.91

[a] All these values were calculated by the compiler.

AUXILIARY INFORMATION

METHOD/APPARATUS/PROCEDURE:	SOURCE AND PURITY OF MATERIALS:
Self-constructed high temperature solubility bomb with sampler ensuring the sampling at the operating temperature. The time of equilibration is not specified. Phosphate determinations were made by a colorimetric method using aminonaphthol-sulfonic acid (1). Hydroxide was determined by titration to the methyl red end-point (2 equivalents/1 mol of phosphate present being deducted).	Merck chemically pure Na$_3$PO$_4$·12H$_2$O was used. The actual phosphate content of this material was determined by analysis (the results not given - compiler). If necessary, the dodecahydrate was dried at 120°C to give approximately the monohydrate or was re-crystallized at 250°C to give anhydrous salt. NaOH - nothing specified.

	ESTIMATED ERROR:
	Phosphate determination: the error not greater than 1%. Nothing else given.

	REFERENCES:
	1. Fiske, C.H.; Subbarow, J.T. J. Biol. Chem. 1925, 66, 375.

COMPONENTS:	ORIGINAL MEASUREMENTS:
(1) Disodium hydrogenphosphate; Na_2HPO_4; [7558-79-4] (2) Sodium hydroxide; NaOH; [1310-73-2] (3) Water; H_2O; [7732-18-5]	Ravich, M.I.; Shcherbakova, L.G. *Izv. Sektora Fiz. Khim. Analiza, Inst. Obsch. Neorg. Khim. Akad. Nauk SSSR* 1955, *26*, 248-58.

VARIABLES:	PREPARED BY:
Composition at 523, 573 and 638 K.	J. Eysseltová

EXPERIMENTAL VALUES:

Part 1. Composition of the saturated liquid phase in the Na_2HPO_4-NaOH-H_2O system.

g ion/100 g soln		g ion/100 g ion		mass%[a]			mol/kg[a]	
PO_4^{3-}	Na^+	PO_4^{3-}	Na^+	Na_2HPO_4	NaOH	H_2O	Na_2HPO_4	NaOH
				temp = 638 K				
0.0092	0.0185	33.3	66.7	1.31	0.004	98.67	0.09	0.001
0.0044	0.0089	32.9	67.1	0.62	0.004	99.37	0.04	0.001
0.0025	0.0055	31.1	68.9	0.36	0.02	99.62	0.02	0.005
0.0035	0.0081	30.4	69.6	0.50	0.04	99.45	0.04	0.01
0.0036	0.0095	27.8	72.2	0.51	0.09	99.40	0.04	0.02
0.0020	0.0058	25.8	74.2	0.28	0.07	99.64	0.02	0.02
0.0022	0.0081	21.6	78.4	0.31	0.15	99.54	0.02	0.04
0.0016	0.0076	17.4	82.6	0.23	0.18	99.60	0.02	0.04
0.0017	0.0116	12.5	87.5	0.24	0.33	99.43	0.02	0.08
0.0010	0.0277	3.6	96.4	0.14	1.03	98.83	0.01	0.26
				temp = 573 K				
0.0202	0.0542	27.1	72.9	2.88	0.55	96.58	0.21	0.14
0.0228	0.0621	26.8	73.2	3.24	0.66	96.10	0.24	0.17
0.0185	0.0524	26.1	73.9	2.63	0.62	96.75	0.19	0.16
0.0166	0.0525	24.0	74.0	2.36	0.77	96.87	0.17	0.20
0.0102	0.0468[b]	17.9	82.1	1.45	1.06	97.49	0.10	0.27
0.0067	0.0535	11.2	88.8	0.95	1.60	97.44	0.07	0.41
0.0057	0.0812	6.5	93.5	0.81	2.80	96.40	0.06	0.72
0.0067	0.1112	5.7	94.3	0.95	3.91	95.13	0.07	1.03

(continued next page)

AUXILIARY INFORMATION

METHOD/APPARATUS/PROCEDURE:	SOURCE AND PURITY OF MATERIALS:
The apparatus has been described previously (1). Samples of the saturated liquid phase were removed after 1 to 2 hours of equilibration, filtered through a silver wire mat, and analyzed volumetrically (2). The samples were also analyzed gravimetrically with phosphate being determined as $Mg_2P_2O_7$ and sodium as sodium zincuranylacetate.	Chemically pure Na_2HPO_4 was recrystallized. The NaOH was supplied by a solution, about 50%, of chemically pure NaOH. It contained less than 0.1% Na_2CO_3.

	ESTIMATED ERROR:
	No indication is given.

REFERENCES:

1. Ravich, M.I.; Borovaya, F.E.; Luk'yanova, E.I.; Elenevskaya, V.M. *Izv. Sektora Fiz.-Khim. Analiza, Inst. Obsch. Neorg. Khim. Akad. Nauk SSSR* 1954, *24*, 280.
2. *Reaktivu Neorganicheskie. Sbornik Tekhnicheskikh Uslovíu (Inorganic Reactives Technical Conditions)*, Standartgiz, 1951. p. 141 (TU MKhP. 1963-49).

COMPONENTS	ORIGINAL MEASUREMENTS
(1) Disodium hydrogenphosphate; Na$_2$HPO$_4$; [7558-79-4]	Ravich, M.I.; Shcherbakova, L.G.
(2) Sodium hydroxide; NaOH; [1310-73-2]	*Izv. Sektora Fiz. Khim. Analiza, Inst. Obsch. Neorg. Khim. Akad. Nauk SSSR* 1955, 26, 248-58.
(3) Water; H$_2$O; [7732-18-5]	

EXPERIMENTAL VALUES cont'd.

g ion/100 g soln		g ion/100 g ion			mass%[a]			mol/kg[a]	
PO$_4^{3-}$	Na$^+$	PO$_4^{3-}$	Na$^+$	Na$_2$HPO$_4$	Na$_2$HPO$_4$	NaOH	H$_2$O	Na$_2$HPO$_4$	NaOH

temp = 523 K

0.0463	0.1666	21.8	78.2	6.58	2.96	90.46	0.51	0.82	

[a] These values were calculated by the compiler.

[b] This equilibrium was established by starting with a supersaturated solution.

Part 2. Composition of saturated solutions for systems in which the beginning P/Na ratio is 1/3.

g ion/100 g soln		g ion/100 g ion			mass%[a]			mol/kg[a]	
PO$_4^{3-}$	Na$^+$	PO$_4^{3-}$	Na$^+$	Na$_2$HPO$_4$	NaOH	H$_2$O	Na$_2$HPO$_4$	NaOH	

temp = 638 K

0.0014	0.0073	16.2	83.8	0.20	0.18	99.62	0.01	0.04
0.0010	0.0079	11.8	88.2	0.14	0.14	99.62	0.01	0.06
0.0010	0.0082	11.4	88.6	0.14	0.25	99.61	0.01	0.06
0.0008	0.0086	8.3	91.7	0.11	0.28	99.61	0.01	0.07

temp = 573 K

0.0147	0.0475	23.6	76.4	2.09	0.72	97.19	0.15	0.19
0.0146	0.0473	23.6	76.4	2.07	0.72	97.20	0.15	0.19
0.0132	0.0485	21.4	78.6	1.88	0.88	97.24	0.14	0.23
0.0123	0.0483	20.3	79.7	1.75	0.95	97.30	0.13	0.24
0.0112	0.0446	20.1	79.9[b]	1.59	0.89	97.52	0.11	0.26
0.0118	0.0491	19.4	80.6	1.68	1.02	97.30	0.12	0.26
0.0098	0.0507	16.2	83.8	1.39	1.24	97.36	0.10	0.32

temp = 523 K

0.0542	0.1641	24.8	75.2	7.70	2.23	90.07	0.60	0.62
0.0484	0.1527	24.0	76.0	6.88	2.24	90.88	0.53	0.61
0.0480	0.2547	23.7	76.3	6.82	2.35	90.83	0.53	0.65

[a] These values were calculated by the compiler.

[b] Equilibrium was established by starting with a supersaturated solution.

COMPONENTS:	ORIGINAL MEASUREMENTS:
(1) Sodium dihydrogenphosphate; NaH$_2$PO$_4$; [7558-80-7] (2) Phosphoric acid; H$_3$PO$_4$; [7664-38-2] (3) Water; H$_2$O, [7732-18-5]	Lilich, L.S.; Vanjusheva, L.N.; Chernykh, L.V. *Zh. Neorg. Khim.* 1971, 16, 2782-9.

VARIABLES:	PREPARED BY:
Composition and temperature.	J. Eysseltová

EXPERIMENTAL VALUES:

Solubility in the NaH$_2$PO$_4$–H$_3$PO$_4$–H$_2$O system.

NaH$_2$PO$_4$		H$_3$PO$_4$		H$_2$O	
mass%	mol/kg	mass%	mol/kg	mass%[a]	solid phase

temp. = 0°C.

mass%	mol/kg	mass%	mol/kg	mass%	solid phase
38.7	5.26	----	----	62.3	NaH$_2$PO$_4$·2H$_2$O
37.00	5.30	5.00	0.90	58.0	"
33.20	5.3	14.50	2.8	52.3	"
33.0	5.8	19.6	4.2	47.4	"
33.6	6.9	25.7	6.5	40.7	"
34.5	7.8	28.8	8.0	36.7	"
36.9	10.5	33.9	11.9	29.2	"
39.2	13.6	36.9	15.7	23.9	"
37.9	13.6	38.8	17.0	23.3	NaH$_2$PO$_4$·H$_3$PO$_4$
29.7	10.1	45.8	19.0	24.5	"
20.4	7.2	55.9	24.1	23.7	"
17.8	6.8	60.4	28.3	21.8	"
8.1	3.7	73.9	41.9	18.0	"
5.4	2.8	78.6	49.9	16.0	"
3.9	2.2	81.2	55.6	14.9	"

(continued next page)

AUXILIARY INFORMATION

METHOD/APPARATUS/PROCEDURE:	SOURCE AND PURITY OF MATERIALS:
The isothermal method was used with equilibrium being reached in 10-12 hours. Phosphoric acid was determined alkalimetrically, the sum of H$_3$PO$_4$ and NaH$_2$PO$_4$ was determined alkalimetrically after ion exchange. The composition of the solid phases was determined by Schreinemakers' method. In the starting materials, H$_3$PO$_4$ was determined gravimetrically and alkalimetrically using bromcresolgreen as indicator. NaH$_2$PO$_4$ was determined alkalimetrically after ion exchange using ionex KU-2.	Chemically pure 90% H$_3$PO$_4$ was used. The NaH$_2$PO$_4$ was dried at 80-100°C.
	ESTIMATED ERROR: The analyses had a precision of ±0.8% relatively. The temperature control was: 0 ± 0.1°C; 25 ± 0.05°C; 50 ± 0.1°C.
	REFERENCES:

COMPONENTS:	ORIGINAL MEASUREMENTS:
(1) Sodium dihydrogenphosphate; NaH$_2$PO$_4$; [7558-80-7]	Lilich, L.S.; Vanjusheva, L.N.; Chernykh, L.V.
(2) Phosphoric acid; H$_3$PO$_4$; [7664-38-2]	_Zh. Neorg. Khim._ 1971, _16_, 2782-9.
(3) Water; H$_2$O; [7732-18-5]	

EXPERIMENTAL VALUES cont'd:

Solubility in the NaH$_2$PO$_4$-H$_3$PO$_4$-H$_2$O system.

NaH$_2$PO$_4$		H$_3$PO$_4$		H$_2$O	
mass%	mol/kg	mass%	mol/kg	mass%[a]	solid phase

temp. = 25°C.

mass%	mol/kg	mass%	mol/kg	mass%	solid phase
49.4	8.13	----	----	50.6	NaH$_2$PO$_4$·2H$_2$O
48.8	8.1	1.1	0.2	50.1	"
45.5	7.8	5.7	1.2	48.8	"
43.8	8.5	13.3	3.2	42.9	"
44.4	9.8	18.0	4.9	37.6	"
44.5	10.6	20.5	6.0	30.5	"
45.7	13.4	25.8	9.3	29.5	NaH$_2$PO$_4$
45.5	14.6	28.6	11.2	25.9	"
44.8	16.2	32.1	14.2	23.1	"
44.6	18.2	34.9	17.4	20.5	"
43.2	18.9	37.8	20.2	19.0	"
40.1	18.6	39.0	21.1	20.9	NaH$_2$PO$_4$+ NaH$_2$PO$_4$·H$_3$PO$_4$
41.5	19.3	40.6	23.2	17.9	NaH$_2$PO$_4$·H$_3$PO$_4$
35.1	15.9	46.5	25.8	18.4	"
31.7	14.3	49.8	27.4	18.5	"
23.2	10.3	57.9	31.4	18.9	"
11.7	6.3	73.0	46.5	15.3	"

temp. = 50°C.

mass%	mol/kg	mass%	mol/kg	mass%	solid phase
62.6	13.95	----	----	37.4	NaH$_2$PO$_4$
60.5	13.9	3.3	0.9	36.2	"
54.3	14.5	14.6	4.8	31.1	"
52.0	14.7	18.5	6.4	29.5	"
49.3	15.5	24.2	9.3	26.5	"
45.2	19.0	35.0	18.0	19.8	"
46.9	16.0	28.7	12.0	24.4	"
43.8	20.2	38.1	21.5	18.1	"
43.7	21.7	39.6	24.0	19.7	"
41.5	23.4	43.7	29.0	14.8	NaH$_2$PO$_4$+ NaH$_2$PO$_4$·H$_3$PO$_4$
40.5	22.5	44.5	29.1	15.0	NaH$_2$PO$_4$·H$_3$PO$_4$
37.2	20.4	47.4	32.0	15.4	"
30.2	17.3	55.3	38.8	14.5	"
22.6	14.8	64.7	52.0	12.7	"
16.6	11.9	71.7	62.5	11.7	"
13.6	10.4	75.4	70.4	11.0	"

[a]These values were calculated by the compiler.

COMPONENTS:	ORIGINAL MEASUREMENTS:
(1) Disodium hydrogenphosphate; Na$_2$HPO$_4$; [7558-79-4] (2) Sodium hydroxide; NaOH; [1310-73-2] (3) Water; H$_2$O; [7732-18-5]	Broadbent, D.; Lewis, G.G.; Wetton, E.A.M. J. Chem. Soc., Dalton Trans. 1977, 464-8.

VARIABLES:	PREPARED BY:
Composition at 573 and 524 K.	J. Eysseltová

EXPERIMENTAL VALUES:

Solubility in the Na$_2$HPO$_4$–NaOH–H$_2$O system.

Na$_2$O mass%	P$_2$O$_5$ mass%	Na$_2$O/P$_2$O$_5$ mol ratio	Na$_2$HPO$_4$[a] mass%	mol/kg	NaOH[a] mass%	mol/kg	solid phase Na$_2$O/P$_2$O$_5$ mol ratio
\multicolumn{8}{c}{temp = 573 K}								
1.70	0.43	9.05	0.86	0.062	1.71	0.44	2.82
1.40	0.60	5.34	1.20	0.086	1.13	0.29	2.81
1.62	1.32	2.80	2.64	0.19	0.60	0.16	2.73
2.00	1.97	2.33	3.34	0.24	0.70	0.18	2.76
2.29	2.43	2.15	4.87	0.36	0.22	0.056	2.75
3.36	3.59	2.14	7.19	0.55	0.29	0.078	2.76
3.36	3.59	2.14	7.19	0.55	0.29	0.078	2.66
\multicolumn{8}{c}{temp = 524 K}								
6.51	2.15	6.94	4.31	0.34	5.98	1.67	2.75
4.55	3.15	3.31	6.31	0.48	2.32	0.64	2.68
10.5	9.99	2.41	20.01	1.81	2.29	0.74	2.64

[a] These values were calculated by the compiler.

Note: Other data are given in the article but they are presented in graphical form only.

AUXILIARY INFORMATION

METHOD/APPARATUS/PROCEDURE:	SOURCE AND PURITY OF MATERIALS:
The furnace was made from a cylindrical block of aluminum alloy fitted with electric heaters mounted so that it could be rocked. The autoclaves, filters, valves and tubing were made of stainless steel Type 316. Temperatures were measured with chromel-alumel thermocouples connected to a Comark "Electronic Thermometer." Sodium was determined by means of a specific-ion electrode (1). Phosphate was determined colorimetrically by the molybdate-vanadate method. A Unicam SP 1800 spectrometer was used.	The chemicals were of AnalaR quality. The water was deionized and had a sodium content less than 2×10^{-7} mol dm^{-3}.
	ESTIMATED ERROR:
	The temperatures had an accuracy of ± 0.5 K. Most of the experimental points are accurate to ± 5-6%. In the more concentrated solutions, the errors are as high as 8-10%.
	REFERENCES:
	1. Webber, H.M.; Wilson, A.L. Analyst 1969, 94, 209.

COMPONENTS:	ORIGINAL MEASUREMENTS:
(1) Disodium hydrogenphosphate; Na$_2$HPO$_4$; [7558-79-4] (2) Phosphoric acid; H$_3$PO$_4$; [7664-38-2] (3) Water; H$_2$O; [7732-18-5]	Beremzhanov, B.A.; Savich, R.F.; Kunanbaeva, G.S. *Prikl. Teor. Khim.* <u>1978</u>, 8-14.

VARIABLES:	PREPARED BY:
Composition at 25°C.	J. Eysseltová

EXPERIMENTAL VALUES:

Solubility in the Na$_2$HPO$_4$–H$_3$PO$_4$–H$_2$O system at 25°C.

Na$_2$HPO$_4$			H$_3$PO$_4$				refr.	solid
mass%	mol%	mol/kg[a]	mass%	mol%	mol/kg[a]	pH	index	phase
0.66	0.13	0.08	44.22	12.82	8.19	----	1.438	Na$_2$HPO$_4$
2.26	0.34	0.20	17.71	3.89	2.26	1.00	1.452	"
4.19	0.63	0.36	13.60	2.98	1.72	1.14	1.460	"
5.42	0.79	0.45	10.48	2.22	1.27	1.47	1.469	"
6.96	0.99	0.56	5.74	1.18	0.67	2.64	1.484	"
8.61	1.26	0.71	6.57	1.39	0.79	2.23	1.478	"
10.27	1.46	0.82	2.00	0.41	0.23	5.12	1.508	"
12.00	1.70	0.96	----	----	----	9.93	1.520	"

[a] The mol/kg H$_2$O values were calculated by the compiler.

AUXILIARY INFORMATION

METHOD/APPARATUS/PROCEDURE:	SOURCE AND PURITY OF MATERIALS:
Solutions of phosphoric were saturated with Na$_2$HPO$_4$. Equilibrium was established in three days. Sodium was determined using flame photometry, phosphorus was determined gravimetrically. No further details are given.	No information is given.
	ESTIMATED ERROR: No details are given.
	REFERENCES:

COMPONENTS:	ORIGINAL MEASUREMENTS:
(1) Sodium dihydrogenphosphate; NaH_2PO_4; [7558-80-7] (2) Sodium hydroxide; NaOH; [1310-73-2] (3) Water; H_2O; [7732-18-5]	Marshall, W.L. J. Chem. Eng. Data <u>1982</u>, 27, 175-80.

VARIABLES:	PREPARED BY:
Five Na/P ratios.	J. Eysseltová

EXPERIMENTAL VALUES:

Immiscibility and liquid-vapor critical phenomena for aqueous
sodium phosphate solutions.

solute stoichiometry			immiscibility boundary		critical phenomenon	
Na/PO_4 ratio	mass%	mol/kg[a]	t[b]/°C	phase[c]	t/°C	mass%[d]
1	4.96	0.43	376.6 ± 0.2	L_2	383.4 ± 0.4	2.5
1	9.98	0.92	375.0 ± 0.5	L_2	383.7 ± 0.2	2.5
1	20.0	2.08	374.0 ± 0.2	L_2	383.5 ± 0.2	2.5
1	30.1	3.59	376.5 ± 0.3	L^x	384[d]	2.5
1	34[d]	4.29	384[d]	L_1	e	e
1	40.0	4.29	e	e[1]	e	e
1	50.0	8.33	e	e	e	e
1	60.1	12.55	e	e	e	e
1.20	4.55[b]	0.38	350[b]	L_2		2
1.20	5.13	0.43	347.7 ± 0.5	L_2	378.4 ± 0.5	2
1.20	10.0	0.89	340.3 ± 0.4	L_2	380.9 ± 0.5	2
1.20	20.1	2.01	340.7 ± 0.3	L_2	382.9 ± 0.5	2
1.20	30.3	3.47	345.7 ± 0.1	L_2	383.5 ± 0.6	2
1.20	39.8	5.27	353.5 ± 0.1	L^x	380 ± 2	2
1.20	40.6[b]	5.45	350[b]	L_1		2
1.20	50.3	8.07	369.0 ± 0.5	L_1	380[d] ± 2	2
1.20	55[d]	9.75	382[d]	L_1	382[d]	2
1.20	60.1	12.01	e	e[1]	e	e

(continued next page)

AUXILIARY INFORMATION

METHOD/APPARATUS/PROCEDURE:	SOURCE AND PURITY OF MATERIALS:
The synthetic method was used. A chromel-alumel thermocouple was used with a digital readout unit. The experimental details are described in ref. (1).	Analytical reagent grade Na_2HPO_4 and trisodium phosphate hydrate and ACS grade $NaH_2PO_4 \cdot H_2O$ were used.

<table>
<tr><td></td><td>

ESTIMATED ERROR:

The temperature at which immiscibility occurs had a precision of ± 0.1 K and an accuracy of 0.5-1.0 K. The critical temperature had a precision of ± 0.1-0.2 K and an accuracy of 1.0-1.5 K.
</td></tr>
</table>

REFERENCES:
1. Marshall, W.L.; Hall, C.E.; Mesmer, R.E. J. Inorg. Nucl. Chem. <u>1981</u>, 43, 449.
2. Wetton, E.A.M. Power Industry Research <u>1981</u>, 1, 151.

(continued next page)

COMPONENTS:	ORIGINAL MEASUREMENTS:
(1) Sodium dihydrogenphosphate; NaH$_2$PO$_4$; [7558-80-7] (2) Sodium hydroxide; NaOH; [1310-73-2] (3) Water; H$_2$O; [7732-18-5]	Marshall, W.L. J. Chem. Eng. Data 1982, 27, 175-80.

EXPERIMENTAL VALUES cont'd:

Immiscibility and liquid-vapor critical phenomena for aqueous sodium phosphate solutions.

solute stoichiometry			immiscibility boundary		critical phenomenon	
Na/PO$_4$ ratio	mass%	mol/kg[a]	t[b]/°C	phase[c]	t/°C	mass%[d]
1.50	2.05[f]	0.16	350[f]	L_2		1
1.50	4.99	0.40	329.9 ± 0.1	L_2	378.4 ± 0.1	1
1.50	9.59	0.80	319.0 ± 0.1	L_2	378.8 ± 0.1	1
1.50	20.0	1.89	310.3 ± 0.1	L_2	379.3 ± 0.2	1
1.50	30.3	3.29	310.1 ± 0.1	L_2	378.8 ± 0.1	1
1.50	40.1	5.07	313.1 ± 0.1	L_x	379.3 ± 0.1	1
1.50	50.2	7.64	319.9 ± 0.3	L_1	379.3 ± 0.1	1
1.50	60.2[f]	11.46	343.2 ± 0.3	L_1		1
1.50	63.1[f]	12.96	350[f]	L_1		1
1.50	67[d]	15.4	379[d]	L_1	379[d]	1
2.00	0.8[g]	0.06	365[g]	L_2		0.5
2.00	1.14[f]	0.08	350[f]	L_2		0.5
2.00	3.0[h]	0.2	324[h]	L_2		0.5
2.00	5.0[i]	0.37	321.3 ± 0.4	L_2	374.7 ± 0.5	0.5
2.00	10.0[i]	0.78	300[i]	L_2		0.5
2.00	10.3[h]	0.81	305.2 ± 0.1	L_2	374.7 ± 0.5	0.5
2.00	12.4[h]	1.00	300[h]	L_2		0.5
2.00	20.0	1.76	293.6 ± 0.2	L_2	375.4 ± 0.5	0.5
2.00	30.0	3.01	290.6 ± 0.2	L_2	375.1 ± 0.3	0.5
2.00	40.0[i]	4.69	290.6 ± 0.4	L_2	374.7 ± 0.5	0.5
2.00	57.3[i]	9.44	300[i]	L_x		0.5
2.00	72[f]	18.09	350[f]	L_1		0.5
2.00	74[d]	20.02	375[d]	L_1	375[d]	0.5
2.16	4.85	0.35	301 ± 1	L_2		
2.16	10.0	0.76	291 ± 1	L_2		
2.16	20.0	1.71	282 ± 1	L_2		
2.16	28.9	2.77	281 ± 1	L_2		
2.16	37.5	4.10	279 ± 1	L_2		

[a] The mol/kg H$_2$O values were calculated by the compiler.

[b] Lower boundary of observation (appearance of second liquid phase with rising temperature).

[c] L_1 = dilute liquid phase; L_2 = concentrated liquid phase; L_x = liquid phase near the consolute solution composition (where composition L_1 = composition L_2).

[d] Values at the upper temperature limit of immiscibility, determined graphically.

[e] No second liquid or critical phenomenon is observed at temperatures up to 410°C.

[f] From the plots of ref. (2).

[g] From the plots in ref. (3).

[h] From the plots in ref. (4).

[i] From the plots in ref. (5).

REFERENCES cont'd:
3. Ravich, M.I.; Shcherbakova, L.G. Izv. Sekt. Fiz.-Khim. Anal., Inst. Obshch. Neorg. Khim., Akad. Nauk SSSR 1955, 26, 248
4. Panson, A.J.; Economy, G. Liu, C.-T.; Bulischeck, T.S.; Lindsay, W.T., Jr. J. Electrochem. Soc. 1975, 122, 915.
5. Broadbent, D.; Lewis, G.G.; Wetton, E.A.M. J. Chem. Soc., Dalton Trans. 1977, 464.

COMPONENTS:	ORIGINAL MEASUREMENTS:
(1) Trisodium phosphate; Na_3PO_4; [7601-54-9]	D'Ans, J.; Schreiner, O.
(2) Phosphoric acid; H_3PO_4; [7664-38-2]	Z. Anorg. Chem. 1911, 75, 95-102.
(3) Sodium hydroxide; NaOH; [1310-73-2]	
(4) Water; H_2O; [7732-18-5]	

VARIABLES:	PREPARED BY:
One temperature: 25°C	J. Eysseltová
Composition	

EXPERIMENTAL VALUES:

Solubility in the system: Na_3PO_4–NaOH–H_3PO_4–H_2O at 25°C.

		Na_3PO_4[b]		NaOH[b]		H_3PO_4[b]		solid[c]
C_{Na^+}[a]	$C_{PO_4^{3-}}$[a]	mass%	mol/kg	mass%	mol/kg	mass%	mol/kg	phase
4.28	0.040	0.66	0.05	17.07	5.18	----	----	A
3.24	0.183	3.00	0.22	12.72	3.77	----	----	"
2.24	0.752	12.26	0.85	----	----	0.05	0.01	"
2.73	1.08	14.94	1.09	----	----	1.67	0.20	"[d]
3.48	1.33	19.04	1.46	----	----	1.67	0.21	A + B
2.62	1.09	14.33	1.04	----	----	2.12	0.26	B
1.56	0.78	8.54	0.58	----	----	2.55	0.29	"
2.38	1.60	13.02	1.00	----	----	7.90	1.02	"
3.18	2.24	17.40	1.49	----	----	11.56	1.66	"
4.65	3.55	25.44	2.82	----	----	19.60	3.64	"
5.63	3.87	30.80	3.77	----	----	19.53	4.01	"[d]
6.31	4.63	34.52	5.16	----	----	24.76	6.20	C[d]
6.76	4.88	36.99	6.04	----	----	25.74	7.05	"
7.31	5.55	40.00	8.25	----	----	30.51	10.56	metastable soln
6.76	4.88	36.99	6.04	----	----	25.74	7.05	C + D[e]
6.19	4.68	33.87	5.09	----	----	25.64	6.46	E[d]
6.01	4.67	32.88	4.88	----	----	26.13	6.51	"
5.12	4.36	28.01	3.70	----	----	26.00	5.77	"
4.81	4.22	26.32	3.33	----	----	25.64	5.45	"
4.36	4.08	23.86	2.88	----	----	25.74	5.21	"
4.06	4.03	22.21	2.62	----	----	26.23	5.19	"

(continued next page)

AUXILIARY INFORMATION

METHOD/APPARATUS/PROCEDURE:	SOURCE AND PURITY OF MATERIALS:
Isothermal method. Analytical methods: H_3PO_4 was precipitated as $NH_4MgPO_4 \cdot 6H_2O$ and weighed as $Mg_2P_2O_7$. Na^+ was determined as Na_2SO_4 after removing of H_3PO_4 with the aid of lead method.	Commercial materials, pure, recrystallized before use.
	ESTIMATED ERROR:
	Temperature: precision ± 0.05 K
	Nothing else given.
	REFERENCES:

COMPONENTS:	ORIGINAL MEASUREMENTS:
(1) Trisodium phosphate, Na$_3$PO$_4$; [7601-54-9]	D'Ans, J.; Schreiner, O.
(2) Phosphoric acid; H$_3$PO$_4$; [7664-38-2]	Z. Anorg. Chem. <u>1911</u>, 75, 95-102.
(3) Sodium hydroxide; NaOH; [1310-73-2]	
(4) Water; H$_2$O; [7732-18-5]	

EXPERIMENTAL VALUES cont'd:

Solubility in the system: Na$_3$PO$_4$–NaOH–H$_3$PO$_4$–H$_2$O at 25°C.

		Na$_3$PO$_4$[b]		NaOH[b]		H$_3$PO$_4$[b]		
c_{Na^+}[a]	$c_{PO_4^{3-}}$[a]	mass%	mol/kg	mass%	mol/kg	mass%	mol/kg	solid phase[c]
4.19	4.38	22.92	2.91	----	----	29.24	6.24	E
4.32	4.96	23.64	3.43	----	----	34.50	8.41	"
4.65	5.89	25.44	4.83	----	----	42.53	13.55	"
4.88	6.40	26.70	6.12	----	----	46.78	18.00	"[d]

[a] These concentrations are expressed as mol/kg of solution.

[b] All these values were calculated by the compiler.

[c] The solid phases are: A = Na$_3$PO$_4$·12H$_2$O; B = Na$_2$HPO$_4$·12H$_2$O; C = Na$_2$HPO$_4$·7H$_2$O;

D = Na$_2$HPO$_4$·2H$_2$O; E = NaH$_2$PO$_4$·2H$_2$O.

[d] These solid phases were analyzed.

[e] The compiler considers this to be an obvious error. It should be C + E.

COMPONENTS:	ORIGINAL MEASUREMENTS:
(1) Trisodium phosphate; Na_3PO_4; [7601-54-9]	Menzel, H.; v. Sahr, E.
(2) Phosphoric acid; H_3PO_4; [7664-38-2]	*Z. Elektrochem.* 1937, *2*, 104-19.
(3) Sodium hydroxide; NaOH; [1310-73-2]	
(4) Water; H_2O; [7732-18-5]	

VARIABLES:	PREPARED BY:
One temperature: 20°C	J. Eysseltová
Composition	

EXPERIMENTAL VALUES:

Composition of saturated solutions of the Na₂O-P₂O₅-H₂O system at 20°C.

	Na_2O	P_2O_5		Na_3PO_4 [b]		NaOH [b]		H_3PO_4 [b]		solid phase		
N_o^a	mass%	mass%	N_1^a	mass%	mol/kg	mass%	mol/kg	mass%	mol/kg	Na_2O:	P_2O_5:	H_2O
2.00	3.06	3.51	2.00	5.41	0.35	----	----	1.62	0.18	2.00	1	25.0
2.12	3.79	3.97	2.19	6.70	0.44	----	----	1.49	0.16	2.00	1	25.09
2.30	6.01	5.51	2.50	10.63	0.74	----	----	1.27	0.15	2.01	1	24.64
2.40	7.24	6.38	2.60	12.80	0.90	----	----	1.18	0.14	2.02	1	25.85
2.50	8.44	7.25	2.67	14.92	1.08	----	----	1.11	0.14	2.12	1	24.48
2.60	8.43	7.25	2.67	14.91	1.08	----	----	1.12	0.14	2.33	1	21.52
2.70	8.45	7.27	2.67	14.94	1.08	----	----	1.13	0.14	3.09	1	23.67
2.80	8.32	7.09	2.69	14.71	1.06	----	----	1.02	0.12	3.11	1	25.27
2.85	7.35	6.06	2.78	13.00	0.92	----	----	0.62	0.07	3.13	1	23.73
2.90	7.45	6.11	2.79	13.17	0.93	----	----	0.58	0.07	3.14	1	25.41
3.00	6.61	5.22	2.90	11.69	0.81	----	----	0.24	0.03	3.18	1	24.82
3.05	6.17	4.79	2.95	10.91	0.74	----	----	0.11	0.01	3.19	1	25.28
3.10	5.34	4.03	3.03	9.33	0.63	0.08	0.02	----	----	3.19	1	24.73
3.20	4.80	3.46	3.18	8.01	0.53	0.34	0.09	----	----	3.21	1	24.32
3.30	4.53	3.11	3.34	7.20	0.48	0.59	0.16	----	----	3.22	1	23.97
3.50	4.26	2.65	3.68	6.14	0.40	1.02	0.27	----	----	3.22	1	23.73

(continued next page)

AUXILIARY INFORMATION

METHOD/APPARATUS/PROCEDURE:	SOURCE AND PURITY OF MATERIALS:
The components were brought into solution at an elevated temperature. After reaching 20°C each system was equilibrated for 2 or 3 days. The liquid phase was then analyzed and reanalyzed after another 2 or 3 days. The solid phase was separated by a Schott filter and then either washed with ice water or filtered under a pressure of CO_2. It was then dried on a porous plate in an empty desiccator. The samples were titrated with 0.5 mol dm⁻³ HCl using dimethyl yellow as indicator. The indicator was then destroyed by boiling with Br_2 water. The samples were then titrated with 0.5 mol dm⁻³ NaOH using thymolphthalein as indicator. Water was determined by difference.	Na_2HPO_4 was from Sorensen-Kahlbaum, Merck. The NaOH was carbonate-free. The water was conductivity water.
	ESTIMATED ERROR:
	Temperature was constant to within ± 0.1 K.
	REFERENCES:

COMPONENTS:	ORIGINAL MEASUREMENTS:
(1) Trisodium phosphate; Na$_3$PO$_4$; [7601-54-9]	Menzel, H., v. Sahr, E.
(2) Phosphoric acid; H$_3$PO$_4$; [7664-38-2]	Z. Elektrochem. 1937, 2, 104-19.
(3) Sodium hydroxide; NaOH; [1310-73-2]	
(4) Water; H$_2$O; [7732-18-5]	

EXPERIMENTAL VALUES cont'd:

Composition of saturated solutions of the Na$_2$O-P$_2$O$_5$-H$_2$O system at 20°C.

N_o [a]	Na$_2$O mass%	P$_2$O$_5$ mass%	N_1 [a]	Na$_3$PO$_4$ [b] mass%	mol/kg	NaOH [b] mass%	mol/kg	H$_3$PO$_4$ [b] mass%	mol/kg	solid phase Na$_2$O:P$_2$O$_5$:H$_2$O
3.70	4.13	2.38	3.97	5.51	0.36	1.31	0.35	----	----	
3.75	4.07	1.54	6.05	3.56	0.23	2.65	0.71	----	----	3.22 1 23.88
4.00	4.20	1.24	7.75	2.87	0.19	3.32	0.89	----	----	3.23 1 24.15
4.50	4.47	0.95	10.78	2.20	0.14	4.16	1.11	----	----	3.23 1 24.12
5.00	4.77	0.83	13.2	1.92	0.12	4.75	1.27	----	----	3.23 1 23.98
7.00	8.17	0.35	52.8	0.81	0.06	9.95	2.79	----	----	3.23 1 24.11
9.00	13.18	0.20	144.9	0.46	0.03	16.67	5.03	----	----	3.24 1 24.32
	17.86	0.08		0.18	0.01	22.92	7.45	----	----	
	21.73	0.09		0.21	0.02	27.90	9.70	----	----	
	25.39	0.17		0.39	0.04	32.48	12.10	----	----	
	25.85	0.22		0.51	0.05	32.99	12.40	----	----	
	27.79	0.19		0.44	0.04	35.55	13.88	----	----	
	28.15									
	28.36	0.21		0.49	0.05	36.25	14.32	----	----	
	30.23	0.21		0.49	0.05	38.66	15.88	----	----	
	31.07	0.21		0.49	0.05	39.75	16.63	----	----	
	31.18	0.20		0.46	0.05	39.91	16.73	----	----	
	31.64									
	31.97	0.15		0.35	0.04	41.01	17.48	----	----	
	32.58	0.12		0.28	0.03	41.85	18.08	----	----	

[a] N_o is the original ratio of Na$_2$O/P$_2$O$_5$; N_1 is ratio of Na$_2$O/P$_2$O$_5$ in saturated solution.

[b] All these values were calculated by the compiler.

For the composition of the equilibrium solid phases see the Critical Evaluation.

COMPONENTS:	ORIGINAL MEASUREMENTS:
(1) Trisodium phosphate; Na$_3$PO$_4$; [7601-54-9]	Kobe, K.A.; Leipper, A.
(2) Phosphoric acid; H$_3$PO$_4$; [7664-38-2]	
(3) Sodium hydroxide; NaOH; [1310-73-2]	Ind. Eng. Chem. 1940, 32, 198-203.
(4) Water; H$_2$O; [7732-18-5]	

VARIABLES:	PREPARED BY:
Composition at 25°C.	J. Eysseltová

EXPERIMENTAL VALUES:

Composition of saturated solutions of the Na$_2$O–P$_2$O$_5$–H$_2$O system at 25°C.

Na$_2$O	P$_2$O$_5$	Na$_3$PO$_4$ [a]		NaOH [a]		H$_3$PO$_4$ [a]	
mass %	mass %	mass %	mol/kg	mass %	mol/kg	mass %	mol/kg
4.85	1.79	4.15	0.27	3.23	0.87	----	----
4.91	2.38	5.51	0.36	2.31	0.36	----	----
5.06	1.29	2.99	0.20	4.35	1.17	----	----
5.11	2.80	6.48	0.43	1.86	0.51	----	----
7.80	6.30	13.79	0.98	----	----	0.48	0.06
8.56	7.12	15.14	1.10	----	----	0.81	0.10

[a] All these values were calculated by the compiler.

AUXILIARY INFORMATION

METHOD/APPARATUS/PROCEDURE:	SOURCE AND PURITY OF MATERIALS:
Various saturated solutions of Na$_3$PO$_4$ were made up. NaOH was added to some, Na$_2$HPO$_4$ to others. The solutions were rotated in a self-constructed apparatus at 25°C. The amount of solid phase was kept to a minimum. The analyses were done acidimetrically (1).	Baker's C. P. tertiary sodium phosphate was used. According to analysis its composition was Na$_3$PO$_4$·1/7NaOH·12H$_2$O. No other details are given.
	ESTIMATED ERROR:
	Temperature was constant to within ±0.05 K.
	REFERENCES:
	1. Smith, J.H., J. Soc. Chem. Ind. 1917, 36, 415.

COMPONENTS:	ORIGINAL MEASUREMENTS:
(1) Trisodium phosphate; Na$_3$PO$_4$; [7601-54-9] (2) Phosphoric acid; H$_3$PO$_4$; [7664-38-2] (3) Sodium hydroxide; NaOH; [1310-73-2] (4) Water; H$_2$O; [7732-18-5]	Wendrow, B.; Kobe, K.A. *Ind. Eng. Chem.* <u>1952</u>, *44*, 1439-48.
VARIABLES: Composition and temperature.	PREPARED BY: J. Eysseltová

EXPERIMENTAL VALUES:

Composition of saturated solutions of the Na$_2$O–P$_2$O$_5$–H$_2$O system.

Na$_2$O		P$_2$O$_5$		Na$_3$PO$_4$ [a]		NaOH [a]		H$_3$PO$_4$ [a]		solid [b]
mass%	mol%	mass%	mol%	mass%	mol/kg	mass%	mol/kg	mass%	mol/kg	phase
					temp. = 0°C.					
2.23	0.67	1.75	0.23	3.93	0.25	----	----	0.06	0.00	A
2.10	0.62	0.61	0.08	1.41	0.08	1.67	0.43	----	----	A
3.03	0.92	2.25	0.30	5.20	0.33	0.10	0.02	----	----	A
0.754	0.22	0.855	0.11	1.33	0.08	----	----	0.38	0.04	B12
0.746	0.22	0.840	0.11	1.31	0.08	----	----	0.37	0.03	B12
9.28	3.58	21.26	3.58	16.39	1.55	----	----	19.58	3.12	C2
9.36	3.66	21.43	3.66	16.53	1.58	----	----	19.73	3.15	C2
					temp. = 25°C.					
30.35	11.23	0.16	0.025	0.37	0.03	38.90	16.01	----	----	D0.5
28.61	10.44	0.28	0.045	0.64	0.06	36.45	14.48	----	----	D0.5
28.30	10.21	0.31	0.050	0.71	0.06	36.00	14.22	----	----	D0.5 + D6
26.48	9.49	0.28	0.042	0.64	0.06	33.70	12.83	----	----	D6
24.77	8.75	0.30	0.046	0.69	0.06	31.46	11.59	----	----	D6
23.73	8.32	0.29	0.043	0.67	0.05	30.13	10.88	----	----	D6 + A
20.82	7.10	0.22	0.033	0.50	0.04	26.50	9.07	----	----	A
12.31	3.86	0.31	0.043	0.71	0.05	15.13	4.49	----	----	A
5.75	1.75	0.75	0.099	1.73	0.11	6.15	1.67	----	----	A
4.65	1.41	1.40	0.186	3.23	0.21	3.63	0.97	----	----	A
4.61	1.41	2.50	0.335	5.78	0.38	1.72	0.46	----	----	A
4.79	1.48	2.70	0.363	6.24	0.41	1.61	0.43	----	----	A

(continued next page)

AUXILIARY INFORMATION

METHOD/APPARATUS/PROCEDURE:	SOURCE AND PURITY OF MATERIALS:
A standard-type constant temperature bath fitted with automatic controls. Water with ethylene glycol at 0°C, water at 25-60°C and white mineral oil at 80 and 100°C were used as the bath. Self-constructed apparatus for agitation. Analyses: phosphorus was determined according to (1) except for highly alkaline solutions in which the content of P$_2$O$_5$ was 0.8% or less where gravimetric method using magnesium ammonium phosphate was used. Sodium: nothing given, the compiler supposes Smith's method (2) was used. Schreinemakers' method of wet residue was combined with microscopic examination of solid phases.	All chemicals used were C. P. reagent grade. H$_3$PO$_4$, disodium phosphate and monosodium phosphate from J. T. Baker and Co. and disodium phosphate and hemisodium phosphate from Monsanto were used. Merck's sodium hydroxide pellets were used in the preparation of both the samples and the standard NaOH solution.

ESTIMATED ERROR:

Nothing given; the compiler assumes the reproducibility of the analysis to be better than 1%.

REFERENCES:

1. Gerber, A.B.; Miles, P.T. *Ind. Eng. Chem., Anal. Ed.* <u>1941</u>, *13*, 406.

2. Smith, J.H. *J. Soc. Chem. Ind. London* <u>1917</u>, *36*, 420.

COMPONENTS:	ORIGINAL MEASUREMENTS:
(1) Trisodium phosphate; Na₃PO₄, [7601-54-9]	Wendrow, B.; Kobe, K.A.
(2) **Phosphoric acid**; H₃PO₄; [7664-38-2]	*Ind. Eng. Chem.* <u>1952</u>, *44*, 1439-48.
(3) Sodium hydroxide, NaOH; [1310-73-2]	
(4) Water, H₂O; [7732-18-5]	

EXPERIMENTAL VALUES cont'd:

Composition of saturated solutions of the $Na_2O-P_2O_5-H_2O$ system.

Na_2O		P_2O_5		$Na_3PO_4{}^a$		$NaOH^a$		$H_3PO_4{}^a$		Solid b
mass%	mol%	mass%	mol%	mass%	mol/kg	mass%	mol/kg	mass%	mol/kg	phase
7.76	2.53	6.12	0.87	13.70	0.97	----	----	0.27	0.03	A
8.99	3.00	7.50	1.09	15.87	1.16	----	----	0.88	0.10	A
9.91	3.36	8.53	1.26	17.50	1.31	----	----	1.34	0.16	A
10.66	3.67	9.39	1.41	18.82	1.44	----	----	1.73	0.22	A + B12
9.04	3.03	8.20	1.20	15.96	1.18	----	----	1.80	0.22	B12
8.93	2.99	8.01	1.17	15.77	1.16	----	----	1.65	0.20	B12
4.52	1.42	5.23	0.72	7.98	0.54	----	----	2.46	0.28	B12
8.61	3.02	12.95	1.95	15.20	1.21	----	----	8.81	1.18	B12
11.02	4.17	17.59	2.90	19.46	1.74	----	----	12.68	1.90	B12
15.30	6.56	24.58	4.60	27.02	2.98	----	----	17.82	3.29	B12
15.82	6.88	25.05	4.75	27.94	3.14	----	----	17.92	3.38	B12
16.07	7.02	25.32	4.83	28.38	3.22	----	----	18.03	3.43	B12
16.24	7.12	25.37	4.85	28.68	3.27	----	----	17.92	3.42	B12
16.71	7.38	25.69	4.95	29.51	3.41	----	----	17.87	3.46	B12 + B8
16.71	7.45	26.33	5.12	29.51	3.47	----	----	18.75	3.70	B8
16.76	7.54	26.91	5.27	29.60	3.54	----	----	19.50	3.91	B8
17.58	8.12	28.26	5.70	31.05	3.90	----	----	20.50	4.32	B8
17.87	8.31	28.48	5.79	31.56	4.01	----	----	20.50	4.36	B8
19.48	9.61	31.39	6.75	34.40	4.90	----	----	22.82	5.44	B8
20.44	10.40	32.62	7.24	36.10	5.44	----	----	23.51	5.94	B8 + E
19.88	10.04	32.60	7.20	35.11	5.24	----	----	24.07	6.02	E
19.48	9.85	32.93	7.26	34.40	5.15	----	----	24.95	6.26	E
19.08	9.70	33.49	7.44	33.70	5.11	----	----	26.14	6.64	E
18.93	9.60	33.53	7.43	33.43	5.06	----	----	26.36	6.69	E + C2
17.28	8.30	31.20	6.54	30.52	4.17	----	----	24.88	5.69	C2
13.24	5.82	28.06	5.38	23.38	2.75	----	----	24.80	4.88	C2
13.22	5.79	27.84	5.33	23.35	2.72	----	----	24.51	4.79	C2
12.01	5.16	27.50	5.16	21.21	2.41	----	----	25.32	4.83	C2
11.15	4.88	29.50	5.63	19.69	2.33	----	----	28.99	5.76	C2
10.83	4.87	31.65	6.21	19.13	2.39	----	----	32.29	6.78	C2
10.71	4.82	33.94	6.66	18.91	2.53	----	----	35.58	7.98	C2
10.82	5.17	36.23	7.57	19.11	2.75	----	----	38.63	9.32	C2
10.98	5.45	38.60	8.37	19.39	3.03	----	----	41.73	10.95	C2
11.41	5.87	40.51	9.08	20.15	3.41	----	----	43.91	12.47	C2
11.33	5.85	40.95	9.26	20.01	3.44	----	----	44.61	12.86	C1
10.93	5.77	42.70	9.89	19.30	3.53	----	----	47.44	14.56	C1
10.88	5.80	43.37	10.13	19.21	3.61	----	----	48.42	15.27	C1
11.31	6.21	44.85	10.74	19.97	4.05	----	----	50.01	17.01	C1 + C
11.16	6.18	45.39	11.00	19.71	4.08	----	----	50.92	17.69	C
11.08	6.23	45.65	11.18	19.57	4.10	----	----	51.36	18.03	C
10.30	5.92	48.26	12.10	18.19	4.26	----	----	55.78	21.88	C + F
7.25	4.20	51.33	12.99	12.80	3.25	----	----	63.24	26.94	F
2.88	1.73	56.77	14.88	5.08	1.58	----	----	75.35	39.31	F
				temp. = 40°C.						
34.71	13.38	0.12	0.02	0.27	0.03	44.59	20.22	----	----	
29.19	10.75	0.31	0.052	0.71	0.07	37.15	14.94	----	----	D0.5
28.79	10.52	0.38	0.061	0.87	0.08	36.51	14.58	----	----	D0.5
28.37	10.36	0.42	0.067	0.97	0.09	35.90	14.22	----	----	D0.5
27.78	10.11	0.48	0.076	1.11	0.10	35.04	13.72	----	----	D6
27.54	10.00	0.47	0.074	1.08	0.10	34.02	13.11	----	----	D6
26.98	9.73	0.47	0.074	1.08	0.10	34.02	13.11	----	----	D6
25.43	9.06	0.53	0.082	1.22	0.11	31.92	11.94	----	----	D6
24.42	8.61	0.56	0.086	1.29	0.11	30.57	11.51	----	----	D6
23.80	8.36	0.61	0.093	1.41	0.12	29.68	10.77	----	----	D6
23.36	8.20	0.66	0.101	1.52	0.13	29.03	10.45	----	----	D6 + A
19.85	6.75	0.55	0.082	1.27	0.10	24.69	8.33	----	----	A

(continued next page)

COMPONENTS:	ORIGINAL MEASUREMENTS:
(1) Trisodium phosphate; Na_3PO_4; [7601-54-9]	Wendrow, B; Kobe, K.A.
(2) Phosphoric acid; H_3PO_4, [7664-38-2]	*Ind. Eng. Chem.* 1952, 44, 1439-48.
(3) Sodium hydroxide; NaOH; [1310-73-2]	
(4) Water; H_2O; [7732-18-5]	

EXPERIMENTAL VALUES cont'd:

Composition of saturated solutions of the Na_2O-P_2O_5-H_2O system.

Na_2O		P_2O_5		Na_3PO_4[a]		NaOH[a]		H_3PO_4[a]		Solid[b]
mass%	mol%	mass%	mol%	mass%	mol/kg	mass%	mol/kg	mass%	mol/kg	phase
15.64	5.13	0.55	0.079	1.27	0.09	19.25	6.05	----	----	A
15.32	5.02	0.59	0.084	1.36	0.10	18.77	5.87	----	----	A
9.30	2.93	1.17	0.16	2.70	0.18	10.02	2.87	----	----	A
8.59	2.70	1.54	0.21	3.56	0.24	8.48	2.41	----	----	A
7.63	2.41	3.18	0.44	7.35	0.50	4.47	1.26	----	----	A
7.54	2.40	3.73	0.52	8.62	0.59	3.42	0.97	----	----	A
8.97	2.96	6.30	0.91	14.56	1.05	0.92	0.27	----	----	A
12.21	4.28	9.60	1.47	21.56	1.68	----	----	0.39	0.05	A
12.38	4.26	9.81	1.48	21.86	1.71	----	----	0.50	0.06	A
13.88	5.02	11.31	1.79	24.51	2.00	----	----	1.00	0.13	A + D8
14.16	5.16	11.76	1.87	25.01	2.06	----	----	1.32	0.18	D8
14.33	5.26	12.13	1.94	25.31	2.11	----	----	1.65	0.23	D8
15.00	5.57	12.73	2.06	26.49	2.25	----	----	1.78	0.25	D8
15.59	5.90	13.94	2.30	27.53	2.40	----	----	2.83	0.41	D8
17.15	6.78	16.52	2.84	30.29	2.84	----	----	4.75	0.74	D8
17.72	7.10	17.31	3.03	31.31	3.00	----	----	5.24	0.84	D8
19.27	8.05	19.18	3.50	34.03	3.46	----	----	6.19	1.05	D8 + B7
18.78	7.78	18.93	3.42	33.17	3.34	----	----	6.34	1.07	B7
16.05	6.36	17.52	3.03	28.35	2.68	----	----	7.28	1.15	B7
15.10	5.92	17.33	2.96	26.67	2.48	----	----	8.02	1.25	B7
14.71	5.73	17.09	2.91	25.98	2.40	----	----	8.10	1.25	B7
15.34	6.10	18.32	3.18	27.09	2.58	----	----	9.14	1.46	B7
16.14	6.63	20.24	3.65	28.50	2.86	----	----	10.95	1.84	B7
17.87	7.83	24.05	4.60	31.56	3.55	----	----	14.38	2.71	B7
18.36	8.25	25.58	5.02	32.41	3.82	----	----	15.99	3.16	B7
19.18	8.80	26.43	5.30	33.87	4.14	----	----	16.29	3.33	B7 + C2
19.39	8.96	26.72	5.39	34.25	4.23	----	----	16.47	3.41	C2
19.33	8.96	26.99	5.46	34.14	4.25	----	----	16.91	3.52	C2
19.54	9.13	27.41	5.59	34.51	4.36	----	----	17.27	3.65	C2
19.55	9.19	27.82	5.71	34.53	4.41	----	----	17.82	3.81	C2
19.90	9.68	29.94	6.36	35.15	4.81	----	----	20.38	4.67	C2
20.44	10.20	31.23	6.80	36.10	5.20	----	----	21.59	5.21	C2
20.63	10.44	32.00	7.07	36.44	5.40	----	----	22.46	5.57	C2
20.74	10.56	32.34	7.19	36.63	5.50	----	----	22.81	5.74	C2
20.82	10.65	32.53	7.27	36.77	5.56	----	----	22.99	5.83	C2
21.34	11.20	33.79	7.75	37.69	6.02	----	----	24.18	6.47	C2
21.12	11.22	34.73	8.06	37.30	6.14	----	----	25.71	7.09	E
20.68	11.12	35.79	8.40	36.52	6.21	----	----	27.64	7.87	E
20.32	10.98	36.26	8.53	35.89	6.17	----	----	28.66	8.25	E
20.44	11.16	37.01	8.83	36.10	6.40	----	----	29.57	8.79	E + C1
19.46	10.31	35.90	8.31	34.37	5.72	----	----	29.07	8.11	C1
18.95	9.94	35.71	8.18	33.47	5.48	----	----	29.35	8.05	C1
18.76	9.79	35.52	8.10	33.13	5.37	----	----	29.28	7.96	C1
17.56	8.90	34.60	7.66	31.01	4.76	----	----	29.28	7.52	C1
17.09	8.60	34.50	7.58	30.18	4.57	----	----	29.63	7.52	C1
16.92	8.48	34.39	7.53	29.88	4.50	----	----	29.66	7.48	C1
15.77	7.73	33.78	7.23	27.85	4.03	----	----	30.03	7.27	C1
14.66	7.07	33.62	7.08	25.89	3.65	----	----	30.98	7.33	C1
13.96	6.84	35.31	7.56	24.65	3.63	----	----	34.05	8.41	C1
13.61	6.78	36.71	7.99	24.04	3.69	----	----	36.35	9.36	C1
13.07	6.61	38.05	8.40	23.08	3.68	----	----	38.77	10.37	C1
12.37	6.56	41.70	9.66	21.85	3.96	----	----	44.55	13.53	C1
12.53	6.72	42.27	9.90	22.13	4.12	----	----	45.17	14.09	C1
12.25	6.59	42.73	10.04	21.63	4.08	----	----	46.10	14.58	C
11.77	6.54	45.10	10.95	20.79	4.31	----	----	49.87	17.35	C
11.44	6.60	46.38	11.68	20.20	4.42	----	----	51.99	19.08	C
10.92	6.38	48.72	12.44	19.28	4.71	----	----	55.77	22.81	C
10.66	6.51	51.35	13.69	18.82	5.33	----	----	59.67	28.33	C

(continued next page)

COMPONENTS:	ORIGINAL MEASUREMENTS:
(1) Trisodium phosphate; Na_3PO_4; [7601-54-9]	Wendrow, B.; Kobe, K.A.
(2) Phosphoric acid; H_3PO_4, [7664-38-2]	*Ind. Eng. Chem.* 1952, *44*, 1439-48.
(3) Sodium hydroxide; NaOH; [1310-73-2]	
(4) Water; H_2O; [7732-18-5]	

EXPERIMENTAL VALUES cont'd:

Composition of saturated solutions of the Na_2O-P_2O_5-H_2O system.

Na_2O		P_2O_5		Na_3PO_4[a]		NaOH[a]		H_3PO_4[a]		Solid[b]
mass%	mol%	mass%	mol%	mass%	mol/kg	mass%	mol/kg	mass%	mol/kg	phase
10.33	6.36	52.07	14.00	18.24	5.36	----	----	61.01	30.02	F
7.39	4.48	53.66	14.22	13.05	3.85	----	----	66.30	32.78	F
5.10	3.09	55.39	14.64	9.00	2.76	----	----	71.11	36.49	F
2.68	1.71	60.24	16.80	4.73	1.93	----	----	80.35	54.00	F

temp. = 60°C.

Na_2O		P_2O_5		Na_3PO_4		NaOH		H_3PO_4		Solid
mass%	mol%	mass%	mol%	mass%	mol/kg	mass%	mol/kg	mass%	mol/kg	phase
34.39	13.22	0.094	0.016	0.21	0.02	44.22	19.90	----	----	D
30.14	11.16	0.136	0.022	0.31	0.03	38.67	15.84	----	----	D
28.72	10.52	0.31	0.050	0.71	0.06	36.54	14.56	----	----	D
28.29	10.34	0.52	0.082	1.20	0.11	35.63	14.10	----	----	D + DO.5
26.25	9.46	0.90	0.14	2.08	0.19	32.36	12.34	----	----	DO.5
25.68	9.24	1.19	0.19	2.75	0.25	31.13	11.77	----	----	DO.5
25.01	8.97	1.37	0.21	3.16	0.28	29.96	11.20	----	----	DO.5
23.76	8.44	1.54	0.24	3.56	0.31	28.06	10.26	----	----	DO.5
20.60	7.17	2.16	0.33	4.99	0.42	22.93	7.95	----	----	DO.5
19.72	6.84	2.47	0.37	5.71	0.47	21.27	7.28	----	----	A
18.71	6.43	2.55	0.38	5.89	0.48	19.83	6.67	----	----	A
15.79	5.30	2.59	0.38	5.98	0.46	16.00	5.12	----	----	A
14.71	4.02	3.05	0.45	7.05	0.54	13.83	4.37	----	----	A
13.71	4.62	4.18	0.61	9.66	0.73	10.62	3.33	----	----	A
13.28	4.58	7.41	1.11	17.13	1.33	4.61	1.47	----	----	A
15.91	5.84	11.02	1.76	25.48	2.13	1.90	0.65	----	----	A
17.15	6.46	12.24	2.01	28.30	2.45	1.44	0.51	----	----	A + D8
17.29	6.65	12.71	2.13	29.39	2.56	0.82	0.29	----	----	D8
17.79	6.80	13.01	2.17	30.08	2.65	0.96	0.35	----	----	D8
19.20	7.70	15.94	2.79	33.91	3.21	----	----	1.79	0.28	D8
21.73	9.10	20.63	3.95	38.38	4.17	----	----	5.60	1.02	D8
22.13	9.75	21.04	4.05	39.09	4.31	----	----	5.74	1.06	D8
22.48	10.05	21.65	4.21	39.70	4.47	----	----	6.22	1.17	D8
22.62	10.18	22.06	4.33	39.95	4.55	----	----	6.63	1.26	D8 + B2
22.12	9.90	21.85	4.25	39.07	4.40	----	----	6.87	1.29	B2
20.86	9.56	21.82	4.16	36.84	4.08	----	----	8.16	1.51	B2
20.68	9.48	21.99	4.19	36.52	4.05	----	----	8.58	1.59	B2
19.26	8.34	22.11	4.17	34.02	3.71	----	----	10.24	1.87	B2
19.69	8.93	25.13	4.97	34.78	4.13	----	----	13.96	2.78	B2
19.95	9.31	27.09	5.52	35.23	4.43	----	----	16.39	3.45	B2
20.58	10.14	30.32	6.52	36.35	5.09	----	----	20.19	4.74	B2
21.38	10.96	32.29	7.23	37.76	5.72	----	----	22.07	5.60	B2
21.55	11.18	32.90	7.45	38.06	5.91	----	----	22.73	5.91	B2
21.81	11.44	33.43	7.66	38.52	6.13	----	----	23.19	6.18	B2 + G
21.79	11.45	33.53	7.70	38.48	6.14	----	----	23.35	6.24	
21.79	11.45	33.60	7.71	38.48	6.16	----	----	23.44	6.28	G
21.66	11.39	33.69	7.74	38.26	6.12	----	----	23.70	6.36	G
21.37	11.35	34.52	8.01	37.74	6.20	----	----	25.16	6.92	G
21.30	11.31	34.67	8.05	37.62	6.20	----	----	25.44	7.02	G
20.93	11.33	36.17	8.55	36.97	6.41	----	----	27.90	8.10	G
21.02	11.53	36.78	8.81	37.12	6.61	----	----	28.64	8.54	G
20.79	11.49	37.52	9.06	36.72	6.70	----	----	29.91	9.14	G
20.74	11.61	38.20	9.32	36.63	6.87	----	----	30.90	9.71	G + C
18.92	10.07	36.76	8.55	33.42	5.69	----	----	30.83	8.80	C
17.50	9.09	36.21	8.20	30.91	5.01	----	----	31.56	8.58	C
17.16	8.86	36.16	8.15	30.31	4.88	----	----	31.85	8.59	C
16.83	8.65	36.13	8.09	29.72	4.75	----	----	32.16	8.61	C
16.11	8.04	36.89	8.28	28.45	4.61	----	----	33.97	9.22	C
15.08	7.77	37.72	8.49	26.63	4.36	----	----	36.20	9.94	C
13.54	7.12	40.29	9.27	23.91	4.19	----	----	41.37	12.16	C
12.41	6.72	43.02	10.18	21.92	4.20	----	----	46.33	14.89	C

(continued next page)

COMPONENTS:	ORIGINAL MEASUREMENTS:
(1) Trisodium phosphate; Na$_3$PO$_4$; [7601-54-9]	Wendrow, B.; Kobe, K.A.
(2) Phosphoric acid; H$_3$PO$_4$; [7664-38-2]	*Ind. Eng. Chem.* <u>1952</u>, 44, 1439-48.
(3) Sodium hydroxide; NaOH; [1310-73-2]	
(4) Water; H$_2$O; [7732-18-5]	

EXPERIMENTAL VALUES cont'd:

Composition of saturated solutions of the Na$_2$O-P$_2$O$_5$-H$_2$O system.

Na$_2$O		P$_2$O$_5$		Na$_3$PO$_4$[a]		NaOH[a]		H$_3$PO$_4$[a]		Solid[b]
mass%	mol%	mass%	mol%	mass%	mol/kg	mass%	mol/kg	mass%	mol/kg	phase
11.43	6.82	49.48	12.90	20.18	5.22	----	----	56.28	24.41	C
11.39	6.89	50.32	13.29	20.11	5.47	----	----	57.48	26.19	C
11.19	6.90	51.52	13.87	19.76	5.76	----	----	59.35	29.00	C
10.89	7.02	54.09	15.23	19.23	6.67	----	----	63.21	36.76	C
8.91	5.62	54.53	15.02	15.73	5.22	----	----	65.91	36.65	F
7.59	4.77	55.46	15.23	13.40	4.53	----	----	68.58	38.86	F
6.01	3.73	56.11	15.22	10.61	3.54	----	----	71.14	39.80	F
3.63	2.83	59.01	16.30	6.41	2.45	----	----	77.65	49.73	F
					temp. = 80°C.					
36.39	14.30	0.140	0.024	0.32	0.03	46.73	22.06	----	----	D
32.11	12.10	0.210	0.034	0.48	0.05	41.09	17.58	----	----	D
30.64	11.42	0.314	0.051	0.72	0.07	39.01	16.18	----	----	D
30.28	11.26	0.373	0.061	0.86	0.08	38.45	15.84	----	----	D
29.85	11.06	0.514	0.083	1.18	0.11	37.65	15.39	----	----	D0.5
28.03	10.24	0.605	0.096	1.39	0.13	35.15	13.85	----	----	D0.5
27.72	10.07	0.617	0.098	1.42	0.13	34.73	13.60	----	----	D0.5
25.60	9.25	1.65	0.26	3.81	0.35	30.25	11.47	----	----	D0.5
25.24	9.12	1.79	0.28	4.13	0.38	29.55	11.14	----	----	D0.5
24.83	9.11	2.60	0.41	6.01	0.55	27.65	10.42	----	----	D0.5
23.87	8.75	4.29	0.69	9.92	0.90	23.55	8.85	----	----	D0.5 + D6
21.61	8.11	8.35	1.37	19.31	1.75	13.77	5.14	----	----	D6
21.35	8.35	12.15	2.08	28.09	2.63	7.01	2.70	----	----	D6
23.20	9.98	18.20	3.41	40.98	4.28	----	----	0.69	0.12	D6
24.51	11.07	20.96	4.13	43.29	4.92	----	----	3.13	0.59	D6
22.94	10.50	23.29	4.66	40.52	4.79	----	----	8.00	1.58	B2
20.59	9.25	23.56	4.61	36.37	4.19	----	----	10.84	2.09	B2
					temp. = 100°C.					
38.36	15.36	0.41	0.072	0.94	0.11	48.81	24.29	----	----	D
30.05	11.17	0.80	0.13	1.85	0.18	37.43	15.41	----	----	D
26.19	9.48	1.32	0.21	3.07	0.28	31.55	12.06	----	----	D
23.79	8.70	4.17	0.67	9.64	0.88	23.65	8.86	----	----	D
24.37	9.46	9.26	1.57	21.41	2.07	15.79	6.29	----	----	D
24.26	9.45	9.53	1.62	22.03	2.13	15.20	6.05	----	----	D
25.98	10.76	13.18	2.38	30.48	3.18	11.24	4.82	----	----	D
25.43	10.50	13.41	2.42	31.01	3.21	10.15	4.31	----	----	D0.5
24.97	10.42	14.00	2.66	33.76	3.50	7.54	3.21	----	----	D0.5
25.18	10.66	15.62	2.89	36.12	3.80	6.09	2.63	----	----	D0.5
25.76	11.18	16.93	3.20	39.15	4.24	4.62	2.05	----	----	D0.5
26.51	11.74	17.79	3.43	41.14	4.58	4.13	1.89	----	----	D0.5
26.75	12.03	18.87	3.70	43.63	4.94	2.62	1.22	----	----	D6
26.90	11.86	19.29	3.71	44.61	5.10	2.10	0.98	----	----	D6
27.89	13.22	21.77	4.51	49.26	5.99	----	----	0.68	0.14	D6
28.31	13.69	23.01	4.86	50.00	6.34	----	----	1.95	0.41	D6
27.48	13.30	23.46	4.95	48.54	6.16	----	----	3.45	0.73	B
25.73	12.35	24.29	5.08	45.44	5.75	----	----	6.44	1.36	B
24.69	11.73	24.59	5.10	43.61	5.48	----	----	7.95	1.67	B
23.53	11.08	24.84	5.12	41.56	5.17	----	----	9.52	1.98	B
22.13	10.35	25.56	5.22	39.09	4.86	----	----	11.98	2.50	B
23.39	12.56	33.64	7.90	41.31	6.82	----	----	21.81	6.03	B
24.34	13.93	36.61	9.15	42.99	8.16	----	----	24.91	7.92	B + E
23.07	13.31	38.09	9.57	40.75	8.02	----	----	28.29	9.33	E
22.54	13.14	39.21	10.00	39.81	8.14	----	----	30.40	10.41	E
22.33	13.18	40.00	10.32	39.44	8.33	----	----	31.71	11.22	E
22.40	13.31	42.73	11.64	39.56	9.63	----	----	35.41	14.44	E + C

(continued next page)

COMPONENTS:	ORIGINAL MEASUREMENTS:
(1) Trisodium phosphate; Na$_3$PO$_4$; [7601-54-9]	Wendrow, B.; Kobe, K.A.
(2) Phosphoric acid; H$_3$PO$_4$; [7664-38-2]	*Ind. Eng. Chem.* <u>1952</u>, *44*, 1439-48.
(3) Sodium hydroxide; NaOH; [1310-73-2]	
(4) Water; H$_2$O; [7732-18-5]	

EXPERIMENTAL VALUES cont'd:

Composition of saturated solutions of the Na$_2$O-P$_2$O$_5$-H$_2$O system.

Na$_2$O		P$_2$O$_5$		Na$_3$PO$_4$[a]		NaOH[a]		H$_3$PO$_4$[a]		solid[b] phase
mass%	mol%	mass%	mol%	mass%	mol/kg	mass%	mol/kg	mass%	mol/kg	
21.48	13.00	41.91	11.04	37.94	8.62	----	----	35.24	13.41	C
19.87	11.55	41.25	10.50	35.09	7.40	----	----	36.03	12.73	C
19.16	11.07	41.40	10.46	33.84	7.06	----	----	36.98	12.93	C
18.36	10.51	41.72	10.46	32.43	6.74	----	----	38.27	13.32	C
18.31	10.53	41.92	10.53	32.34	6.78	----	----	38.60	13.55	C
16.71	9.57	42.90	10.74	29.51	6.23	----	----	41.63	14.72	C
14.63	8.93	48.22	12.88	25.84	6.85	----	----	51.17	22.72	C
13.14	9.95	53.04	18.33	23.21	13.48	----	----	66.30	64.51	C
12.32	8.80	57.95	18.05	21.76	11.84	----	----	67.04	61.10	F
10.70	7.55	58.68	18.03	18.90	10.15	----	----	69.75	62.74	F
9.86	7.06	60.08	18.74	17.41	10.60	----	----	72.57	73.98	F
9.37	6.86	61.42	19.63	16.55	11.85	----	----	74.93	89.86	F

[a] All these values were calculated by the compiler.

[b] The solid phases are: A = 4(Na$_3$PO$_4$·12H$_2$O)·NaOH;

  B = Na$_2$HPO$_4$; B2 = Na$_2$HPO$_4$·2H$_2$O; B7 = Na$_2$HPO$_4$·7H$_2$O; B8 = Na$_2$HPO$_4$·8H$_2$O;
       B12 = Na$_2$HPO$_4$·12H$_2$O;

  C = NaH$_2$PO$_4$; C1 = NaH$_2$PO$_4$·H$_2$O; C2 = NaH$_2$PO$_4$·2H$_2$O;

  D = Na$_3$PO$_4$; D0.5 = Na$_3$PO$_4$·0.5H$_2$O; D6 = Na$_3$PO$_4$·6H$_2$O; D8 = Na$_3$PO$_4$·8H$_2$O;

  E = Na$_2$HPO$_4$·2NaH$_2$PO$_4$·2H$_2$O;

  F = NaH$_2$PO$_4$·H$_3$PO$_4$;

  G = Na$_2$HPO$_4$·NaH$_2$PO$_4$.

COMPONENTS:	ORIGINAL MEASUREMENTS:
(1) Disodium hydrogenphosphate; Na$_2$HPO$_4$; [7558-79-4] (2) Phosphoric acid; H$_3$PO$_4$; [7664-38-2] (3) Sodium hydroxide; NaOH; [1310-73-2] (4) Water; H$_2$O; [7732-18-5]	Panson, A.J.; Economy, G.; Liu, Chia-sun; Bulischeck, T.S.; Lindsay Jr., W.T. J. Electrochem. Soc. 1975, 122, 915-8.
VARIABLES:	PREPARED BY:
Composition at 548, 573 and 597 K.	J. Eysseltová

EXPERIMENTAL VALUES:

Solubility in the Na$_2$HPO$_4$–H$_3$PO$_4$–NaOH–H$_2$O system.

Na/PO$_4$ ratio		PO$_4$	Na$_2$HPO$_4$[a]	H$_3$PO$_4$[a]	NaOH[a]
solution	solid phase	mol/kg	mass%	mass%	mass%
		temp = 548 K			
2.0	---	>8.0			
2.10	---	>2.5			
2.15	2.13	3.79	34.88	---	0.39
2.15	2.23	3.26	31.54	---	0.41
2.24	2.87	1.29	15.37	---	0.80
2.25	2.86	0.64	8.26	---	0.91
2.32	2.73	0.74	9.41	---	1.14
2.57	2.84	0.42	5.52	---	2.11
2.69	2.76	0.41	5.37	---	2.54
3.24	2.83	0.26	3.40	---	4.56
		temp = 573 K			
1.58	1.62	5.73	27.66	13.80	---
1.60	1.60	5.68	28.38	13.04	---
1.65	1.65	6.95	34.25	12.67	---
1.75	1.84	2.94	22.62	5.20	---
1.81	1.91	0.94	9.62	1.55	---
1.92	1.99	1.06	12.09	0.72	---
1.97	2.04	0.85	10.47	0.22	---
2.00	2.22	0.51	6.76	0.00	0.00
2.10	2.37	0.64	8.31	----	0.36
2.11	2.23	0.68	8.78	----	0.40
2.13	2.56	0.33	4.46	----	0.49
2.13	2.63	0.44	5.86	----	0.49

(continued next page)

AUXILIARY INFORMATION

METHOD/APPARATUS/PROCEDURE:	SOURCE AND PURITY OF MATERIALS:
The experiments were carried out in two-liter autoclaves of stainless steel (A.I.S.I. Type 316) and Inconel Alloy 600 fitted with internal sampling tubes and filters. Three different experimental procedures were used: nonisothermal procedures, isothermal procedures with saturation approached from below, and experiments in which saturation was approached by evaporation at approximately isothermal conditions. Samples of solution were removed periodically from the autoclave and analyzed by a potentiometric acid-base titration.	No information is given.
	ESTIMATED ERROR:
	No information is given.
	REFERENCES:

COMPONENTS:	ORIGINAL MEASUREMENTS:
(1) Disodium hydrogenphosphate; Na$_2$HPO$_4$; [7558-79-4] (2) Phosphoric acid; H$_3$PO$_4$; [7664-38-2] (3) Sodium hydroxide; NaOH; [1310-73-2] (4) Water; H$_2$O; [7732-18-5]	Panson, A.J.; Economy, G.; Liu, Chia-sun; Bulischeck, T.S.; Lindsay Jrs., W.T. J. *Electrochem. Soc.* 1975, *122*, 915-8.

EXPERIMENTAL VALUES cont'd:

Solubility in the Na$_2$HPO$_4$–H$_3$PO$_4$–NaOH–H$_2$O system.

Na/PO$_4$ ratio solution	Na/PO$_4$ ratio solid phase	PO$_4$ mol/kg	Na$_2$HPO$_4$[a] mass%	H$_3$PO$_4$[a] mass%	NaOH[a] mass%
		temp = 573 K cont'd.			
2.14	2.28	0.47	6.23	---	0.52
2.14	2.44	0.55	7.21	---	0.52
2.14	2.35	0.66	8.53	---	0.51
2.16	2.33	0.67	8.64	---	0.58
2.18	2.55	0.38	5.09	---	0.68
2.19	2.58	0.43	5.72	---	0.71
2.46	2.82	0.25	3.37	---	1.74
2.89	2.75	0.14	1.89	---	3.37
		temp = 597 K			
1.48	1.71	2.08	11.37	8.49	---
1.55	1.79	0.70	5.04	2.84	---
1.56	1.66	3.63	19.99	10.83	---
1.63	1.70	3.86	23.27	9.42	---
1.71	1.90	0.38	3.66	1.03	---
1.89	2.03	0.28	3.41	0.29	---
1.93	2.16	0.33	4.17	0.22	---
1.97	2.01	0.34	4.47	0.10	---
1.97	2.14	0.19	2.55	0.05	---
1.98	2.27	0.26	3.49	0.05	---
2.08	2.29	0.24	3.29	----	0.31
2.13	2.37	0.26	3.55	----	0.50
2.14	2.54	0.17	2.35	----	0.54
2.15	2.44	0.20	2.75	----	0.58
2.17	2.48	0.19	2.61	----	0.66
2.32	2.42	0.07	0.97	----	1.25
2.54	2.76	0.11	1.51	----	2.08
3.04	2.74	0.08	1.08	----	3.95
3.28	2.72	0.06	0.80	----	4.83

[a]These values were calculated by the compiler.

COMPONENTS:	EVALUATOR:
(1) Sodium dihydrogenphosphate; NaH_2PO_4; [7558-80-7] (2) Water; H_2O; [7732-18-5]	J. Eysseltová and J. Makovička Charles University Prague, Czechoslovakia May 1985

CRITICAL EVALUATION:

THE BINARY SYSTEM

The isothermal method has been used to determine the solubility of sodium dihydro-genphosphate in water (1, 2). The solubility has also been reported as a limiting condition in the study of several multicomponent systems (3-12). The nature of the equilibrium solid phases was studied in detail by Imadsu (2). He reported the existence of the following solid phases: NaH_2PO_4 [7558-80-7]; $NaH_2PO_4 \cdot H_2O$ [10049-21-5]; $NaH_2PO_4 \cdot 2H_2O$ [13472-35-0]; and $NaH_2PO_4 \cdot 4H_2O$ [101056-45-5]. The transition temperature of the anhydrous salt to the monohydrate was estimated to be 330.5 K while that of the monohydrate to the dihydrate was estimated to be 313.9 K. The transition between the dihydrate and the tetrahydrate was not determined but was considered to be below 273 K (2).

Evaluation Procedure. All the data were examined and evaluated by using the method described by Cohen-Adad (13). Only experimentally obtained data were evaluated. Data obtained from smoothing equations or by extrapolation were excluded from consideration. The data calculated on the basis of sodium determination (1) and the 298 K values reported by others (3, 4) are clearly incorrect and were not included. Some data (4) were reported as mol dm^{-3} and could not be used here because no density information was given and the values could not be recalculated in terms of mole fraction. All the other data are consistent with each other and were evaluated together.

The data were fitted to equation [1], suggested by Cohen-Adad (13).

$$\ln(x/x_o) = A \cdot (1/T - 1/T_o) + B \cdot \ln(T/T_o) + C \cdot (T - T_o) \qquad [1]$$

A, B and C are adjustable parameters. No attempt was made to give a physical meaning to them. The system is too complex for that, as is shown by reported activity coefficient data (14, 15). x_o is a reference mole fraction at temperature T_o. Its choice is arbitrary. The evaluators used the following two criteria for making their selection.

1. x_o was chosen as the mean value of the experimental data of more than one study. Furthermore, the standard deviation did not exceed the experimental uncertainty in obtaining the data.
2. T_o was chosen near the middle of the temperature range in which the hydrate exists rather than at or near a transition point of one hydrate into another.

Each datum selected was given a weight equal to the number of independent deter-minations of the value. In some reports (5, 7, 10, 12) a given value appears to be reported more than once. In such cases the value was given a weight of one. Then an iterative method analagous to that described by others (16) was used. It was necessary to know the experimental uncertainty of the values before the next iteration could be made. However, such information was included in only two reports (2, 4). Therefore, the evaluators tried to estimate the experimental uncertainty on the basis of the data that were included in the report. It appears that for the isothermal studies (1-4, 6-9, 11, 12) the precision is 0.1 to 0.5% while for data derived from polythermal studies (5, 10) it is about 1%. The selection conditions that were used are given in equation [2]. x_j and T_j are the coordinates of the experimental point j in terms of mole fraction and

$$\left| \frac{x_j - x(T_j)}{x(T_j)} \right| \leq 0.015 \qquad [2]$$

temperature. $x(T_j)$ is the calculated mole fraction. Data points that did not meet the conditions of equation [2] were eliminated before the next iteration.

The value of coefficients A, B and C of equation [1] were determined by using a non-linear regression with the experimental points selected as described above. For the regression, equation [1] was put in the form of equation [3]. The calculation was stopped

$$x = x_o \cdot \exp[A(1/T - 1/T_o) + B \cdot \ln(T/T_o) + C \cdot (T - T_o)] \qquad [3]$$

when steady values were obtained for A, B and C.

The solubility results are summarized in Table I. During the iteration procedure all the data in refs (6-8, 12) and most of the data in ref (1) were eliminated. Table II

(continued next page)

COMPONENTS:	EVALUATOR:
(1) Sodium dihydrogenphosphate; NaH_2PO_4; [7558-80-7] (2) Water; H_2O; [7732-18-5]	J. Eysseltová and J. Makovička Charles University Prague, Czechoslovakia May 1985

CRITICAL EVALUATION: (cont'd)

Table I. Solubility of NaH_2PO_4 in water.

T/K	mass%	ref.	weight init/final	T/K	mass%	ref.	weight init/final
				$NaH_2PO_4 \cdot 2H_2O$			
263.3	32.4	5	1/1	298.2	48.97	1	1/1
263.3	30.4	10	1/0	298.2	48.62	2	2/2
266.2	33.6	10	1/1	298.2	48.69	9	1/1
268.2	35.4	12	1/0	298.2	48.03	8	1/0
268.9	34.6	10	1/1	298.2	48.47	11	1/0
273.2	36.1	1	1/0	299.2	49.16	2	1/1
273.2	36.4	5	1/1	299.2	49.17	2	1/1
273.2	36.25	11	1/1	300.2	49.80	2	2/2
273.2	37.6	12	1/0	301.2	50.43	2	1/1
273.3	36.64	2	1/1	301.2	50.41	2	1/1
274.2	37.13	2	1/1	303.2	51.55	2	1/1
274.2	37.15	2	1/1	303.2	51.57	2	1/1
276.2	38.06	2	1/1	303.2	51.2	5	1/1
276.2	38.08	2	1/1	306.2	52.15	2	1/1
278.2	38.95	2	1/1	306.2	52.12	2	1/1
278.2	38.96	2	1/1	307.2	53.93	2	1/1
283.2	41.12	2	1/1	307.2	53.96	2	1/1
283.2	41.14	2	1/1	308.2	54.63	2	1/1
283.2	40.5	5	1/0	308.2	54.64	2	1/1
283.2	42.2	12	1/0	308.2	54.79	3	1/1
288.2	43.42	2	1/1	308.7	53.65	1	1/0
288.2	43.41	2	1/1	313.2	56.41	1	1/0
293.2	46.01	2	1/1	313.2	58.02	2	1/0
293.2	46.00	2	1/1	313.2	58.00	2	1/0
293.2	45.30	5	1/0	313.2	56.31	7	1/0
293.2	46.60	12	1/0				
				$NaH_2PO_4 \cdot H_2O$			
314.2	58.76	2	1/1	323.2	61.16	6	1/0
314.2	58.78	2	1/1	323.2	60.58	8	1/0
315.2	58.98	2	1/1	325.2	62.11	2	1/1
315.2	58.99	2	1/1	325.2	62.09	2	1/0
317.2	57.97	1	1/0	328.2	63.85	1	1/0
318.2	59.71	2	2/2	328.2	63.09	2	1/1
323.2	61.81	1	1/1	328.2	63.07	2	1/1
323.2	61.32	2	1/1	329.2	63.41	2	1/1
323.2	61.34	2	1/1	329.2	63.39	2	1/1
				NaH_2PO_4			
331.2	65.53	1	1/0	338.2	65.89	1	1/0
331.2	63.94	2	1/0	342.2	65.54	2	1/1
331.2	63.92	2	1/0	342.2	65.55	2	1/1
333.2	64.20	2	2/2	343.2	66.25	1	1/0
334.2	65.77	1	1/0	348.2	67.21	1	1/0
335.2	64.44	2	1/1	348.2	66.57	11	1/1
335.2	64.48	2	1/1	353.2	67.44	2	1/0
338.2	64.92	2	1/1	353.2	67.48	2	1/1
338.2	64.90	2	1/1	356.2	69.13	1	1/1

(continued next page)

COMPONENTS:	EVALUATOR:
(1) Sodium dihydrogenphosphate; NaH_2PO_4; [7558-80-7] (2) Water; H_2O; [7732-18-5]	J. Eysseltová and J. Makovička Charles University Prague, Czechoslovakia May 1985

CRITICAL EVALUATION: (cont'd)

Table II. Values for the parameters in equation [1].

	dihydrate		monohydrate		anhydrous	
Parameter	value	σ^a	value	σ^a	value	σ^a
A	-3.52×10^4	100	1.596×10^6	5000	-6.29×10^5	3000
B	-253	1	9940	40	-3700	10
C	0.472	0.002	-15.47	0.07	5.45	0.02
x_o	0.12454		0.19467		0.21720	
T_o	298.2		323.2		338.2	

a These are the standard deviations for the respective parameter.

Table III. Solubility data calculated by equation [1].

T/K	mol fraction	mol/kg	mass%

$NaH_2PO_4 \cdot 2H_2O$

T/K	mol fraction	mol/kg	mass%
273.2	0.079491	4.80	36.54
278.2	0.086735	5.28	38.77
283.2	0.094636	5.81	41.07
288.2	0.10340	6.41	43.47
293.2	0.11327	7.10	46.00
298.2	0.12454	7.90	48.68
303.2	0.13759	8.86	51.54
308.2	0.15287	10.02	54.61
313.2	0.17095	11.46	57.89
313.7 a	0.17299	11.62	58.24

$NaH_2PO_4 \cdot H_2O$

T/K	mol fraction	mol/kg	mass%
313.7	0.17299	11.62	58.24
315.2	0.17505	11.79	58.59
317.2	0.17893	12.11	59.23
319.2	0.18376	12.51	60.02
321.2	0.18915	12.96	60.87
323.2	0.19467	13.43	61.71
325.2	0.19981	13.87	62.48
327.2	0.20412	14.25	63.10
329.2 b	0.20707	14.51	63.52
331.2	0.20815	14.60	63.67

NaH_2PO_4

T/K	mol fraction	mol/kg	mass%
333.2	0.19865	13.77	62.30
338.2	0.21720	14.72	63.85
343.2	0.22437	16.07	65.86
348.2	0.23358	16.93	67.02
353.2	0.24767	18.29	68.70
358.2	0.27010	20.56	71.16

a The dihydrate to monohydrate transition temperature.

b The monohydrate to anhydrous salt transition temperature.

The above transition temperature values were determined graphically by the evaluators.

(continued next page)

COMPONENTS:	EVALUATOR:
(1) Sodium dihydrogenphosphate; NaH_2PO_4; [7558-80-7] (2) Water; H_2O; [7732-18-5]	J. Eysseltová and J. Makovička Charles University Prague, Czechoslovakia May 1985

CRITICAL EVALUATION: (cont'd)

gives a summary of the values for the parameters of equation [1]. In Table III are given some solubility values obtained by the use of equation [1] and the parameters given in Table II. In Table III, the values for $NaH_2PO_4 \cdot 2H_2O$ are recommended values. For the dihydrate and the anhydrous salt the values given are tentative values because only the data of Imadsu (2) survived the iteration procedure.

References

1. Apfel, O. Dissertation, Technical University, Darmstadt 1911.
2. Imadsu, A. *Mem. Col. Sci. Emp. (Kyoto)* 1911-12, *3*, 257.
3. Beremzhanov, B.A.; Savich, R.F. Kunanbaev, G.S. *Khim. Khim. Tekhnol., (Alma Ata)* 1977, *22*, 15.
4. Ferroni, G.; Galea, J.; Antonetti, G. *Bull. Soc. Chim. Fr.* 1974, *12 (Pt. 1)*, 273.
5. Shpunt, S.J. *Zh. Prikl. Khim.* 1940, *13*, 19.
6. Kol'ba, V.I. Zhikharev, M.I.; Sukhanov, L.P.; *Zh. Neorg. Khim.* 1981, *26*, 828.
7. Khallieva, Sh. D. *Izv. Akad. Nauk Turkm. SSR, Ser. Fiz-Tekhn., Khim. Geol. Nauk* 1977, *3*, 125.
8. Girich, T.E.; Gulyamov, Yu. M. *Vopr. Khim. Khim. Tekhnol.* 1979, *57*, 54.
9. Lilich, L.S.; Alekseeva, E.A. *Zh. Neorg. Khim.* 1969, *14*, 1655.
10. Shpunt, S.J.; *Zh. Prikl. Khim.* 1940, *13*, 9.
11. Brunisholz, G.; Bodmer, M. *Helv. Chim. Acta* 1963, *46*, 289, 2575.
12. Babenko, A.M.; Vorob'eva, T.A. *Zh. Prikl. Khim.* 1976, *49*, 1502.
13. Cohen-Adad, R. *Pure Appl. Chem.* 1985, *57*, 255.
14. Platford, R.F. *J. Chem. Eng. Data* 1974, *19*, 166.
15. Scatchard, G.; Breckenridge, R.C. *J. Phys. Chem.* 1954, *58*, 596.
16. Tenu, R.; Counioux, J.J.; Cohen-Adad, R. *8th International CODATA Conference, Jachanka, Poland* 1982.

COMPONENTS:	ORIGINAL MEASUREMENTS:
(1) Sodium dihydrogenphosphate; NaH_2PO_4; [7558-80-7] (2) Water; H_2O; [7732-18-5]	Apfel, O. Dissertation, Technical University, Darmstadt, <u>1911</u>.

VARIABLES:	PREPARED BY:
Composition and temperature.	J. Eysseltová

EXPERIMENTAL VALUES:

Composition of saturated solutions in the NaH_2PO_4-H_2O system.

$t/°C$	PO_4^{3-} c^a	Na^+ c^a	NaH_2PO_4[b] mass%	NaH_2PO_4[b] mol/kg	solid phase
0	3.01		36.13	4.71	
0		2.84	34.08	4.31	
25	4.08		48.97	8.00	$NaH_2PO_4 \cdot 2H_2O$
35.5	4.47		53.65	9.64	"
40	4.70		56.41	10.78	"
44	4.83		57.97	11.49	"
44[c]	5.06		60.73	12.89	$NaH_2PO_4 \cdot H_2O$
50	5.15		61.81	13.48	$NaH_2PO_4 \cdot 2H_2O$
50		5.26	63.13	14.27	
55	5.32		63.85	14.72	$NaH_2PO_4 \cdot H_2O$
58	5.46		65.53	15.84	"
61	5.48		65.77	16.01	NaH_2PO_4
65	5.49		65.89	16.10	"
70	5.52		66.25	16.36	"
75	5.60		67.21	17.08	"
83	5.76		69.13	18.66	"

[a] These concentrations are expressed as mol/1000 g soln.

[b] These values were calculated by the compiler.

[c] This was a metastable equilibrium.

AUXILIARY INFORMATION

METHOD/APPARATUS/PROCEDURE:	SOURCE AND PURITY OF MATERIALS:
All the experiments were performed in a water thermostat. Equilibrium was ascertained by repeated analysis of the liquid phase, which was separated from the solid phase by filtration through a mat of platinum wires. Phosphate was determined gravimetrically as $Mg_2P_2O_7$. Sodium was determined as Na_2SO_4 after phosphoric acid had been removed as lead phosphate.	No information is given.
	ESTIMATED ERROR: No information is given.
	REFERENCES:

COMPONENTS:	ORIGINAL MEASUREMENTS:
(1) Sodium dihydrogenphosphate; NaH_2PO_4; [7558-80-7] (2) Water; H_2O; [7732-18-5]	Imadsu, A. *Mem. Col. Sci. Emp.* (*Kyoto*) <u>1911-12</u>, 3, 257-63.
VARIABLES:	PREPARED BY:
Composition and temperature.	J. Eysseltová

EXPERIMENTAL VALUES:

Solubility in the NaH_2PO_4-H_2O system.

g/100 g H_2O

$t/°C$		mean	mass%[a]	mol/kg[a]
0.10	57.84	57.86	36.64	4.82
	57.87		36.66	4.82
1.00	59.10	59.08	37.15	4.92
	59.06		37.13	4.92
3.00	61.45	61.47	38.06	5.12
	61.49		38.08	5.12
5.00	63.80	63.82	38.95	5.32
	63.84		38.96	5.32
10.00	69.85	69.87	41.12	5.82
	69.89		41.14	5.82
15.00	76.74	76.72	43.42	6.39
	76.70		43.41	6.39
20.00	85.23	85.21	46.01	7.10
	85.18		46.00	7.08
25.00	94.62	94.63	48.62	7.88
	94.63		48.62	7.88
26.00	96.70	96.73	49.16	8.06
	96.75		49.17	8.06
27.00	99.20	99.20	49.80	8.26
	99.19		49.80	8.26
28.00	101.75	101.71	50.43	8.48
	101.67		50.41	8.47

(continued next page)

AUXILIARY INFORMATION

METHOD/APPARATUS/PROCEDURE:	SOURCE AND PURITY OF MATERIALS:
Equilibrium was approached from both under-saturation and supersaturation. Care was taken during sampling to insure the absence of solid particles. Samples of solution were weighed, evaporated to dryness and heated strongly to convert the solid into metaphosphate. The concentration of the solution was calculated from the weight of metaphosphate formed.	NaH_2PO_4 was prepared by adding H_3PO_4 to ordinary sodium phosphate until the solution gave no precipitate with $BaCl_2$. The solution was then evaporated until crystals formed. These crystals were re-crystallized. Subsequent analysis showed that the crystals were free from ordinary impurities.
	ESTIMATED ERROR:
	The temperature was kept constant to within 0.03 K (below 40°C), 0.05 K (between 40 and 60°C), 0.1 K (between 60 and 80°C), and 0.15 K (above 80°C). Duplicate analyses agreed within ±0.1%.
	REFERENCES:

COMPONENTS:	ORIGINAL MEASUREMENTS:
(1) Sodium dihydrogenphosphate; NaH_2PO_4; [7558-80-7] (2) Water; H_2O, [7732-18-5]	Imadsu, A. *Mem. Col. Sci. Emp. (Kyoto)* <u>1911-12</u>, 3, 257-63.

EXPERIMENTAL VALUES cont'd:

Solubility in the $NaH_2PO_4-H_2O$ system.

g/100 g H_2O

$t/°C$		mean	mass%[a]	mol/kg[a]
30.00	106.40	106.45	51.55	8.86
	106.50		51.57	8.87
31.00	108.99	108.93	52.15	9.08
	108.87		52.12	9.07
33.00	114.38	114.31	53.35	9.53
	114.23		53.32	9.52
34.00	117.08	117.14	53.93	9.76
	117.20		53.96	9.76
35.00	120.42	120.44	54.63	10.03
	120.45		54.64	10.04
37.00	126.82	126.76	55.91	10.57
	126.70		55.89	10.56
40.00	138.22	138.16	58.02	11.52
	138.10		58.00	11.51
40.20	139.12	139.06	58.18	11.59
	139.00		58.16	11.58
40.55	140.95	140.83	58.50	11.74
	140.70		58.45	11.72
41.00	142.50	142.55	58.76	11.87
	142.60		58.78	11.88
42.00	143.80	143.83	58.98	11.98
	143.85		58.99	11.99
45.00	148.19	148.20	59.71	12.35
	148.20		59.71	12.35
50.00	158.55	158.61	61.32	13.21
	158.67		61.34	13.22
52.00	163.91	163.84	62.11	13.66
	163.76		62.09	13.64
55.00	170.93	170.85	63.09	14.24
	170.77		63.07	14.23
56.00	173.15	173.23	63.39	14.43
	173.30		63.41	14.44
57.00	175.87	175.81	63.75	14.65
	175.74		63.73	14.64
58.00	177.33	177.24	63.94	14.78
	177.14		63.92	14.76
60.0	179.31	179.33	64.20	14.94
	179.34		64.20	14.94
62.0	181.20	181.35	64.44	15.10
	181.50		64.48	15.12
65.0	185.06	184.99	64.92	15.42
	184.92		64.90	15.41
69.0	190.17	190.24	65.54	15.84
	190.31		65.55	15.86
80.0	207.08	207.29	67.44	17.25
	207.50		67.48	17.29
90.0	225.17	225.31	69.25	18.76
	225.45		69.27	18.78
99.1	246.20	146.56	71.11	20.51
	246.92		71.17	20.57

[a] These values were calculated by the compiler.

Examination of the equilibrium solid phases showed the presence of the anhydrous salt, the monohydrate, dihydrate and tetrahydrate. The transition points of anhydrous salt and monohydrate and of monohydrate and dihydrate were estimated to be 57.4°C and 40.8°C, respectively.

COMPONENTS:	ORIGINAL MEASUREMENTS:
(1) Sodium metaphosphate; $NaPO_3$; [10361-03-2] (2) Water; H_2O; [7732-18-5]	Morey, G.W. J. Am. Chem. Soc. <u>1953</u>, 75, 5794-7.

VARIABLES:	PREPARED BY:
Temperature and composition.	J. Eysseltová

EXPERIMENTAL VALUES:

Solubility of $NaPO_3$ in water.

concentration

wt. fraction	mass%[a]	mol/kg[a]	$t/°C.$	primary phase
0.70	70	22.88	147	NaH_2PO_4
0.739	73.9	27.77	159	"
0.765	76.5	31.92	210	$Na_2H_2P_2O_7$
0.78	78	34.77	235	"
0.794	79.4	37.80	256	"
0.849	84.9	55.14	305	"
0.92	92	112.8	348	"
0.93	93	130.3	402	$NaPO_3II$
0.96	96	235.4	517	$NaPO_3I$

[a] These values were calculated by the compiler.

AUXILIARY INFORMATION

METHOD/APPARATUS/PROCEDURE:	SOURCE AND PURITY OF MATERIALS:
The solubilities below 400°C were made in sealed glass tubes rotating in an oven which was provided with an automatic temperature control (1). Runs above 400°C were also made in sealed glass tubes but in an ordinary furnace without continuous rotation. The tubes were inverted several times to make sure that equilibrium was obtained. Temperatures were determined with a Pt-Pt90Rh10 thermocouple, the bare junction of which was within a few mm of the middle of the tube. The glass tubes were Corning 702 glass.	$NaPO_3 \cdot H_2O$ was obtained from Ontario Research Foundation, but the purity is not specified.

ESTIMATED ERROR:
No information is given but the compiler estimates the accuracy of the temperature measurement to be within ± 1°C.

REFERENCES:
1. Kracek, F.C.; Morey, G.W.; Merwin, H.E. Am. J. Sci. <u>1938</u>, 35A, 143.

COMPONENTS:	EVALUATOR:
(1) Sodium dihydrogenphosphate; NaH_2PO_4; [7553-80-7] (2) Water; H_2O; [7732-18-5]	J. Eysseltová Charles University Prague, Czechoslovakia May 1985

CRITICAL EVALUATION:

The values reported for the solubilities in multicomponent systems containing sodium dihydrogenphosphate will be presented, and, where possible, evaluated in this chapter.

Ternary systems with two saturating components

Seven such systems have been studied, but in several of these reports, there are insufficient data to permit a critical evaluation to be made. There was no evidence for the formation of solid solutions and/or solid ternary compounds in any of these systems. Critical evaluations will be made for the following four systems.

1. The NaH_2PO_4-$NaBO_2$-H_2O system. This system has been studied at 298 and 308 K (1). The appearance of $Na_2B_4O_7 \cdot 10H_2O$ [61028-24-8] as one of the solid phases in this system indicates that the system cannot be treated as a ternary one, but should be considered as a part of the Na_2O-B_2O_3-P_2O_5-H_2O system.
2. The NaH_2PO_4-$NaNO_3$-H_2O system. Solubility data for this system have been obtained at 273, 283, 293 and 303 K (2), and at 323 K (3). The solubility data for the isotherms at 273, 283 and 293 K can be described by equation [1], derived by Kirgintsev (4). In this equation, m_1 is the molality of the salt, m_o is the molality of the same salt in a

$$\log(m_1/m_o) = (-1/\alpha)\log y_1 \qquad [1]$$

saturated aqueous solution under the same conditions, y_1 is the solute mole fraction of the component, and α is an adjustable parameter. In an ideal solution $\alpha = da_w/dm$ where a_w is the activity of the water and m is the molality of the component. The values of $(-1/\alpha)$ for this system are given in Table I.

Table I. Parameters of equation [1] for the NaH_2PO_4-$NaNO_3$-H_2O system.

	NaH_2PO_4		$NaNO_3$	
T/K	$-1/\alpha$	σ^a	$-1/\alpha$	σ^a
273	0.68	0.02	1.03	0.06
283	0.70	0.02	0.98	0.03
293	0.68	0.03	0.88	0.06

[a] the standard deviation of the $(-1/\alpha)$ value

The data at 303 K (2) and at 323 K (3) do not give a constant value for α. This is possibly due to some interaction between the two salts at these higher temperatures, whereas in the derivation of equation [1] it was assumed that there is no such interaction. Figure 1 is a summary of the solubility data obtained experimentally and by the use of equation [1].
3. The NaH_2PO_4-$NaCl$-H_2O system. Solubility data for this system were reported for 273, 298 and 348 K (5), at 298 and 323 K (6), and at 313 K (7). Of all these data only the 298 K isotherms (5, 6) can be compared directly, Figure 2. The solubility of NaH_2PO_4 can be expressed by equation [2]. Neither the solubility of NaH_2PO_4 at other temperatures nor the solubility of $NaCl$ can be expressed by equation [1]. Attempts to use

$$\log m_1 = \log 7.7 - (1.1 \pm 0.1) \log y_1 \qquad [2]$$

equation [1] gave values for α that either varied or had standard deviations of about 50%.
4. The NaH_2PO_4-KH_2PO_4-H_2O system. Babenko and Vorob'eva (8) present solubility data from which it is possible to construct a polytherm and make comparison with data obtained by others for solutions simultaneously saturated with two solids (5, 7). This is done on Figure 3. The agreement is fairly good. However, equation [1] could not be used for this system because it was impossible to obtain a constant value for α. This is likely due to the fact that equation [1] was derived on the basis of simplifying assumptions and this system is too complex for such assumptions to be valid. A more precise model is needed but this may require additional parameters. At present there are too few data available to calculate values for additional parameters.

Three other ternary systems have been studied but only a limited amount of experimental data is available. Therefore, no evaluation of these data can be made.

(continued next page)

COMPONENTS:	EVALUATOR:
(1) Sodium dihydrogenphosphate; NaH_2PO_4; [7558-80-7] (2) Water; H_2O; [7732-18-5]	J. Eysseltová Charles University Prague, Czechoslovakia May 1985

CRITICAL EVALUATION:

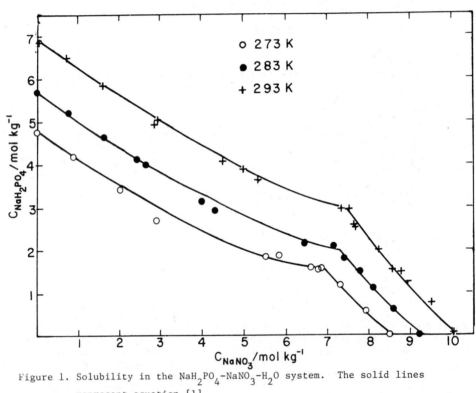

Figure 1. Solubility in the NaH_2PO_4-$NaNO_3$-H_2O system. The solid lines represent equation [1].

COMPONENTS:	EVALUATOR:
(1) Sodium dihydrogenphosphate; NaH_2PO_4; [7558-80-7] (2) Water; H_2O; [7732-18-5]	J. Eysseltová Charles University Prague, Czechoslovakia May 1985

CRITICAL EVALUATION:

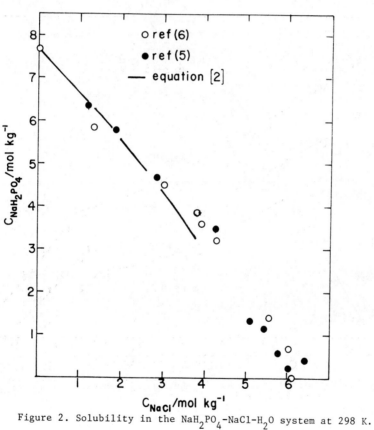

Figure 2. Solubility in the NaH_2PO_4-NaCl-H_2O system at 298 K.

COMPONENTS:	EVALUATOR:
(1) Sodium dihydrogenphosphate; NaH_2PO_4; [7558-80-7] (2) Water; H_2O; [7732-18-5]	J. Eysseltová Charles University Prague, Czechoslovakia May 1985

CRITICAL EVALUATION:

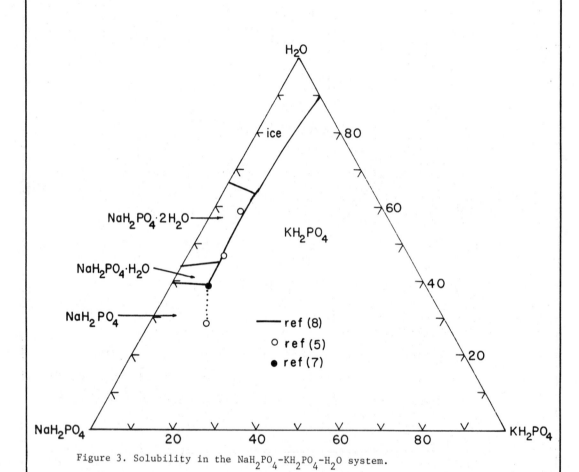

Figure 3. Solubility in the NaH_2PO_4-KH_2PO_4-H_2O system.

COMPONENTS:	EVALUATOR:
(1) Sodium dihydrogenphosphate; NaH_2PO_4; [7558-80-7] (2) Water; H_2O; [7732-18-5]	J. Eysseltová Charles University Prague, Czechoslovakia May 1985

CRITICAL EVALUATION: (cont'd)

These other systems are: $NaH_2PO_4-Na_2SO_4-H_2O$ studied at 298 K (9); $NaH_2PO_4-NaClO_4-H_2O$ studied at 298 K (10); and $NaH_2PO_4-NH_4H_2PO_4-H_2O$, studied at 263 to 303 K (11).

Systems having an organic component

Solubility data for the NaH_2PO_4-acetone-H_2O system and for two sections through the $NaH_2PO_4-NaClO_4$-acetone-H_2O system at 298 K have been reported (12). Layer formation was observed in all the studies. Not enough data are available to make a critical evaluation of the work. However, the value reported for the solubility of $NaH_2PO_4 \cdot 2H_2O$ in water at 298 K is in error by about +20%.

Quaternary systems

Solubility data have been reported for four quaternary systems: (a) a section through the $NaH_2PO_4-NH_4H_2PO_4-(NH_4)_2HPO_4-H_2O$ system at 262 to 343 K (13); (b) three isotherms of the $Na^+, K^+||H_2PO_4^-, NO_3^--H_2O$ system (14); (c) the $Na^+, NH_4^+||H_2PO_4^-, Cl^--H_2O$ system at 298 K (15); and (d) the $Na^+, K^+||H_2PO_4^-, Cl^--H_2O$ system at 298 K (15), at 313 K (7), and at 273, 298, 323 and 373 K (16).

The 298 K solubility isotherm for the $Na^+, K^+||H_2PO_4^-, Cl^--H_2O$ system is given on Figure 4. The values given for the boundary ternary systems are those reported for the eutonic solutions in the $NaH_2PO_4-NaCl-H_2O$ system (5, 6), the $NaH_2PO_4-KH_2PO_4-H_2O$ system (5), the $KH_2PO_4-KCl-H_2O$ system (17-19), and the $KCl-NaCl-H_2O$ system (20). There appears to be a systematic error in the data of Brunisholz and Bodmer (16). The phosphate content is too large. Therefore, the data of Solov'ev, et al. (15) are preferred.

References

1. Beremzhanov, B.A.; Savich, R.F.; Kunanbaev, G.S. *Khim. Khim. Tekhnol. (Alma Ata)* <u>1977</u>, *22*, 15.
2. Shpunt, S.J. *Zh. Prikl. Khim.* <u>1940</u>, *13*, 19.
3. Kol'ba, V.I.; Zhikharev, M.I.; Sukhanov, L.P. *Zh. Neorg. Khim.* <u>1981</u>, *26*, 828.
4. Kirgintsev, A.N. *Izv. Akad. Nauk SSSR, Ser. Khim. Nauk* <u>1965</u>, *8*, 1591.
5. Brunisholz, G.; Bodmer, M. *Helv. Chim. Acta* <u>1963</u>, *46*, 288, 2566.
6. Girich, T.E.; Gulyamov, Yu. M.; Ganz, S.N.; Miroshina, O.S. *Vopr. Khim. Khim. Tekhnol.* <u>1979</u>, *57*, 58.
7. Khallieva, Sh. D. *Izv. Akad. Nauk Terkm. SSSR, Ser. Fiz.-Tekh., Khim. Geol. Nauk* <u>1977</u>, *3*, 125.
8. Babenko, A.M.; Vorob'eva, T.A. *Zh. Prikl. Khim (Leningrad)* <u>1975</u>, *48*, 2437.
9. Apfel, O. Dissertation, Technical University, Darmstadt <u>1911</u>.
10. Lilich, L.S.; Alekseeva, E.A. *Zh. Neorg. Khim.* <u>1969</u>, *14*, 1655.
11. Shpunt, S.J. *Zh. Prikl. Khim.* <u>1940</u>, *13*, 9.
12. Ferroni, G.; Galea, J. Antonetti, G. *Bull. Soc. Chim. Fr* <u>1974</u>, 12 (Pt. 1), 273.
13. Babenko, A.M.; Vorob'eva, T.A. *Zh. Prikl. Khim.* <u>1976</u>, *49*, 1502.
14. Girich, T.E.; Gulyamov, Yu. M. *Vopr. Khim. Khim. Tekhnol.* <u>1979</u>, *57*, 54.
15. Solov'ev, A.P.; Balashova, E.F.; Verendyakina, N.A.; Zyzina, L.F. *Resp. Sb. Nauch. Tr.-Yaroslav. Gos. Ped. In-t.* <u>1978</u>, *169*, 79.
16. Brunisholz, G.; Bodmer, M. *Helv. Chim. Acta* <u>1963</u>, *46*, 289, 2575.
17. Krasil'shtschikov, A.I. *Izv. In-ta Fiz.-Khim. Anal.* <u>1933</u>, *6*, 159.
18. Filipescu, L. *Rev. Chim. (Bucharest)* <u>1971</u>, *22*, 533.
19. Mraz, R.; Srb, V.; Tichy, D.; Vosolsobe, J. *Chem. Prum.* <u>1976</u>, *26*, 511.
20. Seidell, A. *Solubilities of Inorganic and Metal Inorganic Compounds*, D. Van Nostrand Co., New York <u>1953</u>.

COMPONENTS:	EVALUATOR:
(1) Sodium dihydrogenphosphate; NaH_2PO_4; [7558-80-7] (2) Water; H_2O; [7732-18-5]	J. Eysseltová Charles University Prague, Czechoslovakia May 1985

CRITICAL EVALUATION:

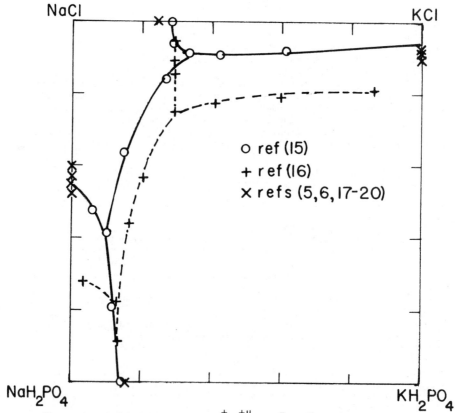

Figure 4. Solubility in the $Na^+, K^+ \| H_2PO_4^-, Cl^-$ -H_2O system at 298 K.

COMPONENTS:	ORIGINAL MEASUREMENTS:
(1) Sodium dihydrogenphosphate; NaH_2PO_4; [7558-80-7] (2) Disodium sulfate; Na_2SO_4; [7757-82-6] (3) Water; H_2O, [7732-18-5]	Apfel, O. Dissertation, Technical University, Darmstadt <u>1911</u>.

VARIABLES:	PREPARED BY:
Composition at 25°C.	J. Eysseltová

EXPERIMENTAL VALUES:

Composition of saturated solutions in the NaH_2PO_4-Na_2SO_4-H_2O system at 25°C.

PO_4^{3-}	SO_4^{2-}	$NaH_2PO_4{}^a$		$Na_2SO_4{}^a$		$H_2O{}^a$
mol/1000 g soln	mol/1000 g soln	mass%	mol/kg	mass%	mol/kg	mass%
4.08	----	48.97	8.00	----	----	51.03
3.92	0.11	47.05	7.63	1.56	0.21	51.39
3.82	0.26	45.85	7.57	3.69	0.52	50.46
3.58	0.45	42.97	7.07	6.39	0.89	50.64
3.27	0.71	39.25	6.45	10.08	1.40	50.67
3.29	0.72	39.49	6.54	10.23	1.43	50.28

a The mass% and mol/kg H_2O values were calculated by the compiler.

AUXILIARY INFORMATION

METHOD/APPARATUS/PROCEDURE:	SOURCE AND PURITY OF MATERIALS:
Equilibrium was reached isothermally. Equilibrium was ascertained by repeated analysis of the liquid phase. The solid and liquid phases were separated from each other by filtration through a mat of platinum wires. Phosphate content was determined gravimetrically as $Mg_2P_2O_7$. The sulfate content was determined gravimetrically as $BaSO_4$. Sodium content was determined as Na_2SO_4 after removing the phosphate and sulfate as $Pb_3(PO_4)_2$ and $PbSO_4$.	No information is given.

	ESTIMATED ERROR:
	No information is given.

	REFERENCES:

COMPONENTS:	ORIGINAL MEASUREMENTS:
(1) Sodium dihydrogenphosphate; NaH_2PO_4; [7558-80-7] (2) Ammonium dihydrogenphosphate; $NH_4H_2PO_4$; [7722-76-1] (3) Water; H_2O; [7732-18-5]	Shpunt, S.J. *Zh. Prikl. Khim.* <u>1940</u>, *13*, 9-18.

VARIABLES:	PREPARED BY:
Temperature and composition.	J. Eysseltová

EXPERIMENTAL VALUES:

Part 1. Crystallization temperatures on sections
of the NaH_2PO_4-$NH_4H_2PO_4$-H_2O system.

NaH_2PO_4		$NH_4H_2PO_4$			
mass%	mol/kg^a	mass%	mol/kg^a	$t/°C$	solid$_b$ phase
			Section I		
7.7	0.70	----	----	-2.1	A
7.4	0.70	4.8	0.48	-3.3	"
7.0	0.70	9.1	0.94	-4.5	"
6.8	0.71	13.0	1.41	-5.5	"
6.5	0.70	16.7	1.89	-0.6	B
6.2	0.70	20.0	2.36	8.5	"
6.0	0.70	23.1	2.83	17.1	"
5.8	0.71	25.9	3.30	25.0	"
5.6	0.71	28.6	3.78	30.8	"
			Section II		
15.4	1.52	----	----	-4.5	A
15.0	1.52	2.9	0.31	-5.0	"
14.6	1.53	5.7	0.62	-5.6	"
14.2	1.53	8.3	0.93	-6.1	"
13.9	1.54	10.7	1.23	-6.7	"
13.2	1.54	15.2	1.84	1.4	B
12.8	1.52	17.3	2.15	8.9	"
12.5	1.53	19.3	2.46	13.6	"
12.2	1.53	21.3	2.78	19.8	"
12.0	1.53	23.1	3.09	24.5	"
11.7	1.54	24.8	3.39	30.1	"

(continued next page)

AUXILIARY INFORMATION

METHOD/APPARATUS/PROCEDURE:	SOURCE AND PURITY OF MATERIALS:
A standard visual polythermic method and the isothermal method were used but no details are given. The P_2O_5 content was determined by a standard method described in the "NIUIF materials" but no reference is given. The ammonia content was determined by the Kjeldahl method. The sodium ion content was probably determined by difference-compiler.	No information is given.
	ESTIMATED ERROR: The temperature was controlled to within ±0.2 K.
	REFERENCES:

COMPONENTS:	ORIGINAL MEASUREMENTS:
(1) Sodium dihydrogenphosphate; NaH_2PO_4; [7558-80-7]	Shpunt, S.J.
(2) Ammonium dihydrogenphosphate; $NH_4H_2PO_4$; [7722-76-1]	*Zh. Prikl. Khim.* <u>1940</u>, *13*, 9-18.
(3) Water; H_2O; [7732-18-5]	

EXPERIMENTAL VALUES cont'd.

Part 1. Crystallization temperatures on sections of the NaH_2PO_4-$NH_4H_2PO_4$-H_2O system.

NaH_2PO_4		$NH_4H_2PO_4$			
mass%	mol/kg^a	mass%	mol/kg^a	t/°C	solid phase^b
		Section III			
23.1	2.50	----	----	-6.6	A
22.7	2.51	2.0	0.23	-7.1	"
22.2	2.50	3.8	0.45	-7.6	"
21.8	2.50	5.7	0.68	-8.0	"
21.4	2.50	7.4	0.90	-8.5	"
21.0	2.50	9.1	1.13	-9.0	"
20.3	2.50	12.2	1.57	-2.2	B
19.9	2.50	13.8	1.81	3.5	"
19.6	2.50	15.2	2.03	7.9	"
19.2	2.52	17.3	2.37	13.1	"
18.7	2.51	19.3	2.70	19.0	"
18.1	2.50	21.3	3.06	26.1	"
17.7	2.49	23.1	3.39	30.1	"
		Section IV			
34.6	4.41	----	----	-2.7	C
33.9	4.41	2.0	0.27	-2.6	"
33.2	4.39	3.8	0.52	-2.5	"
32.6	4.40	5.7	0.80	-2.6	"
32.0	4.40	7.4	1.06	-2.5	"
31.5	4.42	9.1	1.33	-2.9	"
30.9	4.41	10.7	1.59	1.9	B
30.4	4.41	12.2	1.85	9.0	"
29.9	4.42	13.8	2.13	13.9	"
29.4	4.42	15.2	2.38	18.5	"
28.4	4.41	18.0	2.92	27.6	"
		Section V			
38.5	5.22	----	----	7.1	C
37.8	5.23	2.0	0.29	6.8	"
37.1	5.23	3.8	0.56	6.9	"
36.4	5.24	5.7	0.86	7.2	"
35.8	5.25	7.4	1.13	7.1	"
35.2	5.26	9.1	1.42	7.1	"
34.5	5.24	10.7	1.70	7.1	B + C
33.9	5.24	12.2	1.97	12.1	B
33.3	5.24	13.8	2.27	17.9	"
32.8	5.26	15.2	2.54	22.8	"
32.2	5.25	16.7	2.84	27.3	"
		Section VI			
42.3	6.11	----	----	14.5	C
41.4	6.09	2.0	0.31	14.4	"
39.9	6.11	5.7	0.91	14.7	"
38.4	6.09	9.1	1.51	14.6	"
37.1	6.10	12.2	2.09	14.5	"
36.4	6.09	13.8	2.41	20.9	B
35.9	6.12	15.2	2.70	26.6	"
35.2	6.10	16.7	3.02	31.1	"

(continued next page)

COMPONENTS:	ORIGINAL MEASUREMENTS:
(1) Sodium dihydrogenphosphate; NaH_2PO_4; [7558-80-7] (2) Ammonium dihydrogenphosphate; $NH_4H_2PO_4$; [7722-76-1] (3) Water, H_2O; [7732-18-5]	Shpunt, S.J. *Zh. Prikl. Khim.* 1940, *13*, 9-18

EXPERIMENTAL VALUES cont'd.

Part 1. Crystallization temperatures on sections
of the $NaH_2PO_4-NH_4H_2PO_4-H_2O$ system.

NaH_2PO_4		$NH_4H_2PO_4$			
mass%	mol/kg[a]	mass%	mol/kg[a]	$t/°C$	solid[b] phase
		Section VII			
24.9	2.91	3.8	0.46	-8.4	A
26.6	3.18	3.7	0.46	-8.8	"
28.2	3.44	3.6	0.46	-9.7	"
31.1	3.96	3.5	0.46	-8.2	C
32.5	4.22	3.4	0.46	-5.3	"
34.1	4.54	3.3	0.46	-1.2	"
36.5	5.04	3.2	0.46	3.4	"
38.5	5.49	3.1	0.46	7.8	"
41.9	6.32	2.9	0.46	15.6	"
44.8	7.12	2.8	0.46	20.8	"
46.1	7.50	2.7	0.46	24.3	"
		Section VIII			
19.9	2.30	8.0	0.96	-7.9	A
22.0	2.61	7.8	0.96	-8.7	"
23.8	2.89	7.6	0.96	-9.3	"
25.6	3.18	7.4	0.96	-9.9	"
27.3	3.47	7.2	0.96	-10.7	A + C
28.8	3.74	7.1	0.96	-8.0	C
31.7	4.29	6.8	0.96	-3.3	"
34.1	4.79	6.6	0.97	1.5	"
36.5	5.33	6.4	0.97	6.5	"
38.5	5.80	6.2	0.97	10.7	"
40.2	6.22	6.0	0.97	14.7	"
41.9	6.68	5.8	0.96	18.3	"
44.8	7.52	5.6	0.98	23.5	"
47.3	8.31	5.3	0.97	28.2	"

Part 2. Solutions coexisting with two solid phases.

	NaH_2PO_4		$NH_4H_2PO_4$		H_2O	
$t/°C$	mass%	mol/kg[a]	mass%	mol/kg[a]	mass%	solid[b] phase
-4.3	----	----	16.7	1.74	83.3	A + B
-9.9	32.4	3.99	----	----	67.6	A + C
-6.0	6.6	0.70	14.7	1.62	78.7	A + B
-7.1	13.6	1.53	12.4	1.46	74.0	"
-9.1	21.0	2.53	9.8	1.23	69.2	"
-2.8	31.3	4.39	9.3	1.36	59.4	B + C
7.1	34.4	5.22	10.7	1.69	54.9	"
14.7	37.3	6.09	11.7	1.99	51.0	"
-10.2	30.0	3.75	3.4	0.44	66.6	A + C
-10.7	27.4	3.49	7.2	0.96	65.4	"

(continued next page)

COMPONENTS:	ORIGINAL MEASUREMENTS:
(1) Sodium dihydrogenphosphate; NaH_2PO_4; [7558-80-7] (2) Ammonium dihydrogenphosphate; $NH_4H_2PO_4$; [7722-76-1] (3) Water; H_2O; [7732-18-5]	Shpunt, S.J. *Zh. Prikl. Khim.* 1940, *13*, 9-18.

EXPERIMENTAL VALUES cont'd.

Part 3. Solubility isotherms.

$NH_4H_2PO_4$			NaH_2PO_4			H_2O		solid phase[b]
mass%	c^c	mol/kg^a	mass%	c^c	mol/kg^a	mass%	c^c	
temp. = -9.9°C								
----	----	----	32.4	100.0	3.99	67.6	1391	A + C
3.6	11.7	0.46	28.6	88.3	3.51	67.8	1400	A
7.5	24.0	0.96	24.6	76.0	3.02	67.9	1398	"
9.2	29.6	1.18	22.8	70.4	2.79	68.0	1402	A + B
8.5	23.8	1.17	28.4	76.2	3.75	63.1	1129	B + C
3.4	10.4	0.45	30.4	89.6	3.83	66.2	1302	C
temp. = -7°C								
----	----	----	24.2	100.0	2.66	75.8	2092	A
4.0	16.5	0.46	21.0	83.5	2.33	75.0	1988	"
8.4	33.5	0.98	17.4	66.5	1.95	74.2	1893	"
12.3	48.1	1.45	13.8	51.9	1.56	73.9	1851	A + B
10.6	34.6	1.34	20.8	65.4	2.53	68.6	1436	B
8.8	23.7	1.24	29.4	76.3	3.96	61.8	1068	B + C
7.0	19.8	0.96	29.7	80.2	3.91	63.3	1140	C
3.4	10.0	0.45	31.5	90.0	4.03	65.1	1239	"
----	----	----	33.6	100.0	4.22	66.4	1261	"
temp. = -4.3°C								
----	----	----	15.1	100.0	1.48	84.9	3745	A
8.8	56.8	0.91	7.0	43.2	0.69	84.2	3484	"
16.8	100.0	1.76	----	----	----	83.2	3184	A + B
15.3	70.9	1.70	6.6	29.1	0.70	78.1	2309	B
13.3	50.8	1.58	13.5	49.2	1.54	73.2	1782	"
11.4	36.4	1.46	20.6	63.6	2.52	68.0	1396	"
9.1	23.8	1.30	30.3	76.2	4.16	60.6	1015	B + C
6.9	18.9	0.96	31.0	81.1	4.16	62.1	1084	C
3.3	9.5	0.45	32.8	90.5	4.28	63.9	1175$_d$	"
----	----	----	34.6	100.0	4.41	65.4	----^{d}	"
temp. = 0°C								
18.4	100.0	1.96	----	----	----	81.6	2833	B
16.9	73.4	1.92	6.4	26.6	0.70	76.7	2127	"
16.06	65.0	1.86	9.02	35.0	1.00	74.92	1941	"
14.7	54.0	1.77	13.1	46.0	1.51	72.2	1697	"
13.11	42.6	1.66	18.46	57.4	2.25	68.43	1420	"
12.9	40.1	1.67	20.0	59.9	2.48	67.1	1333	"
10.53	28.5	1.48	27.47	71.5	3.69	62.0	1077	"
10.0	25.1	1.48	31.1	74.9	4.40	58.9	944	"
9.64	24.1	1.43	31.69	75.9	4.50	58.67	937	B + C
9.8	24.3	1.46	31.7	75.7	4.51	58.5	931	"
6.7	17.4	0.97	33.2	82.8	4.60	60.1	999	C
5.8	15.3	0.83	33.5	84.7	4.60	60.7	1024	"
3.2	8.8	0.45	34.8	91.2	4.68	62.0	1083	"
2.62	7.3	0.36	34.84	92.7	4.64	62.54	1109	"
----	----	----	36.4	100.0	4.77	63.6	1150	"

(continued next page)

COMPONENTS:	ORIGINAL MEASUREMENTS:
(1) Sodium dihydrogenphosphate; NaH_2PO_4; [7558-80-7]	Shpunt, S.J.
(2) Ammonium dihydrogenphosphate; $NH_4H_2PO_4$; [7722-76-1]	*Zh. Prikl. Khim.* <u>1940</u>, *13*, 9-18.
(3) Water; H_2O; [7732-18-5]	

EXPERIMENTAL VALUES cont'd.

Part 3. Solubility isotherms.

$NH_4H_2PO_4$			NaH_2PO_4			H_2O		solid[b]
mass%	c^c	mol/kg^a	mass%	c^c	mol/kg^a	mass%	c^c	phase

temp. = +10°C

mass%	c^c	mol/kg^a	mass%	c^c	mol/kg^a	mass%	c^c	phase
21.8	100.0	2.42	----	----	----	78.2	2288	B
20.5	77.6	2.43	6.2	22.4	0.70	73.3	1766	"
18.0	59.7	2.26	12.7	40.3	1.53	69.3	1470	"
16.1	46.4	2.16	19.4	53.6	2.51	64.5	1196	"
12.7	30.5	1.93	30.2	69.5	4.41	57.1	875	"
11.5	26.0	1.84	34.1	74.0	5.22	54.4	788	"
11.1	24.6	1.81	35.5	75.4	5.54	53.4	757	B + C
6.2	14.6	0.96	38.0	85.4	5.67	55.8	838	C
3.0	7.3	0.45	39.5	92.7	5.72	57.5	900	"
----	----	----	40.5	100.0	5.67	59.5	961	"

temp. = +20°C

mass%	c^c	mol/kg^a	mass%	c^c	mol/kg^a	mass%	c^c	phase
25.9	100.0	3.04	----	----	----	74.1	1824	B
24.2	81.4	3.00	5.7	18.6	0.68	70.1	1508	"
21.4	64.9	2.80	12.1	35.1	1.52	66.5	1293	"
19.5	51.8	2.75	18.9	48.2	2.56	61.6	1048	"
15.8	36.1	2.50	29.2	63.9	4.42	55.0	806	"
14.5	31.5	2.40	32.9	68.5	5.21	52.6	730	"
13.4	27.6	2.33	36.7	72.4	6.13	49.9	658	"
12.4	24.9	2.22	39.1	75.1	6.72	48.5	621	B + C
5.7	12.2	0.96	42.9	87.8	6.95	51.4	701	C
2.8	6.2	0.45	44.2	93.8	6.95	53.0	750	"
----	----	----	45.3	100.0	6.90	54.7	789	"

temp. = +30°C

mass%	c^c	mol/kg^a	mass%	c^c	mol/kg^a	mass%	c^c	phase
30.2	100.0	3.76	----	----	----	69.8	1477	B
28.0	84.0	3.66	5.5	16.0	0.70	66.5	1278	"
26.10	73.6	3.54	9.77	26.4	1.27	64.13	1156	"
25.2	69.5	3.46	11.5	30.5	1.51	63.3	1116	"
23.0	56.9	3.40	18.2	43.1	2.58	58.8	930	"
20.0	45.6	3.16	24.97	54.4	3.78	55.03	802	"
18.6	40.7	3.04	28.2	59.3	4.42	53.2	745	"
17.3	36.1	2.96	31.9	63.9	5.23	50.8	678	"
16.2	32.3	2.91	35.4	67.7	6.09	48.4	617	"
15.78	31.6	2.82	35.62	68.4	6.11	48.60	622	"
13.30	24.9	2.58	41.89	75.1	7.79	44.81	536	B + C
13.20	24.9	2.54	41.60	75.1	7.67	45.20	544	"
9.55	18.5	1.78	43.75	81.5	7.80	46.70	580	C
6.26	12.3	1.15	46.50	87.7	8.20	47.24	594	"
5.2	10.2	0.96	47.8	89.8	8.47	47.0	588	"
3.42	6.8	0.62	48.55	93.2	8.42	48.03	615	"
2.5	5.0	0.45	49.0	95.0	8.42	48.5	627	"
----	----	----	51.2	100.0	8.74	48.8	628	"

[a] The mol/kg H_2O values were calculated by the compiler.

[b] The solid phases are: A = ice; B = $NH_4H_2PO_4$; C = $NaH_2PO_4 \cdot 2H_2O$.

[c] The concentration units are: mol/100 mol of solute.

[d] The compiler calculates this missing value to be 1259.

COMPONENTS:	ORIGINAL MEASUREMENTS:
(1) Sodium dihydrogenphosphate; NaH_2PO_4; [7558-80-7] (2) Sodium nitrate; $NaNO_3$; [7631-99-4] (3) Water; H_2O; [7732-18-5]	Shpunt, S.J. *Zh. Prikl. Khim.* <u>1940</u>, *13*, 19-28.

VARIABLES:	PREPARED BY:
Temperature and composition.	J. Eysseltová

EXPERIMENTAL VALUES:

Part 1. Composition of the relevant sections.

I. 46.9% $NaNO_3$ + 4.8% NaH_2PO_4 + 48.2% H_2O, water added.
II. 43.8% $NaNO_3$ + 8.5% NaH_2PO_4 + 47.7% H_2O, water added.
III. 42.0% $NaNO_3$ + 11.0% NaH_2PO_4 + 47.0% H_2O, water added.
IV. 39.0% $NaNO_3$ + 13.2% NaH_2PO_4 + 47.8% H_2O, water added.
V. 38.0% $NaNO_3$ + 18.0% NaH_2PO_4 + 44.0% H_2O, water added.
VI. 28.7% $NaNO_3$ + 27.4% NaH_2PO_4 + 43.9% H_2O, water added.
VII. 26.4% $NaNO_3$ + 29.0% NaH_2PO_4 + 41.6% H_2O, water added.
VIII. 16.0% $NaNO_3$ + 38.0% NaH_2PO_4 + 46.0% H_2O, water added.
IX. 12.0% $NaNO_3$ + 88.0% H_2O, NaH_2PO_4 added.
X. 6.0% $NaNO_3$ + 94.0% H_2O, NaH_2PO_4 added.

Part 2. Crystallization temperatures.

Section I

NaH_2PO_4		$NaNO_3$			NaH_2PO_4		$NaNO_3$		
mass%	mol/kg^a	mass%	mol/kg^a	t/°C	mass%	mol/kg^a	mass%	mol/kg^a	t/°C
4.7	0.78	46.2	11.07	36.5	4.0	0.58	38.8	7.98	2.2
4.5	0.73	44.1	10.09	26.1	3.8	0.53	37.3	7.45	-4.5
4.3	0.67	42.2	9.28	17.8	3.7	0.51	35.9	6.99	-10.2
4.2	0.63	40.4	8.58	10.4	3.6	0.49	34.7	6.62	-15.4
3.4	0.45	33.4	6.22	-18.4	1.6	0.16	15.4	2.18	-6.8
3.2	0.41	31.2	5.60	-16.1	1.4	0.14	13.3	1.83	-5.8
3.0	0.37	29.3	5.09	-14.8	1.1	0.10	10.4	1.38	-4.4
2.3	0.26	22.6	3.54	-10.4	0.9	0.08	8.6	1.12	-3.7
2.0	0.21	18.3	2.70	-8.2					

(continued next page)

AUXILIARY INFORMATION

METHOD/APPARATUS/PROCEDURE:	SOURCE AND PURITY OF MATERIALS:
No information is given.	No information is given.
	ESTIMATED ERROR:
	No information is given.
	REFERENCES:

COMPONENTS:	ORIGINAL MEASUREMENTS:
(1) Sodium dihydrogenphosphate; NaH_2PO_4; [7558-80-7]	Shpunt, S.J.
(2) Sodium nitrate; $NaNO_3$; [7631-99-4]	*Zh. Prikl. Khim.* <u>1940</u>, *13*, 19-28.
(3) Water; H_2O; [7732-18-5]	

EXPERIMENTAL VALUES cont'd:

Part 2. Crystallization temperatures.

Section II

NaH_2PO_4		$NaNO_3$			NaH_2PO_4		$NaNO_3$		
mass%	mol/kg^a	mass%	mol/kg^a	$t/°C$	mass%	mol/kg^a	mass%	mol/kg^a	$t/°C$
8.2	1.39	42.7	10.23	35.3	5.8	0.75	29.9	5.47	-17.4
7.5	1.17	39.1	8.61	17.8	5.4	0.68	28.1	4.97	-16.0
7.2	1.08	37.5	7.98	9.8	4.1	0.46	21.5	3.40	-10.9
6.9	1.01	36.0	7.42	1.4	3.3	0.35	17.4	2.58	-8.4
6.8	0.98	35.2	7.14	-2.3	2.8	0.28	14.6	2.08	-6.9
6.6	0.93	34.4	6.86	-6.0	2.4	0.24	12.6	1.74	-6.0
6.5	0.90	33.6	6.61	-9.7	1.9	0.18	9.9	1.32	-4.5
6.3	0.86	32.9	6.37	-13.6	1.6	0.15	8.0	1.04	-3.8
6.2	0.84	32.0	6.09	-15.8	1.3	0.12	6.9	0.89	-3.0
6.0	0.79	30.9	5.76	-18.2					

Section III

NaH_2PO_4		$NaNO_3$			NaH_2PO_4		$NaNO_3$		
mass%	mol/kg^a	mass%	mol/kg^a	$t/°C$	mass%	mol/kg^a	mass%	mol/kg^a	$t/°C$
10.6	1.81	40.6	9.79	34.3	8.0	1.08	30.4	5.80	-16.4
10.2	1.67	38.8	8.95	24.7	5.8	0.67	22.3	3.65	-12.2
9.8	1.54	37.2	8.26	16.3	4.7	0.51	17.8	2.70	-9.1
9.4	1.43	35.7	7.65	9.2	3.9	0.40	14.8	2.14	-7.4
9.0	1.32	34.3	7.12	0.8	3.3	0.33	12.7	1.78	-6.3
8.7	1.24	33.1	6.69	-6.0	2.9	0.28	11.1	1.52	-5.4
8.4	1.17	31.9	6.29	-11.2	2.3	0.22	8.9	1.18	-4.3
8.2	1.13	31.4	6.12	-13.2	1.9	0.17	7.4	0.96	-3.5
8.1	1.11	31.0	5.99	-15.2					

Section IV

NaH_2PO_4		$NaNO_3$			NaH_2PO_4		$NaNO_3$		
mass%	mol/kg^a	mass%	mol/kg^a	$t/°C$	mass%	mol/kg^a	mass%	mol/kg^a	$t/°C$
12.7	2.12	37.5	8.86	27.7	9.0	1.16	26.6	4.86	-18.0
12.1	1.94	35.8	8.08	19.0	8.5	1.06	25.0	4.42	-16.4
11.6	1.79	34.3	7.46	11.1	8.0	0.97	23.6	4.06	-14.0
11.1	1.65	32.9	6.91	3.5	6.2	0.68	18.3	2.85	-10.1
10.7	1.56	32.2	6.63	0.4	5.0	0.52	14.9	2.19	-8.1
10.2	1.42	30.2	5.96	-5.3	4.3	0.43	12.6	1.78	-6.5
10.0	1.38	29.6	5.77	-7.2	3.3	0.32	9.6	1.30	-4.9
9.7	1.31	28.5	5.42	-13.0					

Section V

NaH_2PO_4		$NaNO_3$			NaH_2PO_4		$NaNO_3$		
mass%	mol/kg^a	mass%	mol/kg^a	$t/°C$	mass%	mol/kg^a	mass%	mol/kg^a	$t/°C$
16.5	2.83	34.9	8.45	30.7	12.3	1.66	25.9	4.93	-9.8
16.1	2.69	34.0	8.02	26.5	11.6	1.51	24.4	4.48	-14.3
15.9	2.62	33.5	7.79	22.7	11.3	1.45	23.7	4.29	-16.8
15.8	2.58	33.2	7.66	20.7	11.0	1.39	23.1	4.12	-16.0
15.5	2.49	32.7	7.43	18.0	10.4	1.28	21.9	3.80	-14.5
15.3	2.43	32.3	7.25	15.5	9.9	1.19	20.9	3.55	-13.5
14.8	2.29	31.3	6.83	13.2	9.4	1.11	19.9	3.31	-12.6
14.5	2.20	30.5	6.52	10.6	8.6	0.98	18.2	2.92	-11.2
14.0	2.07	29.6	6.17	8.0	7.1	0.76	15.0	2.26	-8.9
13.9	2.04	29.2	6.04	6.3	6.0	0.62	12.8	1.85	-7.3
13.5	1.94	28.4	5.75	3.7	5.3	0.53	11.8	1.67	-6.3
13.1	1.84	27.5	5.45	-1.1	4.2	0.40	8.9	1.20	-4.9
12.7	1.75	26.7	5.18	-4.8	3.5	0.33	7.3	0.96	-3.8

Section VI

NaH_2PO_4		$NaNO_3$			NaH_2PO_4		$NaNO_3$		
mass%	mol/kg^a	mass%	mol/kg^a	$t/°C$	mass%	mol/kg^a	mass%	mol/kg^a	$t/°C$
34.9	7.88	28.2	8.99	38.5	30.7	5.75	24.8	6.56	24.1
34.3	7.52	27.7	8.58	34.4	29.6	5.30	23.9	6.05	20.5
33.8	7.24	27.3	8.26	31.6	27.6	4.61	22.5	5.30	15.9
33.2	6.92	26.8	7.88	29.5	25.8	4.02	20.8	4.58	6.8
31.9	6.27	25.7	7.13	26.2					

(continued next page)

COMPONENTS:	ORIGINAL MEASUREMENTS
(1) Sodium dihydrogenphosphate; NaH$_2$PO$_4$; [7558-80-7]	Shpunt, S.J.
(2) Sodium nitrate; NaNO$_3$; [7631-99-4]	*Zh. Prikl. Khim.* 1940, *13*, 19-28.
(3) Water; H$_2$O; [7732-18-5]	

EXPERIMENTAL VALUES cont'd:

Part 2. Crystallization temperatures.

Section VII

NaH$_2$PO$_4$		NaNO$_3$			NaH$_2$PO$_4$		NaNO$_3$		
mass%	mol/kg^a	mass%	mol/kg^a	t/°C	mass%	mol/kg^a	mass%	mol/kg^a	t/°C
26.7	4.57	24.6	5.94	28.8	15.8	1.89	14.6	2.47	-12.9
25.7	4.23	23.7	5.51	24.7	15.1	1.77	13.9	2.30	-11.9
23.8	3.66	22.0	4.78	18.7	12.2	1.33	11.3	1.74	-9.1
22.2	3.22	20.4	4.18	11.9	10.3	1.07	9.5	1.39	-7.4
20.8	2.89	19.2	3.76	6.2	8.9	0.89	8.2	1.16	-6.1
20.0	2.71	18.5	3.54	2.5	7.8	0.76	7.2	1.00	-5.5
19.3	2.56	17.8	3.33	-1.2	7.0	0.67	6.5	0.88	-4.9
18.4	2.37	17.0	3.10	0.0	5.8	0.54	5.3	0.70	-3.9
17.5	2.20	16.1	2.85	-11.1	4.9	0.45	4.5	0.58	-3.2
16.6	2.03	15.3	2.64	-13.7					

Section VIII

38.0	6.88	16.0	4.09	33.6	21.3	2.55	9.0	1.52	-11.7
37.0	6.50	15.6	3.87	31.7	20.3	2.37	8.4	1.39	-10.8
35.4	5.93	14.9	3.53	28.2	18.4	2.08	7.8	1.24	-9.4
33.9	5.45	14.3	3.25	24.7	15.1	1.60	6.3	0.94	-7.2
31.4	4.72	13.2	2.80	17.6	12.8	1.30	5.4	0.78	-5.9
29.1	4.14	12.3	2.47	11.2	11.0	1.08	4.6	0.64	-4.9
27.8	3.83	11.7	2.28	7.8	8.7	0.83	3.7	0.50	-3.9
26.0	3.44	11.0	2.05	0.8	7.2	0.67	3.0	0.39	-3.2
24.5	3.13	10.3	1.86	-4.2	5.3	0.48	2.2	0.28	-2.4
23.8	3.00	10.0	1.78	-7.0	4.2	0.37	1.8	0.22	-1.8
22.5	2.76	9.5	1.64	-12.7					

Section IX

0	0.0	12.0	1.60	-4.8	34.1	4.90	7.9	1.60	13.3
7.0	0.71	11.2	1.61	-6.6	36.4	5.42	7.6	1.60	16.6
12.7	1.40	11.5	1.78	-8.4	38.4	5.91	7.5	1.63	20.5
17.7	2.04	9.9	1.61	-10.2	39.4	6.16	7.3	1.61	22.2
22.0	2.67	9.4	1.61	-11.6	40.2	6.36	7.1	1.58	23.7
25.6	3.26	8.9	1.60	-6.4	41.1	6.60	7.0	1.59	25.3
27.3	3.55	8.7	1.60	-2.8	41.9	6.82	6.9	1.59	26.3
28.8	3.83	8.6	1.62	-0.1	43.4	7.26	6.8	1.61	28.7
31.7	4.39	8.2	1.60	7.1	45.1	7.78	6.6	1.61	29.8

Section X

----	----	6.0	0.75	-2.2	36.4	5.07	3.8	0.75	9.2
7.0	0.67	5.6	0.75	-4.1	38.4	5.52	3.7	0.75	13.4
12.7	1.29	5.2	0.74	-5.7	39.4	5.76	3.6	0.74	14.9
17.7	1.90	4.9	0.74	-7.2	40.2	5.96	3.6	0.75	16.6
22.0	2.50	4.7	0.75	-8.7	41.1	6.18	3.5	0.74	18.1
25.6	3.05	4.5	0.76	-10.2	41.9	6.38	3.4	0.73	19.5
27.3	3.33	4.4	0.76	-10.9	43.4	6.80	3.4	0.73	22.6
28.8	3.59	4.3	0.76	-8.5	44.8	7.18	3.2	0.72	24.8
30.3	3.85	4.2	0.75	-4.7	45.5	7.39	3.2	0.73	26.2
31.7	4.11	4.1	0.75	-1.8	46.1	7.58	3.2	0.74	27.4
34.1	4.59	4.0	0.76	4.3					

(continued next page)

AMO—F

COMPONENTS:	ORIGINAL MEASUREMENTS:
(1) Sodium dihydrogenphosphate; NaH_2PO_4; [7558-80-7]	Shpunt, S.J.
(2) Sodium nitrate; $NaNO_3$; [7631-99-4]	*Zh. Prikl. Khim.* 1940, *13*,]9-28.
(3) Water; H_2O; [7732-18-5]	

EXPERIMENTAL VALUES cont'd:

Part 3. Solutions coexisting with two equilibrium solid phases.

NaH_2PO_4		$NaNO_3$		H_2O		
mass%	mol/kg[a]	mass%	mol/kg[a]	mass%	$t/°C$	solid phase[c]
----	----	38.4	7.33	61.6	-17.5	A + B
3.6	0.48	34.1	6.44	62.3	-18.4	A + B
6.2	0.84	32.0	6.09	61.8	-18.9	A + B
8.4	1.17	31.8	6.26·	59.8	-12.8	B + C
7.8	1.03	29.4	5.51	62.8	-18.8	A + C
11.2	1.67	32.8	6.89	56.0	2.3	B + C
9.2	1.20	27.2	5.03	63.6	-18.5	A + C
15.5	2.48	32.4	7.32	52.1	15.1	B + C
11.6	1.51	24.4	4.48	64.0	-17.2	A + C
26.0	4.63	27.2	6.84	46.8	29.8	B + C
17.7	2.23	16.2	2.89	66.1	-14.6	A + C
23.0	2.84	9.6	1.68	67.4	-12.8	A + C
23.4	2.89	9.2	1.60	67.4	-12.5	A + C
27.8	3.41	4.3	0.74	67.9	-10.9	A + C

Part 4. Solubility isotherms in the NaH_2PO_4-$NaNO_3$-H_2O system.

NaH_2PO_4			$NaNO_3$			H_2O		solid phase[c]
mass%	mol/kg[a]	M[b]	mass%	mol/kg[a]	M[b]	mass%	M[b]	
			temp. = -17.5°C.					
----	----	---	38.4	7.33	100.0	61.6	758	A + B
3.6	0.48	7.0	34.4	6.53	93.0	62.0	793	B
6.3	0.85	12.1	32.2	6.16	87.9	61.5	793	"
7.6	1.03	14.8	31.0	5.94	85.2	61.4	797	B + C
8.0	1.08	16.0	30.0	5.69	84.0	62.0	820	C
9.4	1.24	19.5	27.5	5.13	80.5	63.1	872	"
10.7	1.39	23.1	25.2	4.62	76.9	64.1	924	A + C
8.9	1.14	19.3	26.3	4.77	80.7	64.8	940	A
7.4	0.96	19.6	28.3	5.18	84.4	64.3	906	"
5.9	0.77	12.2	30.0	5.51	87.8	64.1	886	"
3.4	0.44	6.8	32.8	6.05	93.2	63.8	856	"
			temp. = -14°C.					
----	----	---	39.1	7.55	100.0	60.9	736	B
3.7	0.50	7.0	35.3	6.81	93.0	61.0	759	"
6.4	0.88	12.1	33.0	6.41	87.9	60.6	762	"
8.2	1.13	15.6	31.4	6.12	84.4	60.4	766	B + C
8.3	1.14	15.9	31.2	6.07	84.1	60.5	771	C
9.6	1.28	19.4	28.2	5.33	80.6	62.2	840	"
12.0	1.59	25.3	25.0	4.67	74.7	63.0	888	"
17.8	2.25	43.9	16.2	2.89	56.1	66.0	1082	"
19.4	2.43	49.6	14.0	2.47	50.4	66.6	1133	A + C
17.1	2.11	43.9	15.5	2.70	56.1	67.4	1152	A
10.0	1.21	25.1	21.2	3.62	74.9	68.8	1149	"
7.6	0.90	19.4	22.5	3.79	80.6	69.9	1182	"
6.4	0.77	15.7	24.4	4.15	84.3	69.2	1132	"
5.1	0.61	12.3	25.8	4.39	87.7	69.1	1109	"
2.9	0.35	6.8	28.2	4.82	93.2	68.9	1076	"
----	----	---	31.4	5.38	100.0	68.6	1028	"

(continued next page)

COMPONENTS:	ORIGINAL MEASURMENTS:
(1) Sodium dihydrogenphosphate; NaH_2PO_4; [7558-80-7]	Shpunt, S.J.
(2) Sodium nitrate; $NaNO_3$; [7631-99-4]	*Zh. Prikl. Khim.* <u>1940</u>, *13*, 19-28
(3) Water; H_2O; [7732-18-5]	

EXPERIMENTAL VALUES cont'd:

Part 4. Solubility isotherms in the NaH_2PO_4-$NaNO_3$-H_2O system.

NaH_2PO_4			$NaNO_3$			H_2O		solid[c]
mass%	mol/kg[a]	M[b]	mass%	mol/kg[a]	M[b]	mass%	M[b]	phase[c]
			temp. = -9.9°C.					
----	----	---	39.8	7.78	100.0	60.2	714	B
1.30	0.18	2.3	38.99	7.68	97.7	59.71	707	"
3.8	0.53	6.9	36.2	7.10	93.1	60.0	728	"
6.01	0.85	10.8	34.95	6.96	89.2	59.04	712	"
6.6	0.92	12.1	33.7	6.64	87.9	59.7	736	"
8.5	1.20	15.7	32.3	6.42	84.3	59.2	730	"
9.02	1.28	16.5	32.37	6.50	83.5	58.61	714	B + C
8.8	1.24	16.3	32.0	6.36	83.7	59.2	781	C
10.0	1.37	19.5	29.3	5.68	80.5	60.7	789	"
10.43	1.46	19.8	29.85	5.88	80.2	59.72	757	"
12.3	1.66	25.8	25.8	4.90	74.8	61.9	848	"
15.75	2.08	34.4	21.3	3.98	65.6	62.95	916	"
18.1	2.31	43.9	16.6	2.99	56.1	65.3	1039	"
20.74	2.62	52.4	13.38	2.39	47.6	65.88	1108	"
23.6	2.95	63.1	9.8	1.73	36.9	66.6	1187	"
24.1	3.00	65.3	9.1	1.60	34.7	66.8	1205	"
28.4	3.52	82.3	4.3	0.75	17.7	67.3	1300	"
32.4	3.99	100.0	----	----	----	67.6	1391	A
25.0	2.95	79.7	4.5	0.75	20.3	70.5	1497	"
19.2	2.20	63.0	8.0	1.29	37.0	72.8	1592	"
17.3	1.98	55.2	10.0	1.62	44.8	72.7	1545	"
13.3	1.48	43.8	12.1	1.91	56.2	74.6	1639	"
7.9	0.87	25.1	16.6	2.59	74.9	75.5	1607	"
6.0	0.66	19.5	17.7	2.73	80.5	76.3	1639	"
5.0	0.55	15.8	19.0	2.94	84.2	76.0	1592	"
3.9	0.43	12.1	20.2	3.13	87.9	75.9	1562	"
2.3	0.25	6.9	21.8	3.38	93.1	75.9	1531	"
----	----	----	23.3	3.57	100.0	76.7	1552	"
			temp. = 0°C.					
----	----	----	41.9	8.48	100.0	58.1	654	B
4.02	0.58	6.9	38.67	7.94	93.1	57.31	653	"
4.0	0.58	6.9	38.43	7.85	93.1	57.6	660	"
7.0	0.67	12.2	35.8	7.33	87.8	57.2	666	"
7.98	1.17	13.8	35.24	7.30	86.2	56.78	656	"
9.0	1.32	15.7	34.2	7.08	84.3	56.8	661	"
10.64	1.57	18.7	32.96	6.88	81.3	56.40	657	B + C
10.5	1.54	18.6	32.7	6.77	81.4	56.8	667	"
10.9	1.59	19.6	32.0	6.59	80.4	57.1	679	C
13.17	1.89	24.5	28.81	5.84	75.5	58.02	718	"
13.2	1.86	25.2	27.8	5.54	74.8	59.0	750	"
19.50	2.68	40.9	19.97	3.88	59.1	60.53	847	"
20.0	2.70	43.8	18.2	3.46	56.2	61.8	902	"
25.9	3.41	63.0	10.8	2.01	37.0	63.3	1027	"
26.67	3.54	64.3	10.52	1.97	35.7	62.81	1008	"
31.83	4.18	82.8	4.70	0.87	17.2	63.47	1106	"
32.6	4.28	85.3	4.0	0.74	14.7	63.4	1106	"
36.4	4.77	100.0	----	----	----	63.6	1150	"

(continued next page)

COMPONENTS:	ORIGINAL MEASUREMENTS:
(1) Sodium dihydrogenphosphate; NaH_2PO_4; [7558-80-7] (2) Sodium nitrate; $NaNO_3$; [7631-99-4] (3) Water; H_2O; [7732-18-5]	Shpunt. S.J. *Zh. Prikl. Khim.* <u>1940</u>, *13*, 19-28.

EXPERIMENTAL VALUES cont'd:

Part 4. Solubility isotherms in the NaH_2PO_4-$NaNO_3$-H_2O system.

NaH_2PO_4			$NaNO_3$			H_2O		solid
mass%	mol/kg[a]	M[b]	mass%	mol/kg[a]	M[b]	mass%	M[b]	phase[c]
					temp. = 10°C.			
----	----	---	43.9	9.21	100.0	56.1	604	B
4.2	0.63	6.9	40.5	8.62	93.1	55.3	603	"
7.4	1.12	12.2	37.7	8.09	87.8	54.9	604	"
9.5	1.46	15.8	36.1	7.81	84.2	54.4	600	"
11.6	1.78	19.5	34.1	7.39	80.5	54.3	605	"
13.6	2.11	22.7	32.7	7.16	77.3	53.7	599	B + C
14.4	2.17	25.2	30.3	6.45	74.8	55.3	645	C
20.4	2.92	40.4	21.3	4.30	59.6	58.3	769	"
21.9	3.14	43.8	19.9	4.02	56.2	58.2	778	"
29.0	4.10	63.0	12.1	2.42	37.0	58.9	856	"
33.0	4.66	74.5	8.0	1.60	25.5	59.0	890	"
36.8	5.16	87.2	3.8	0.75	12.8	59.4	939	"
40.5	5.67	100.0	----	----	----	59.5	961	"
					temp. = 20°C.			
----	----	----	46.0	10.02	100.0	54.0	555	B
4.5	0.71	6.8	42.7	9.51	93.2	52.8	545	"
7.8	1.24	12.2	39.8	8.94	87.8	52.4	546	"
8.86	1.41	13.9	38.89	8.76	86.1	52.25	547	"
9.9	1.58	15.6	38.0	8.58	84.4	52.1	546	"
12.4	2.01	19.6	36.1	8.25	80.4	51.5	545	"
15.57	2.54	24.9	33.35	7.68	75.1	51.08	544	"
15.8	2.58	25.2	33.1	7.62	74.8	51.1	545	"
17.79	2.96	28.3	32.05	7.52	71.7	50.16	540	B + C
18.1	2.99	29.0	31.5	7.35	71.0	50.4	542	"
23.0	3.62	40.5	24.0	5.33	59.5	53.0	621	C
24.7	3.90	43.7	22.5	5.01	56.3	52.08	624	"
26.10	4.07	48.4	20.44	4.50	51.6	52.46	625	"
32.6	5.05	62.9	13.6	2.97	37.1	53.8	692	"
32.27	4.94	63.2	13.33	2.88	36.8	54.40	709	"
38.1	5.84	78.3	7.5	1.62	21.7	54.4	745	"
42.2	6.46	89.9	3.4	0.74	10.1	54.4	772	"
45.3	6.90	100.0	----	----	----	54.7	789	"
					temp. = 30°C.			
----	----	----	48.0	10.86	100.0	52.0	490	B
4.7	0.78	6.8	44.8	10.44	93.2	50.5	498	"
6.25	1.05	9.1	43.98	10.40	90.9	49.77	485	"
8.1	1.35	12.1	41.9	9.86	87.9	50.0	496	"
9.10	1.53	13.5	41.31	9.80	86.5	49.59	490	"
10.4	1.74	15.6	39.8	9.40	84.4	49.8	494	"
13.0	2.21	19.5	38.0	9.12	80.5	49.0	490	"
13.31	2.27	19.9	37.77	9.08	80.1	48.92	490	"
16.7	2.87	25.4	34.8	8.44	74.6	48.5	491	"
20.3	3.56	30.7	32.25	8.00	69.3	47.45	482	"
24.99	4.44	38.6	28.10	7.05	61.4	46.91	485	"
26.72	4.80	41.2	26.89	6.82	58.8	46.39	479	B + C
26.0	4.63	40.4	27.2	6.84	59.6	46.8	485	"
27.6	4.87	43.7	25.2	6.28	56.3	47.2	498	C
28.85	5.13	45.7	24.3	6.11	54.3	46.82	493	"
36.5	6.32	62.7	15.4	3.77	37.3	48.10	551	"
37.45	6.56	64.0	14.95	3.70	36.0	47.60	542	"
40.76	7.10	71.7	11.41	2.81	28.3	47.83	561	"
45.0	7.75	82.9	6.6	1.60	17.1	48.4	594	"
46.20	8.10	84.0	6.25	1.55	16.0	47.55	576	"
48.87	8.47	92.0	3.04	0.74	8.0	48.09	603	"
49.1	8.54	92.0	3.0	0.74	8.0	47.9	599	"
51.2	8.74	100.0	----	----	----	48.3	628	"

[a] The mol/kg H_2O values were calculated by the compiler.
[b] The concentration units are: mol/100 mol of solute.
[c] The solid phases are: A = ice; B = $NaNO_3$; C = $NaH_2PO_4 \cdot 2H_2O$.

COMPONENTS:	ORIGINAL MEASUREMENTS:
(1) Sodium dihydrogenphosphate; NaH_2PO_4; [7558-80-7] (2) Sodium chloride; NaCl; [7647-14-5] (3) Water; H_2O; [7732-18-5]	Brunisholz, G.; Bodmer, M. *Helv. Chim. Acta* <u>1963</u>, *46*, 7, 288, 2566-74.

VARIABLES:	PREPARED BY:
Composition and temperature.	J. Eysseltová

EXPERIMENTAL VALUES:

Solubility isotherms in the NaH_2PO_4-NaCl-H_2O system

Na^+ ion%	Cl^- ion%	NaH_2PO_4 mass%	NaH_2PO_4 mol/kg[a]	NaCl mass%	NaCl mol/kg[a]	H_2O n[b]	H_2O mass%[a]	solid phase[c]
				temp. = 0°C.				
91.06	86.56	2.64	0.30	24.88	5.87	818.6	72.47	A
76.31	64.51	8.32	0.99	22.11	5.44	658.7	69.55	A + B
63.31	44.92	13.50	1.59	16.09	3.91	637.9	70.39	B
50.76	26.47	19.97	2.39	10.50	2.58	568.9	69.52	"
43.09	14.98	25.54	3.13	6.57	1.65	502.0	67.87	"
				temp. = 25°C.				
93.26	89.86	1.97	0.22	25.53	6.02	828.1	72.48	A
89.73	84.52	3.30	0.39	26.34	6.40	732.8	70.35	"
84.62	76.97	4.92	0.57	24.02	5.78	739.1	71.05	"
74.50	61.37	9.40	1.13	21.81	5.42	628.1	68.77	"
65.45	48.15	14.42	1.82	19.57	5.07	527.2	66.00	"
52.67	28.95	24.99	3.46	14.87	4.23	379.8	60.12	A + B
44.71	16.87	32.45	4.66	9.62	2.84	329.7	57.91	B
39.75	9.65	38.46	5.77	6.00	1.84	289.9	55.53	"
37.45	5.89	41.64	6.36	3.80	1.19	274.0	54.55	"

(continued next page)

AUXILIARY INFORMATION

METHOD/APPARATUS/PROCEDURE:	SOURCE AND PURITY OF MATERIALS:
At 0 and 25°C the usual techniques were used (1). At 75°C a self-constructed apparatus was used for equilibration and sampling. The dihydrogenphosphate content was determined acidimetrically (after first changing it to H_3PO_4 by ion exchange) using chlorophenol red as indicator. The chloride ion content was determined by titrating potentiometrically with silver nitrate. The sodium ion and water contents were determined by difference.	No information is given.
	ESTIMATED ERROR: No information is given.
	REFERENCES: 1. Flatt, R. *Chimia* <u>1962</u>, *6*, 62.

COMPONENTS:	ORIGINAL MEASUREMENTS:
(1) Sodium dihydrogenphosphate; NaH_2PO_4; [7558-80-7]	Brunisholz, G.; Bodmer, M.
(2) Sodium chloride; NaCl; [7647-14-5]	*Helv. Chim. Acta* <u>1963</u>, *46*, 7, 288, 2566-74.
(3) Water; H_2O; [7732-18-5]	

EXPERIMENTAL VALUES cont'd:

Solubility isotherms in the NaH_2PO_4-NaCl-H_2O system.

Na^+ ion%	Cl^- ion%	NaH_2PO_4 mass%	NaH_2PO_4 mol/kg[a]	NaCl mass%	NaCl mol/kg[a]	H_2O n[b]	H_2O mass%[a]	solid phase[c]
				temp. = 75°C.				
73.29	59.89	10.40	1.29	22.70	5.80	572.9	66.88	A
43.69	15.21	39.75	6.64	10.41	3.57	236.2	49.82	"
36.06	3.79	60.24	13.83	3.46	1.63	128.8	36.28	A + C

[a] These values were calculated by the compiler.

[b] The concentration units are: mol/100 mol solute.

[c] The solid phases are: A = NaCl; B = $NaH_2PO_4 \cdot 2H_2O$; C = NaH_2PO_4.

COMPONENTS:	ORIGINAL MEASUREMENTS:
(1) Sodium dihydrogenphosphate; NaH_2PO_4; [7558-80-7]	Brunisholz, G.; Bodmer, M.
(2) Potassium dihydrogenphosphate; KH_2PO_4; [7778-70-0]	*Helv. Chim. Acta* <u>1963</u>, *46*, 288, 2566-74.
(3) Water; H_2O; [7732-18-5]	

VARIABLES:	PREPARED BY:
Composition and temperature.	J. Eysseltová

EXPERIMENTAL VALUES:

Solubility isotherms in the KH_2PO_4-NaH_2PO_4-H_2O system.

K^+ ion%	Na^+ ion%	H_2O n^b	$KH_2PO_4{}^a$ mass%	mol/kg	$NaH_2PO_4{}^a$ mass%	mol/kg	H_2O mass%	solid phase^c
				temp. = 0°C				
27.03	6.30	1495	11.73	1.00	2.41	0.23	85.85	A
21.62	11.71	1217	11.20	0.98	5.35	0.53	83.42	"
16.47	16.86	941.5	10.56	0.97	9.53	0.99	79.89	"
13.00	20.33	756.9	9.92	0.95	13.68	1.49	76.39	"
5.11	28.22	326.1	6.98	0.87	34.03	4.80	58.97	A + B
2.80	30.53	355.5	3.64	0.43	35.08	4.77	61.26	B
0	33.33	390.7	0.00	0.00	36.25	4.73	63.74	"

(continued next page)

AUXILIARY INFORMATION

METHOD/APPARATUS/PROCEDURE:	SOURCE AND PURITY OF MATERIALS:
The usual techniques (1) were used at 0 and 25°C. At 75°C a self-constructed apparatus was used for equilibration and for sampling. $H_2PO_4^-$ was changed to H_3PO_4 by ion exchange and then titrated acidimetrically using chlorophenol red as indicator. K^+ was determined gravimetrically as $KClO_4$ or as the tetraphenylborate. Na^+ and H_2O were determined by difference.	No information is given.

	ESTIMATED ERROR:
	No information is given.

	REFERENCES:
	1. Flatt, R. *Chimia* <u>1962</u>, *6*, 62.

COMPONENTS	ORIGINAL MEASUREMENTS:
(1) Sodium dihydrogenphosphate; NaH_2PO_4; [7558-80-7]	Brunisholz, G. Bodmer, M.
(2) Potassium dihydrogenphosphate; KH_2PO_4; [7778-70-0]	*Helv. Chim. Acta* <u>1963</u>, *46*, 288, 2566-74.
(3) Water; H_2O; [7732-18-5]	

EXPERIMENTAL VALUES cont'd:

Solubility isotherms in the KH_2PO_4-NaH_2PO_4-H_2O system.

K^+ ion%	Na^+ ion%	H_2O n^b	KH_2PO_4[a] mass%	mol/kg	NaH_2PO_4[a] mass%	mol/kg	H_2O mass%	solid phase[c]
				temp. = 25°C				
30.44	2.89	927.3	19.55	1.82	1.63	0.17	78.80	A
26.20	7.13	808.6	18.79	1.80	4.50	0.48	76.69	"
22.96	10.37	718.3	18.06	1.77	7.19	0.80	74.74	"
18.58	14.75	602.5	16.69	1.71	11.68	1.36	71.61	"
13.29	20.04	457.0	14.53	1.61	19.33	2.43	66.12	"
10.90	22.43	347.0	14.23	1.74	25.83	3.59	59.93	"
5.03	28.30	198.4	8.94	1.40	44.38	7.92	46.66	A + B
4.71	28.62	200.4	8.34	1.30	44.70	7.93	46.94	B
2.56	30.77	211.9	4.43	0.67	47.01	8.06	48.55	"
1.44	31.89	226.2	2.42	0.35	47.28	7.83	50.29	"
1.11	32.22	227.5	1.86	0.27	47.66	7.86	50.47	"
0	33.33	236.2	0.00	0.00	48.47	7.83	51.52	"
				temp. = 75°C				
26.76	6.57	332.5	34.96	4.47	7.57	1.09	57.46	A
18.23	15.10	236.3	29.02	4.28	21.20	3.55	49.76	"
10.53	22.80	149.6	20.88	3.91	39.87	8.46	39.23	"
6.22	27.11	94.1	14.61	3.67	56.15	16.00	29.23	A + C
2.93	30.40	105.6	6.70	1.54	61.33	15.99	31.95	C
0	33.33	111.7	0.00	0.00	66.57	16.57	33.44	"

[a]These values were calculated by the compiler.

[b]The concentration units are: mol H_2O/100 g equiv. of the solute.

[c]The solid phases are: A = KH_2PO_4; B = $NaH_2PO_4 \cdot 2H_2O$; C = NaH_2PO_4.

COMPONENTS:	ORIGINAL MEASUREMENTS:
(1) Sodium dihydrogenphosphate; NaH_2PO_4; [7558-80-7] (2) Sodium perchlorate; $NaClO_4$; [7601-89-0] (3) Water; H_2O; [7732-18-5]	Lilich, L.S.; Alekseeva, E.A.; *Zh. Neorg. Khim.* 1969, *14*, 1655-8.

VARIABLES:	PREPARED BY:
Composition at 25°C.	J. Eysseltová

EXPERIMENTAL VALUES:

Solubility in the NaH_2PO_4-$NaClO_4$-H_2O system at 25°C.

NaH_2PO_4		$NaClO_4$		H_2O	solid$_a$ phase
mass%	mol/kg	mass%	mol/kg	mass%	
48.69	7.91	----	----	51.31	A
46.59	7.53	1.86	0.29	51.55	"
40.84	6.50	6.80	1.06	52.36	"
37.71	5.95	9.50	1.47	52.79	"
34.24	5.38	12.81	1.97	52.95	"
27.99	4.37	18.67	2.86	53.34	"
24.69	3.85	21.51	3.26	53.50	"
22.89	3.55	23.43	3.56	53.68	"
13.89	2.18	33.15	5.11	52.96	"
11.53	1.84	36.41	5.71	52.06	"
8.37	1.37	40.82	6.56	50.81	"
5.23	0.90	46.36	7.82	48.41	"
3.55	0.64	50.11	8.83	46.34	"
1.47	0.31	59.44	12.42	39.09	"
0.66	0.17	67.09	16.99	32.25	B
0.74	0.19	65.97	16.94	32.29	"
0.39	0.10	67.35	17.05	32.26	"

aThe solid phases are: A = $NaH_2PO_4 \cdot 2H_2O$; B = $NaClO_4 \cdot H_2O$.

AUXILIARY INFORMATION

METHOD/APPARATUS/PROCEDURE:	SOURCE AND PURITY OF MATERIALS:
No information is given.	No information is given.
	ESTIMATED ERROR: No information is given.
	REFERENCES:

COMPONENTS:	ORIGINAL MEASUREMENTS:
(1) Sodium dihydrogenphosphate; NaH_2PO_4; [7558-80-7] (2) Potassium dihydrogenphosphate; KH_2PO_4; [7778-70-0] (3) Water; H_2O; [7732-18-5]	Babenko, A.M.; Vorob'eva, T.A. *Zh. Prikl. Khim. (Leningrad)* <u>1975</u>, *48*, 11, 2437-41.

VARIABLES:	PREPARED BY:
Temperature and composition.	J. Eysseltová

EXPERIMENTAL VALUES:

Part 1. Solubility isotherms in the KH_2PO_4-NaH_2PO_4-H_2O system.

NaH_2PO_4		KH_2PO_4		H_2O	solid[b]
mass%	mol/kg[a]	mass%	mol/kg[a]	mass%	phase
		temp. = $-10°C$			
34.2	4.51	2.632	0.31	63.168	A
31.0	4.03	4.83	0.55	64.170	"
28.6	3.67	6.426	0.73	64.974	"
34.7	4.61	2.612	0.31	62.688	B
33.0	4.41	4.69	0.55	62.31	"
29.3	3.79	6.363	0.72	64.337	"
		temp. = $-5°C$			
19.2	1.98	0	0	80.8	A
17.2	1.80	3.312	0.31	79.488	"
14.0	1.46	6.02	0.55	79.980	"
11.4	1.18	7.974	0.73	80.626	"
18.16	2.08	9.2	0.93	72.64	C
27.42	3.57	8.6	0.99	63.98	"
35.4	4.56	0	0	64.6	B
36.2	4.92	2.522	0.31	61.248	"
34.6	4.74	4.578	0.55	60.822	"
30.8	4.08	6.228	0.73	62.976	"

(continued next page)

AUXILIARY INFORMATION

METHOD/APPARATUS/PROCEDURE:	SOURCE AND PURITY OF MATERIALS:
A modified polythermic method was used (1).	Chemically pure or reagent grade dihydrogenphosphates were used. They were recrystallized twice and dried at 105°C. The purity is stated to be near to 100%.
	ESTIMATED ERROR: Nothing is stated.
	REFERENCES: 1. Kaganskii, I.M. *Zavod. Lab.* <u>1967</u>, *1*, 119.

COMPONENTS:	ORIGINAL MEASUREMENTS:
(1) Sodium dihydrogenphosphate; NaH_2PO_4; [7558-80-7]	Babenko, A.M.; Vorob'eva, T.A.
(2) Potassium dihydrogenphosphate; KH_2PO_4; [7778-70-0]	*Zh. Prikl. Khim. (Leningrad)* <u>1975</u>, *48*, 11, 2437-41.
(3) Water; H_2O; [7732-18-5]	

EXPERIMENTAL VALUES cont'd:

Part 1. Solubility isotherms in the $KH_2PO_4-NaH_2PO_4-H_2O$ system.

NaH_2PO_4		KH_2PO_4		H_2O	solid[b]
mass%	mol/kg[a]	mass%	mol/kg[a]	mass%	phase
temp. = 0°C					
0	0	12.3	1.03	87.7	C
8.92	0.92	10.8	0.99	80.280	"
17.98	2.06	10.1	1.03	71.92	"
27.18	3.57	9.4	1.09	63.42	"
37.6	5.02	0	0	62.4	B
37.8	5.27	2.488	0.31	59.712	"
36.0	5.04	4.48	0.55	52.52	"
32.1	4.33	6.111	0.73	61.79	"
temp. = 10°C					
0	0	15.2	1.32	84.8	C
8.72	0.93	12.8	1.20	78.48	"
17.6	2.08	12.0	1.25	70.4	"
26.7	3.57	11.0	1.30	62.3	"
32.868	4.69	8.7	1.09	58.432	"
42.2	6.08	0	0	57.8	B
34.056	4.69	5.4	0.66	60.544	"
41.0	6.03	2.36	0.31	56.64	"
39.4	5.82	4.242	0.55	56.358	"
35.2	4.97	5.832	0.73	58.97	"
temp. = 20°C					
0	0	18.0	1.61	82.0	C
8.5	0.92	15.0	1.44	76.5	"
17.24	2.08	13.8	1.47	68.96	"
26.25	3.57	12.5	1.50	61.25	"
32.4	4.69	10.1	1.29	57.6	"
46.6	7.27	0	0	53.4	B
44.0	6.82	2.24	0.31	53.760	"
42.4	6.59	4.032	0.55	53.568	"
38.2	5.66	5.562	0.73	56.238	"

Part 2. Crystallization temperatures and composition of solutions existing in equilibrium with two or three solid phases.

	NaH_2PO_4		KH_2PO_4		H_2O	solid[b]
t/°C	mass%	mol/kg[a]	mass%	mol/kg[a]	mass%	phase
-2.5	0	0	11.6	0.96	88.4	A + C
-4.2	9.0	0.92	10.0	0.91	81.0	A + C
-6.5	18.2	2.08	9.0	0.91	72.8	A + C
-9.6	27.6	3.57	8.0	0.91	64.4	A + B + C
-8.8	33.5	4.20	0	0	66.5	A + B
-10.1	34.6	4.59	2.616	0.31	62.784	A + B
-10.5	32.8	4.37	4.704	0.55	62.496	A + B
-10.1	29.2	3.78	6.372	0.73	64.428	A + B
7.6	32.940	4.69	8.5	1.07	58.560	B + C
14.6	36.8	5.55	8.0	1.06	55.2	B + C
24.8	41.4	6.82	8.0	1.16	50.6	B + C
33.0	45.5	8.33	9.0	1.45	45.5	B + C + D
40.2	56.0	10.60	0	0	44.0	B + D
47.0	52.0	9.40	1.92	0.31	46.08	B + D
50.0	52.0	9.70	3.36	0.55	44.64	B + D
52.8	49.0	8.80	4.59	0.73	46.41	B + D
19.2	48.5	8.98	6.5	1.06	45.0	B + D

(continued next page)

COMPONENTS:	ORIGINAL MEASUREMENTS:
(1) Sodium dihydrogenphosphate; NaH_2PO_4; [7558-80-7]	Babenko, A.M.; Vorob'eva, T.A.
(2) Potassium dihydrogenphosphate; KH_2PO_4; [7778-70-0]	*Zh. Prikl. Khim.* (*Leningrad*) <u>1975</u>, *48*, 11, 2437-41.
(3) Water; H_2O; [7732-18-5]	

EXPERIMENTAL VALUES cont'd:

Part 2. Crystallization temperatures and composition of solutions existing in equilibrium with two or three solid phases.

	NaH_2PO_4		KH_2PO_4		H_2O	
$t/°C$	mass%	mol/kg[a]	mass%	mol/kg[a]	mass%	solid phase[b]
45.1	52.780	11.51	9.0	1.73	38.220	C + D + E
62.4	55.8	13.59	10.0	2.15	34.2	D + E
57.2	60.8	12.92	0	0	39.2	D + E
65.0	60.0	13.02	1.6	0.31	38.4	D + E
64.0	58.0	12.37	2.94	0.55	39.06	D + E
66.5	56.0	11.96	3.96	0.52	40.04	D + E
49.0	53.45	11.16	7.05	0.97	39.5	D + E

[a] The mol/kg H_2O values were calculated by the compiler.

[b] The solid phases are: A = ice; B = $NaH_2PO_4 \cdot 2H_2O$; C = KH_2PO_4; D = $NaH_2PO_4 \cdot H_2O$; E = NaH_2PO_4.

COMPONENTS:	ORIGINAL MEASUREMENTS:
(1) Sodium dihydrogenphosphate; NaH_2PO_4; [7558-80-7]	Beremzhanov, B.A.; Savich, R.F.; Kunanbaev, G.S.
(2) Sodium borate; $NaBO_2$; [7775-19-1]	*Khim. Khim. Tekhnol. (Alma Alta)* <u>1977</u>, *22*,
(3) Water; H_2O, [7732-18-5]	15-20.

VARIABLES:	PREPARED BY:
Composition at 25 and 35°C.	J. Eysseltová

EXPERIMENTAL VALUES:

Solubility isotherms in the NaH_2PO_4-$NaBO_2$-H_2O system.

B_2O_3		P_2O_5		refr.		NaH_2PO_4[a]		$NaBO_2$[a]		solid[b]
mass%	mol%	mass%	mol%	index	pH	mass%	mol/kg	mass%	mol/kg	phase

temp. = 25°C

mass%	mol%	mass%	mol%	index	pH	mass%	mol/kg	mass%	mol/kg	phase
8.91	0.0125	----	----	1.394	10.80	0.00	0.00	16.84	2.87	A
4.52	0.0226	2.04	0.0002	1.365	10.38	3.45	0.32	8.54	1.46	"
4.01	0.0107	1.25	0.0016	1.365	10.49	2.11	0.20	7.56	1.29	"
1.47	0.0039	3.92	0.0052	1.354	9.69	6.63	0.62	2.78	0.47	"
5.43	0.0147	1.15	0.0015	1.380	10.73	1.94	0.18	10.26	1.75	"
2.52	0.0062	5.23	0.0070	1.362	9.31	8.84	0.83	4.76	0.81	"
2.74	0.0075	5.50	0.0074	1.364	9.20	9.30	0.87	5.18	0.88	"
1.50	0.0041	5.98	0.0081	1.369	8.18	10.11	0.94	2.84	0.48	"
7.00	0.0191	0.99	0.0013	1.372	10.92	1.67	0.16	13.23	2.25	"
3.25	0.0090	5.97	0.0090	1.370	9.05	10.10	0.94	6.14	1.05	A + B
----	-----	26.40	0.0435	1.407	3.03	44.64	4.17	0.00	0.00	C
1.28	0.0038	14.23	0.0212	1.405	4.25	24.06	2.25	2.32	0.40	"
1.02	0.0031	16.06	0.0239	1.400	3.99	27.16	2.54	1.93	0.33	"
0.50	0.0015	18.00	0.0272	1.401	4.01	30.44	2.84	0.94	0.16	"
1.42	0.0041	13.17	0.0198	1.411	4.52	22.27	2.08	2.68	0.46	"
3.69	0.0109	12.10	0.0177	1.418	6.87	20.46	1.91	6.97	1.19	B + C
1.51	0.0041	5.58	0.0075	1.367	8.48	9.44	0.88	2.85	0.49	B
2.40	0.0069	10.81	0.0154	1.400	6.67	18.23	1.71	4.54	0.77	"
1.36	0.0038	8.96	0.0124	1.375	7.50	15.15	1.42	2.57	0.43	"
1.01	0.0028	8.60	0.0118	1.373	7.48	14.54	1.36	1.91	0.33	"

(continued next page)

AUXILIARY INFORMATION

METHOD/APPARATUS/PROCEDURE:	SOURCE AND PURITY OF MATERIALS:
The isothermal method was used. The phases were separated from each other by filtration through a Schott filter. In the analyses, the BO_2^- content was determined by titration with a 0.1 N solution of a base containing mannite, Na^+ was determined by flame photometry, and PO_4^{3-} was determined gravimetrically by precipitation as $NH_4MgPO_4 \cdot 6H_2O$. The precipitating solution contained limonic acid.	The materials were of a chemically pure grade.
	ESTIMATED ERROR:
	No information is given.
	REFERENCES:

COMPONENTS:	ORIGINAL MEASUREMENTS:
(1) Sodium dihydrogenphosphate; NaH_2PO_4; [7558-80-7]	Beremzhanov, B.A.; Savich, R.F.; Kunanbaev, G.S.
(2) Sodium borate; $NaBO_2$; [7775-19-1]	*Khim. Khim. Tekhnol.* (*Alma Alta*) 1977, 22, 15-20.
(3) Water; H_2O; [7732-18-5]	

EXPERIMENTAL VALUES cont'd:

Solubility isotherms in the NaH_2PO_4-$NaBO_2$-H_2O system.

B_2O_3		P_2O_5		refr. index	pH	NaH_2PO_4[a]		$NaBO_2$[a]		solid phase[b]
mass%	mol%	mass%	mol%			mass%	mol/kg	mass%	mol/kg	
				temp. = 35°C						
13.16	0.0375	----	----	1.400	10.36	0.00	0.00	24.87	4.24	A
2.42	0.0065	3.85	0.0051	1.368	10.13	6.51	0.61	4.57	0.78	"
2.51	0.0068	4.00	0.0053	1.371	10.65	6.76	0.63	4.74	0.81	"
1.26	0.0033	1.18	0.0001	1.370	10.36	2.00	0.19	2.38	0.40	"
5.02	0.0134	0.75	0.0010	1.391	11.17	1.27	0.12	9.45	1.62	"
1.49	0.0039	2.59	0.0034	1.365	10.02	4.38	0.41	2.82	0.48	"
2.54	0.0071	7.20	0.0099	1.365	8.19	12.18	1.14	4.80	0.82	B
1.90	0.0054	6.17	0.0083	1.362	9.60	10.43	0.97	3.59	0.61	"
1.83	0.0049	2.51	0.0032	1.367	10.66	4.24	0.40	3.46	0.59	A
5.03	0.0146	8.96	0.0128	1.391	7.51	11.77	1.10	9.51	1.62	B
----	------	32.40	0.0603	1.412	3.02	54.79	5.12	0.00	0.00	C
0.84	0.0025	15.25	0.0596	1.377	3.24	25.79	2.41	1.59	0.27	"
1.20	0.0035	14.16	0.0206	1.380	3.40	23.94	2.24	2.27	0.39	"
1.71	0.0049	11.82	0.0169	1.380	3.63	19.99	1.87	3.23	0.55	"
2.51	0.0073	11.76	0.0169	1.383	3.70	19.87	1.89	4.74	0.81	"
7.48	0.0234	10.10	0.0149	1.413	7.00	17.08	1.60	14.14	2.41	B + C
1.43	0.0043	12.71	0.0183	1.381	3.61	21.49	2.01	2.70	0.46	C
4.20	0.0121	9.20	0.0131	1.385	7.44	15.56	1.45	7.94	1.35	B
2.05	0.0057	7.03	0.0096	1.365	9.10	11.89	1.11	3.87	0.66	"
3.50	0.0099	7.98	0.0111	1.371	7.92	13.49	1.26	6.62	1.13	"

[a] These values were calculated by the compiler from the authors' data.

[b] The solid phases are: A = $NaBO_2 \cdot 4H_2O$; B = $Na_2B_4O_7 \cdot 10H_2O$; C = $NaH_2PO_4 \cdot 2H_2O$.

COMPONENTS:	ORIGINAL MEASUREMENTS:
(1) Sodium dihydrogenphosphate; NaH_2PO_4; [7558-80-7] (2) Potassium dihydrogenphosphate; KH_2PO_4; [7778-70-0] (3) Water: H_2O; [7732-18-5]	Khallieva, Sh.D. *Izv. Akad. Nauk Turkm. SSR, Ser. Fiz.-Tekh., Khim. Geol. Nauk* <u>1977</u>, 3, 125-6.

VARIABLES:	PREPARED BY:
Composition at 40°C.	J. Eysseltová

EXPERIMENTAL VALUES:

Solubility in the KH_2PO_4-NaH_2PO_4-H_2O system at 40°C.

KH_2PO_4		NaH_2PO_4		H_2O	
mass%	mol/kg[a]	mass%	mol/kg[a]	mass%	solid[b] phase
27.12	2.73	----	----	72.88	A
25.05	2.55	2.76	0.32	72.19	"
23.60	2.50	7.10	0.83	69.30	"
20.36	2.32	15.19	1.96	64.45	"
18.33	2.20	20.53	2.79	61.14	"
17.01	2.18	25.74	3.75	57.25	"
15.69	2.19	31.60	5.00	52.71	"
9.79	1.78	49.90	10.31	40.31	A + B
5.19	0.92	53.32	10.71	41.49	B
-----	----	56.31	10.74	43.69	"

[a] The mol/kg H_2O values were calculated by the compiler.

[b] The solid phases are: A = KH_2PO_4; B = $NaH_2PO_4 \cdot H_2O$.

AUXILIARY INFORMATION

METHOD/APPARATUS/PROCEDURE:	SOURCE AND PURITY OF MATERIALS:
The isothermal method was used. The experiments were performed in glass vessels with stirrers. Equilibrium was checked by repeated analysis of the saturated solution. Standard analytical methods were used for the determination of sodium, potassium, and dihydrogenphosphate ions.	Reagent grade materials were used.
	ESTIMATED ERROR: The temperature was constant to within ±0.5 K. No other information is given.
	REFERENCES:

COMPONENTS:	ORIGINAL MEASUREMENTS:
(1) Sodium dihydrogenphosphate; NaH_2PO_4; [7558-80-7] (2) Sodium chloride; NaCl; [7647-14-5] (3) Water; H_2O; [7732-18-5]	Khallieva, Sh.D. *Izv. Akad. Nauk Turkm. SSR, Ser. Fiz.-Tekh., Khim. Geol. Nauk* 1977, 3, 125-6.

VARIABLES:	PREPARED BY:
Composition at 40°C.	J. Eysseltová

EXPERIMENTAL VALUES:

Solubility in the NaH_2PO_4-NaCl-H_2O system at 40°C.

NaH_2PO_4		NaCl		H_2O	solid
mass%	mol/kg[a]	mass%	mol/kg[a]	mass%	phase[b]
56.31	10.74	----	----	43.69	A
47.09	8.38	6.08	2.22	46.83	"
45.25	8.09	8.14	2.99	46.61	"
37.15	6.25	13.31	4.60	49.54	A + B
30.57	4.56	13.60	4.17	55.83	B
16.67	2.16	19.05	5.07	66.28	"
10.22	1.26	21.94	5.53	67.84	"
3.80	0.44	24.90	5.98	71.30	"
----	----	26.54	6.18	73.46	"

[a] The mol/kg H_2O values were calculated by the compiler.

[b] The solid phases are: A = $NaH_2PO_4 \cdot H_2O$; B = NaCl.

AUXILIARY INFORMATION

METHOD/APPARATUS/PROCEDURE:	SOURCE AND PURITY OF MATERIALS:
The isothermal method was used. Equilibrium was ascertained by repeated analysis of the saturated solution. Standard analytical methods were used for the determination of sodium, chloride, and dihydrogenphosphate ions, but no details are given. The water content was probably determined by difference (compiler).	The sodium dihydrogenphosphate and the sodium chloride were of reagent grade quality.
	ESTIMATED ERROR:
	The temperature was controlled to within ±0.5K. No other information is given.
	REFERENCES:

COMPONENTS:	ORIGINAL MEASUREMENTS:
(1) Sodium dihydrogenphosphate; NaH_2PO_4; [7558-80-7]	Girich, T.E.; Gulyamov, Yu.M.; Ganz, S.N.; Miroshina, O.S.
(2) Sodium chloride; NaCl; [7647-14-5]	*Vopr. Khim. Khim. Tekhnol.* 1979, 57, 58-61.
(3) Water; H_2O; [7732-18-5]	

VARIABLES:	PREPARED BY:
Composition at 298 and 323 K.	J. Eysseltová

EXPERIMENTAL VALUES:

Composition and properties of saturated solutions in the NaH_2PO_4-NaCl-H_2O system.

NaH_2PO_4				NaCl				H_2O			density	solid
mass%	mol/kg^a	M^b	c^b	mass%	mol/kg^a	M^b	mass%^{a}	c^b	η/cP	g cm^{-3}	phase^c

temp. = 298 K.

mass%	mol/kg	M	c	mass%	mol/kg	M	mass%	c	η/cP	g cm⁻³	phase
----	----	---	---	26.23	6.08	109.41	73.77	914.0	1.695	1.199	A
5.94	0.71	12.80	10.59	24.49	6.01	108.08	69.61	827.3	2.547	1.231	"
11.46	1.42	25.72	20.46	21.72	5.56	99.98	66.82	795.5	2.724	1.256	"
23.55	3.21	57.79	42.81	15.34	4.29	77.19	61.11	740.8	4.846	1.324	"
23.64	3.22	58.03	43.04	15.25	4.27	76.80	61.11	741.69	5.390	1.331	B
26.00	3.59	64.76	47.94	13.77	3.91	70.32	60.23	740.31	6.574	1.350	"
27.59	3.87	69.81	49.08	13.13	3.79	68.12	59.28	739.57	6.195	1.340	C
31.53	4.51	81.19	60.03	10.23	3.00	54.04	58.24	739.37	6.959	1.352	"
39.40	5.84	105.18	81.39	4.39	1.33	24.05	56.21	773.78	11.35	1.397	"
48.03	7.70	138.63	100.0	----	----	----	51.97	721.35	24.40	1.446	"

(continued next page)

AUXILIARY INFORMATION

METHOD/APPARATUS/PROCEDURE:	SOURCE AND PURITY OF MATERIALS:
The isothermal method was used. The mixtures were equilibrated for 13 hours at 298 K and for 8 hours at 323 K. The phosphate ion content was determined photo-colorimetrically, the sodium ion photometrically and the chloride ion by difference. The composition of the solid phases was determined by the Schreinemakers' method.	The NaCl was of a special purity. Reagent grade NaH_2PO_4 was recrystallized twice before being used.

ESTIMATED ERROR:
Nothing is stated.

REFERENCES:

COMPONENTS:	ORIGINAL MEASUREMENTS:
(1) Sodium dihydrogenphosphate; NaH_2PO_4; [7558-80-7]	Girich, T.E.; Gulyamov, Yu.M.; Ganz, S.N.; Miroshina, O.S.
(2) Sodium chloride; NaCl; [7647-14-5]	*Vopr. Khim. Khim. Tekhnol.* 1979, 57, 58-61.
(3) Water; H_2O; [7732-18-5]	

EXPERIMENTAL VALUES cont'd:

Composition and properties of saturated solutions in the NaH_2PO_4-NaCl-H_2O system.

NaH_2PO_4				NaCl			H_2O			density	solid
mass%	mol/kg[a]	M[b]	c[b]	mass%	mol/kg[a]	M[b]	mass%[a]	c[b]	η/cP	g cm[-3]	phase[c]
					temp. = 323 K.						
----	----	---	---	26.99	6.32	113.75	73.01	879.09	1.502	1.202	A
7.34	0.88	15.91	13.20	23.52	5.82	104.68	69.14	829.07	2.589	1.248	"
13.84	1.76	31.77	24.46	20.83	5.45	98.12	65.33	769.46	3.169	1.273	"
19.43	2.59	46.74	34.19	18.22	5.00	89.96	62.35	731.43	3.064	1.311	"
23.07	3.18	57.29	40.49	16.53	4.68	84.22	60.40	706.67	3.094	1.328	"
27.82	3.99	72.00	48.80	14.23	4.20	75.54	57.95	677.77	4.117	1.452	"
36.16	5.69	102.52	61.70	10.94	3.53	63.63	52.90	601.86	7.597	1.709	"
45.82	8.01	144.30	77.37	6.54	2.34	42.21	47.64	536.18	----	1.921	"
47.78	8.67	156.08	78.68	6.31	2.35	42.30	45.91	504.08	12.734	1.951	D
52.25	9.97	179.53	88.15	4.09	1.60	28.85	43.66	479.88	17.800	1.559	"
53.54	10.35	185.71	88.48	3.39	1.34	24.17	43.07	479.45	----	1.606	E
54.05	10.45	188.18	90.19	2.87	1.13	20.47	43.08	479.28	18.263	1.471	"
60.58	12.80	230.58	100.0	----	----	----	39.42	433.69	23.159	1.718	"

[a] These values were calculated by the compiler.

[b] The concentration units are: M = mol/1000 mol H_2O; c = mol/100 mol solute.

[c] The solid phases are: A = NaCl; B = NaCl + $NaH_2PO_4 \cdot 2H_2O$; C = $NaH_2PO_4 \cdot 2H_2O$; D = NaCl + $NaH_2PO_4 \cdot H_2O$; E = $NaH_2PO_4 \cdot H_2O$.

COMPONENTS:	ORIGINAL MEASUREMENTS:
(1) Sodium dihydrogenphosphate; NaH_2PO_4; [7558-80-7] (2) Sodium nitrate; $NaNO_3$; [7631-99-4] (3) Water; H_2O; [7732-18-5]	Kol'ba, V.I.; Zhikharev, M.I.; Sukhanov, L.P. *Zh. Neorg. Khim.* 1981, *26*, 828-30.

VARIABLES:	PREPARED BY:
Composition at 50°C.	J. Eysseltová

EXPERIMENTAL VALUES:

Solubility isotherm in the NaH_2PO_4-$NaNO_3$-H_2O system at 50°C.

NaH_2PO_4			$NaNO_3^a$		H_2O^a		viscosity	solid
mass%	mol/kg	mol%[b]	mass%	mol/kg	mass%	d/kg m^{-3}	10^6m^2 s^{-1}	phase
61.16	13.12	100	----	----	38.84	1.574	17.784	$NaH_2PO_4 \cdot H_2O$
56.70	12.19	89.93	4.56	1.38	38.74	1.575	15.786	"
54.16	11.41	85.91	6.29	1.88	39.55	1.575	14.172	"
52.69	11.22	82.01	8.19	2.46	39.12	1.574	12.421	"
43.92	9.29	65.10	16.68	4.98	39.40	1.578	9.850	"
37.21	7.79	53.40	23.00	6.80	39.79	1.578	9.810	"
32.75	6.88	45.63	27.60	8.18	39.65	1.579	9.795	eutonic pt.[c]
32.75	6.89	43.63	27.64	8.21	39.61	1.579	9.795	"
30.70	6.24	43.45	28.30	8.12	41.00	1.532	5.499	$NaNO_3$
25.37	4.97	35.89	32.10	8.88	42.53	1.513	4.213	"
19.15	3.55	27.39	35.96	9.42	44.89	1.486	2.982	"
17.88	3.31	25.47	37.06	9.68	45.06	1.482	2.873	"
11.60	2.11	16.19	42.51	10.90	45.89	1.445	1.778	"
4.10	0.72	5.62	48.67	12.12	47.23	1.437	1.571	"
-----	-----	-----	46.80	13.37	46.80	1.429	1.436	"

[a] The mass% and mol/kg H_2O values were calculated by the compiler.

[b] These values are actually mol/100 mol solute values (compiler).

[c] Isothermal invariant point.

AUXILIARY INFORMATION

METHOD/APPARATUS/PROCEDURE:	SOURCE AND PURITY OF MATERIALS:
The isothermal method was used. The mixtures were allowed to equilibrate for 7-8 hours with constant agitation. The $H_2PO_4^-$ content was determined colorimetrically, the sum of the salt content was determined by evaporation to dryness, and the nitrate content was determined by the difference. The composition of the solid phases was determined by the Schreinemakers' method. The viscosity was measured with the aid of an Ostwald viscometer. The density was measured by the use of calibrated 10ml pycnometers.	A pure form of $NaNO_3$ was used and the NaH_2PO_4 was of reagent grade quality.
	ESTIMATED ERROR: No details are given.
	REFERENCES:

COMPONENTS:	ORIGINAL MEASUREMENTS:
(1) Sodium dihydrogenphosphate; NaH_2PO_4; [7558-80-7] (2) Potassium dihydrogenphosphate; KH_2PO_4; [7778-70-0] (3) Sodium chloride; NaCl; [7647-14-5] (4) Potassium chloride; KCl; [7747-40-7] (5) Water; H_2O; [7732-18-5]	Brunishloz, G.; Bodmer, M. *Helv. Chim. Acta* 1963, 46, 7, 289, 2575-86.

VARIABLES:	PREPARED BY:
Temperature and composition.	J. Eysseltová

EXPERIMENTAL VALUES:

Part 1. Solubility in the K^+, $Na^+ \| Cl^-$, $H_2PO_4^-$-H_2O system.

soln. no.	K^+ eq%	H^+ eq%	Cl^- eq%	H_2O concn. [a]	solid phases [b]
			temp. = 0°C		
1	17.65	14.31	78.58	682.5	A + B + C
2	10.40	25.25	62.16	598.7	B + C + D
3	18.93	8.77	86.80	731.1	A + B
4	19.08	7.96	88.08	738.4	"
5	67.33	11.73	82.36	1132	A + C
6	52.28	12.46	81.31	1016	"
7	35.16	12.95	80.64	876.7	"
8	9.59	30.36	54.47	589.1	C + D
9	8.76	38.93	41.57	555.8	"
10	7.07	49.68	25.43	492.3	"
11	5.51	24.28	63.52	627.6	B + D
12	14.92	17.23	74.21	660.4	B + C
13	11.38	22.80	65.80	618.5	"
14	75.45	6.89	88.68	1241	A
15	7.54	15.04	77.42	719.8	B
16	50.20	44.64	33.05	1807	C
17	49.60	44.76	32.96	1797	"
18	60.52	29.38	56.00	1628	"
19	44.64	36.48	46.85	1467	"
20	29.13	58.27	12.60	1396	"
21	58.45	20.36	69.47	1322	"
22	26.01	52.02	21.97	1183	"
23	22.73	45.45	31.82	1028	" (continued next page)

(continued next page)

AUXILIARY INFORMATION

METHOD/APPARATUS/PROCEDURE:	SOURCE AND PURITY OF MATERIALS:
Nothing is stated, but it probably is the same as in ref. (1).	Nothing is stated, but it probably is the same as in ref. (1).
	ESTIMATED ERROR: No information is given.
	REFERENCES: 1. Brunisholz, G.; Bodmer, M. *Helv. Chim. Acta* 1963, 46, 288, 2566-74.

COMPONENTS:	ORIGINAL MEASUREMENTS:
(1) Sodium dihydrogenphosphate; NaH_2PO_4; [7558-80-7] (2) Potassium dihydrogenphosphate; KH_2PO_4; [7778-70-0] (3) Sodium chloride; NaCl; [7647-14-5] (4) Potassium chloride; KCl; [7747-40-7] (5) Water; H_2O; [7732-18-5]	Brunishloz, G.; Bodmer, M. *Helv. Chim. Acta* 1963, 46, 7, 289, 2575-86.

EXPERIMENTAL VALUES cont'd:

Part 1. Solubility in the K^+, $Na^+||Cl^-$, $H_2PO_4^--H_2O$ system.

soln. no.	K^+ eq%	H^+ eq%	Cl^- eq%	H_2O concn.[a]	solid phases[a]
			temp. = 0°C		
24	32.71	19.59	70.64	957.9	C
25	19.41	38.81	41.78	898.7	"
26	13.54	51.95	22.06	747.5	"
27	15.15	30.32	54.53	744.2	"
28	19.08	17.44	73.43	730.6	"
29	3.73	38.51	42.49	598.9	D
			temp. = 25°C		
30	24.60	16.74	74.89	606.1	A + B + C
31	6.38	51.88	22.23	303.6	B + C + D
32	28.23	3.83	94.26	725.3	A + B
33	27.31	7.19	89.20	695.2	"
34	26.56	9.65	85.45	673.3	"
35	25.35	14.52	78.25	627.5	"
36	75.84	12.85	80.73	926.9	A + C
37	69.29	13.28	80.06	889.8	"
38	50.79	13.76	79.30	800.4	"
39	35.16	15.12	77.42	696.8	"
40	5.57	58.93	11.62	258.7	C + D
41	1.48	48.48	27.67	366.2	B + D
42	14.45	28.90	56.65	523.5	B + C
43	10.39	38.49	42.47	443.1	"
44	6.33	51.79	22.57	308.0	"
45	71.18	5.34	92.02	1002	A
46	58.43	6.82	89.82	896.0	"
47	39.70	7.00	89.55	791.0	"
48	4.31	8.61	87.08	794.0	B
49	10.27	20.55	69.18	638.5	"
50	5.07	40.42	39.41	443.4	"
51	57.90	34.84	47.68	1149	C
52	57.29	24.78	62.81	1015	"
53	45.76	34.22	48.71	987.7	"
54	35.72	42.47	36.46	913.9	"
55	29.73	59.45	10.82	881.5	"
56	33.80	31.30	53.20	823.1	"
57	39.91	21.71	67.44	806.1	"
58	26.63	52.67	21.00	786.4	"
59	23.27	46.55	30.18	713.0	"
60	19.87	39.75	40.48	639.4	"
61	16.28	32.57	51.15	564.3	"
62	12.17	51.63	22.43	466.5	"
63	6.39	59.08	11.62	283.7	"
64	2.76	56.71	14.89	296.5	D
			temp. = 50°C		
65	29.62	21.36	68.07	518.5	A + B + C
66	5.01	64.08	3.86	131.9	B + C + E
67	37.76	0	100	702.8	A + B
68	33.65	9.72	85.34	621.1	"
69	19.84	32.95	50.70	444.0	B + C
70	12.81	45.67	31.63	336.0	"
71	6.15	62.03	7.00	162.0	"
72	5.10	64.00	3.89	133.2	"

(continued next page)

COMPONENTS:	ORIGINAL MEASUREMENTS:
(1) Sodium dihydrogenphosphate; NaH_2PO_4; [7558-80-7]	Brunishloz, G.; Bodmer, M.
(2) Potassium dihydrogenphosphate; KH_2PO_4; [7778-70-0]	*Helv. Chim. Acta* 1963, 46, 7, 289, 2575-86.
(3) Sodium chloride; NaCl; [7647-14-5]	
(4) Potassium chloride; KCl; [7747-40-7]	
(5) Water; H_2O; [7732-18-5]	

EXPERIMENTAL VALUES cont'd:

Part 1. Solubility in the K^+, $Na^+||Cl^-$, $H_2PO_4^-$-H_2O system.

soln. no.	K^+ eq%	H^+ eq%	Cl^- eq%	H_2O concn.[a]	solid phases[b]
			temp. = 75°C		
73	31.14	29.73	55.47	402.2	A + B + C
74	6.77	64.85	2.88	102.8	B + C + E
75	39.70	10.85	83.70	556.5	A + B
76	69.77	20.51	69.24	613.0	A + C
77	54.56	23.08	65.46	534.6	"
78	25.72	36.04	46.04	365.3	B + C
79	21.08	42.16	36.76	317.9	"
80	14.97	51.89	22.09	237.0	"
81	9.74	61.15	8.15	149.8	"
82	6.80	64.60	3.02	102.5	"
83	61.31	10.06	84.90	649.7	A
84	10.00	20.00	70.00	603.8	B
85	15.63	31.26	53.11	457.8	"
86	59.39	36.58	45.34	622.9	C
87	38.70	46.08	30.90	503.7	"
88	30.24	60.49	9.27	393.6	"
89	28.42	56.84	14.74	381.9	"
			temp. = 100°C		
90	27.84	43.16	35.40	256.6	A + B + C
91	8.47	65.27	2.10	71.6	B + C + E
92	50.41	0	100	594.2	A + B
93	44.56	11.88	82.22	418.6	"
94	35.98	27.85	58.34	375.8	"
95	24.04	48.08	27.88	223.5	B + C
96	20.20	52.93	20.56	191.7	"
97	15.15	59.27	10.85	138.1	"
98	12.65	61.99	6.89	115.5	"
99	12.10	62.73	5.74	105.6	"
100	9.29	64.91	2.39	76.1	"
101	8.70	65.11	2.30	74.4	"

[a] The concentration units are: mol/100 eq% of solute.

[b] The solid phaes are: A = KCl; B = NaCl; C = KH_2PO_4; D = $NaH_2PO_4 \cdot 2H_2O$; E = NaH_2PO_4.

Part 2. The compiler has calculated the following results from the data in Part 1.

soln. no.	K^+ mass%	K^+ mol/kg	Na^+ mass%	Na^+ mol/kg	Cl^- mass%	Cl^- mol/kg	$H_2PO_4^-$ mass%	$H_2PO_4^-$ mol/kg	H_2O mass%
				temp. = 0°C					
1	3.83	1.44	8.68	5.53	15.46	6.38	3.84	0.58	68.18
2	2.53	0.96	9.19	5.97	13.70	5.75	7.60	1.17	66.98
3	3.88	1.44	8.72	5.49	16.14	6.58	2.23	0.33	69.02
4	3.88	1.44	8.73	5.49	16.25	6.61	2.00	0.30	69.14
5	9.78	3.30	1.78	1.03	10.82	4.03	2.11	0.29	75.52
6	8.30	2.86	3.29	1.93	11.70	4.43	2.45	0.34	74.25
7	6.30	2.23	5.46	3.29	13.10	5.10	2.87	0.41	72.28
8	2.38	0.90	8.76	5.66	12.25	5.12	9.33	1.43	67.27
9	2.30	0.88	8.06	5.23	9.88	4.14	12.67	1.95	67.09
10	2.06	0.80	7.40	4.88	6.71	2.86	17.93	2.80	65.91
11	1.30	0.49	9.75	6.22	13.60	5.61	7.12	1.08	68.22
12	3.33	1.26	8.92	5.71	15.04	6.23	4.78	0.72	67.94
13	2.69	1.02	9.15	5.91	14.11	5.90	6.69	1.02	67.35
14	10.10	3.34	1.39	0.79	10.77	3.96	1.25	0.17	76.49
15	1.59	0.58	9.62	5.98	14.83	5.96	3.94	0.58	70.01

(continued next page)

COMPONENTS:	ORIGINAL MEASUREMENTS:
(1) Sodium dihydrogenphosphate: NaH_2PO_4; [7558-80-7] (2) Potassium dihydrogenphosphate; KH_2PO_4; [7778-70-0] (3) Sodium chloride; NaCl; [7647-14-5] (4) Potassium chloride; KCl; [7747-40-7] (5) Water: H_2O; [7732-18-5]	Brunishloz, G.; Bodmer, M. *Helv. Chim. Acta* 1963, 46, 7, 289, 2575-86.

EXPERIMENTAL VALUES cont'd:

Part 2. The compiler has calculated the following results from the data in Part 1.

soln. no.	K^+ mass%	K^+ mol/kg	Na^+ mass%	Na^+ mol/kg	Cl^- mass%	Cl^- mol/kg	$H_2PO_4^-$ mass%	$H_2PO_4^-$ mol/kg	H_2O mass%
				temp. = 0°C					
16	5.17	1.54	0.31	0.16	3.09	1.01	5.70	0.67	85.72
17	5.14	1.53	0.34	0.17	3.10	1.02	5.74	0.69	85.68
18	6.70	2.06	0.66	0.34	5.62	1.91	4.03	0.50	82.99
19	5.46	1.69	1.36	0.71	5.20	1.77	5.38	0.67	82.61
20	3.82	1.16	0.97	0.50	1.50	0.50	9.47	1.16	84.24
21	7.61	2.46	1.62	0.89	8.20	2.91	3.29	0.43	79.27
22	3.89	1.22	1.93	1.03	2.98	1.03	9.66	1.22	81.53
23	3.79	1.23	3.12	1.72	4.81	1.72	9.40	1.23	78.88
24	5.54	1.90	4.75	2.77	10.86	4.09	4.11	0.57	74.73
25	3.57	1.20	4.52	2.58	6.97	2.58	8.85	1.20	76.09
26	2.93	1.01	4.39	2.56	4.33	1.64	13.94	1.93	74.42
27	3.18	1.13	6.72	4.07	10.37	4.06	7.88	1.13	71.84
28	3.96	1.45	7.72	4.80	13.84	5.57	4.57	0.67	69.91
29	0.93	0.35	8.50	5.36	9.64	3.93	11.90	1.78	69.02
				temp. = 25°C					
30	5.76	2.25	8.08	5.38	15.91	6.85	4.86	0.77	65.38
31	2.50	1.17	9.62	7.64	7.90	4.06	25.20	4.74	54.77
32	5.73	2.16	8.11	5.20	17.36	7.20	0.96	0.15	67.82
33	5.74	2.18	8.10	5.23	17.00	7.11	1.88	0.29	67.28
34	5.73	2.19	8.09	5.26	16.72	7.03	2.60	0.40	66.87
35	5.78	2.24	8.06	5.32	16.18	6.91	4.10	0.64	65.87
36	12.68	4.54	1.11	0.68	12.24	4.83	2.66	0.38	71.31
37	11.98	4.33	1.77	1.09	12.56	4.98	2.85	0.41	70.84
38	9.60	3.52	3.94	2.46	13.59	5.49	3.24	0.48	69.64
39	7.42	2.80	6.17	3.96	14.81	6.16	3.94	0.60	67.67
40	2.43	1.20	9.11	7.62	4.60	2.49	31.89	6.33	51.97
41	0.52	0.22	10.35	7.59	8.82	4.19	21.03	3.66	59.28
42	3.84	1.53	8.86	6.01	13.66	6.00	9.53	1.53	64.10
43	3.14	1.30	9.09	6.41	11.65	5.31	14.39	2.40	61.72
44	2.46	1.14	9.58	7.56	7.96	4.06	24.49	4.66	55.11
45	11.19	3.95	2.17	1.30	13.11	5.09	1.04	0.15	72.49
46	10.05	3.62	3.52	2.15	14.01	5.55	1.45	0.21	70.97
47	7.56	2.79	5.97	3.74	15.47	6.27	1.64	0.24	69.36
48	0.84	0.30	10.03	6.09	15.46	6.08	2.09	0.30	71.57
49	2.37	0.89	9.39	6.02	14.48	6.00	5.88	0.89	67.88
50	1.55	0.64	9.80	6.83	10.93	4.92	15.32	2.53	62.41
51	8.54	2.80	0.63	0.35	6.38	2.30	6.38	0.84	78.06
52	9.20	3.14	1.69	0.98	9.14	3.43	4.94	0.68	75.02
53	7.64	2.57	1.96	1.13	7.38	2.73	7.08	0.96	75.93
54	6.44	2.17	2.31	1.32	5.96	2.21	9.47	1.29	75.82
55	5.66	1.87	1.21	0.68	1.87	0.68	14.03	1.87	77.22
56	6.50	2.28	3.94	2.36	9.27	3.58	7.44	1.05	72.84
57	7.65	2.75	4.33	2.64	11.72	4.64	5.16	0.75	71.14
58	5.49	1.88	2.51	1.46	3.92	1.48	13.46	1.86	74.61
59	5.12	1.81	3.91	2.35	6.02	2.34	12.71	1.81	72.24
60	4.69	1.73	5.60	3.51	8.66	3.51	11.61	1.72	69.44
61	4.14	1.60	7.65	5.04	11.80	5.02	10.28	1.60	66.11
62	3.66	1.45	6.40	4.31	6.11	2.66	19.28	3.08	64.55
63	2.65	1.25	8.43	6.77	4.37	2.27	30.34	5.77	54.21
64	1.12	0.52	9.65	7.59	5.47	2.78	28.50	5.32	55.27
				temp = 50°C					
65	7.69	3.17	7.48	5.25	16.02	7.28	6.85	1.14	61.96
66	3.00	2.11	10.89	13.02	2.10	1.62	47.63	13.50	36.38
67	7.73	2.98	7.49	4.92	18.56	7.88	0	0	66.22
68	7.61	3.01	7.53	5.06	17.49	7.61	2.74	0.44	64.63
69	5.86	2.48	8.19	5.91	13.57	6.33	12.04	2.06	60.34

(continued next page)

COMPONENTS:	ORIGINAL MEASUREMENTS:
(1) Sodium dihydrogenphosphate; NaH_2PO_4; [7558-80-7]	Brunishloz, G.; Bodmer, M.
(2) Potassium dihydrogenphosphate; KH_2PO_4; [7778-70-0]	*Helv. Chim. Acta* 1963, 46, 7, 289, 2575-86.
(3) Sodium chloride; NaCl; [7647-14-5]	
(4) Potassium chloride; KCl; [7747-40-7]	
(5) Water; H_2O; [7732-18-5]	

EXPERIMENTAL VALUES cont'd:

Part 2. The compiler has calculated the following results from the data in Part 1.

soln. no.	K^+ mass%	K^+ mol/kg	Na^+ mass%	Na^+ mol/kg	Cl^- mass%	Cl^- mol/kg	$H_2PO_4^-$ mass%	$H_2PO_4^-$ mol/kg	H_2O mass%
				temp. = 50°C					
70	4.62	2.12	8.81	6.86	10.35	5.22	20.40	3.77	55.82
71	3.37	2.11	10.24	10.91	3.47	2.39	42.10	10.63	40.82
72	3.04	2.13	10.84	12.89	2.10	1.62	47.42	13.36	36.59
				temp. = 75°C					
73	9.54	4.30	7.05	5.40	15.41	7.64	11.28	2.05	56.72
74	4.40	3.66	10.86	15.34	1.70	1.55	52.25	17.50	30.78
75	9.58	3.96	7.02	4.94	18.32	8.33	3.25	0.54	61.83
76	15.65	6.32	1.28	0.88	14.08	6.26	5.70	0.93	63.29
77	13.58	5.67	3.27	2.32	14.78	6.79	7.11	1.20	61.26
78	8.50	3.91	7.43	5.82	13.79	6.98	14.74	2.74	55.55
79	7.67	3.68	7.87	6.42	12.14	6.41	19.04	3.68	53.28
80	6.56	3.51	8.54	7.77	8.78	5.16	28.25	6.09	47.85
81	5.44	3.61	9.55	10.80	4.12	3.01	42.39	11.35	38.49
82	4.42	3.68	10.95	15.50	1.78	1.63	52.16	17.52	30.69
83	13.14	5.24	3.61	2.45	16.50	7.24	2.68	0.43	64.09
84	2.40	0.92	9.86	6.44	15.21	6.42	5.94	0.92	66.59
85	4.54	1.90	9.06	6.44	13.98	6.43	11.25	1.90	61.17
86	13.66	5.30	0.54	0.36	9.46	4.03	10.39	1.63	65.95
87	10.61	4.27	2.45	1.68	7.68	3.40	15.67	2.54	63.58
88	10.07	4.27	1.81	1.31	2.80	1.30	24.98	4.27	60.33
89	9.58	4.13	2.92	2.14	4.50	2.14	23.76	4.13	59.24
				temp. = 100°C					
90	11.20	6.03	6.86	6.28	12.92	7.64	21.49	4.66	47.53
91	6.06	6.57	11.05	20.38	1.36	1.62	57.94	25.32	23.59
92	11.36	4.71	6.57	4.64	20.43	9.32	0	0	61.64
93	12.65	5.91	7.27	5.78	21.17	10.88	4.18	0.79	54.72
94	11.33	5.32	6.70	5.35	16.66	8.60	10.85	2.05	54.47
95	10.53	5.98	7.18	6.93	11.08	6.91	26.13	5.98	45.08
96	9.68	5.85	7.57	7.79	8.94	5.94	31.49	7.67	42.31
97	8.54	6.09	8.48	10.29	5.55	4.35	41.57	11.95	35.85
98	7.71	6.08	9.09	12.20	3.81	3.30	46.96	14.93	32.42
99	7.63	6.36	9.33	13.24	3.28	3.01	49.13	16.53	30.64
100	6.52	6.78	10.66	18.83	1.52	1.74	56.69	23.75	24.60
101	6.16	6.50	10.90	19.56	1.48	1.71	57.21	24.32	24.25

Part 3. The points of simultaneous crystallization of NaCl + KH_2PO_4 + $NaH_2PO_4 \cdot xH_2O$.

soln. no.	t/°C	eq% K^+	eq% H^+	eq% Cl^-	conc H_2O [a]	x
102	0	10.40	25.25	62.16	598.7	2
103	12.5	8.52	36.77	44.93	475.0	2
104	25	6.38	51.88	22.23	303.6	2
105	30	5.47	57.33	14.00	239.0	2
106	35	4.98	61.16	8.33	186.3	2
107	40	4.85	62.60	5.99	163.1	1
108	45	5.00	63.59	4.75	145.7	1
109	50	5.01	64.08	3.86	131.9	0
110	75	6.77	64.85	2.88	102.8	0
111	100	8.47	65.27	2.10	71.6	0
112	36.0	4.90	61.86	7.20	177.5	2+1 [b]
113	48.2	4.90	64.06	3.91	133.9	1+0 [c]

[a] The concentration unit is mol/100 eq% of solute.

[b] An invariant point, the solid phases being NaCl+KH_2PO_4+$NaH_2PO_4 \cdot 2H_2O$+$NaH_2PO_4 \cdot H_2O$.

[c] An invariant point, the solid phases being NaCl+KH_2PO_4+$NaH_2PO_4 \cdot 2H_2O$+NaH_2PO_4.

COMPONENTS:	ORIGINAL MEASUREMENTS:
(1) Sodium dihydrogenphosphate; NaH_2PO_4; [7558-80-7]	Brunishloz, G.; Bodmer, M.
(2) Potassium dihydrogenphosphate; KH_2PO_4; [7778-70-0]	*Helv. Chim. Acta* <u>1963</u>, *46*, 7, 289, 2575-86.
(3) Sodium chloride; NaCl; [7647-14-5]	
(4) Potassium chloride; KCl; [7747-40-7]	
(5) Water; H_2O; [7732-18-5]	

EXPERIMENTAL VALUES cont'd:

Part 4. The compiler has calculated the following values from the data in Part 3.

soln. no.	K^+ mass%	K^+ mol/kg	Na^+ mass%	Na^+ mol/kg	Cl^- mass%	Cl^- mol/kg	$H_2PO_4^-$ mass%	$H_2PO_4^-$ mol/kg	H_2O mass%
102	2.55	0.96	9.19	5.97	13.70	5.75	7.60	1.17	66.98
103	2.46	1.00	9.31	6.40	11.79	5.24	13.18	2.15	63.26
104	2.50	1.17	9.62	7.64	7.90	4.06	25.20	4.74	54.77
105	2.47	1.27	9.89	8.65	5.74	3.25	32.15	6.66	49.74
106	2.57	1.49	10.26	10.10	3.89	2.48	39.07	9.11	44.20
107	2.66	1.65	10.50	11.09	2.98	2.04	42.66	10.67	41.20
108	2.88	1.91	10.64	11.98	2.48	1.81	45.37	12.11	38.63
109	3.00	2.11	10.89	13.02	2.10	1.62	47.63	13.50	36.38
110	4.40	3.66	10.86	15.34	1.70	1.55	52.25	17.50	30.78
111	6.06	6.57	11.05	20.38	1.36	1.62	57.94	25.32	23.59
112	2.59	1.53	10.32	10.40	3.45	2.25	40.51	9.68	43.14
113	2.92	2.03	10.88	12.88	2.11	1.62	47.35	13.29	36.74

COMPONENTS:	ORIGINAL MEASUREMENTS:
(1) Sodium dihydrogenphosphate; NaH_2PO_4; [7558-80-7]	Babenko, A.M.; Vorob'eva, T.A.
(2) Ammonium dihydrogenphosphate; $NH_4H_2PO_4$; [7722-76-1]	$Zh.\ Prikl.\ Khim.$ 1976, 49, 1502-6.
(3) Diammonium hydrogenphosphate; $(NH_4)_2HPO_4$; [7783-28-0]	
(4) Water; H_2O; [7732-18-5]	

VARIABLES:	PREPARED BY:
Temperature and concentration of NaH_2PO_4 in a mixture containing a mol ratio of $NH_4H_2PO_4/(NH_4)_2HPO_4 = 1$.	J. Eysseltová

EXPERIMENTAL VALUES:

Part 1. Points of simultaneous crystallization of two or three solid phases in the $NH_4H_2PO_4-(NH_4)_2HPO_4-NaH_2PO_4-H_2O$ system.

mixture^a		$NH_4H_2PO_4$	$(NH_4)_2HPO_4$	NaH_2PO_4			solid
mass%	mol/kg^b	mass%^{b}	mass%^{b}	mass%	mol/kg^b	$t/°C$	phases^c
38.6	2.54	17.97	20.63	0.00	0.00	-8.6	A + C
32.0	2.11	14.90	17.10	6.8	0.92	-8.4	A + C
20.4	1.29	9.50	10.90	15.92	2.08	-8.4	A + C
9.8	0.63	4.56	5.24	27.06	3.57	-9.4	A + C
21.25	1.35	9.89	11.36	15.0	1.96	-8.0	A + C
16.0	1.01	7.45	8.55	20.0	2.60	-9.0	A + C
6.8	0.45	3.16	3.63	32.0	4.36	-11.0	A + C
6.0	0.43	2.79	3.21	37.6	5.55	+0.5	A + B + C
0.0	0.00	0.0	0.0	33.5	4.20	-8.8	A + B
2.4	0.17	1.12	1.28	40.0	5.79	-12.0	A + B
8.0	0.64	3.72	4.28	41.4	6.82	19.2	B + C
8.0	0.70	3.72	4.28	46.0	8.33	25.2	B + C
9.8	0.96	4.56	5.24	48.708	9.78	35.0	B + C + D
61.0	6.33	28.40	32.60	0.0	0.0	39.0	C + D
56.0	5.72	26.07	29.93	4.4	0.92	45.6	C + D
48.0	4.67	22.34	25.65	10.4	2.08	48.3	C + D
36.0	3.25	16.76	19.24	19.2	3.57	44.6	C + D
28.0	2.62	13.04	14.96	28.8	5.55	50.3	C + D

(continued next page)

AUXILIARY INFORMATION

METHOD/APPARATUS/PROCEDURE:	SOURCE AND PURITY OF MATERIALS:
An improved polythermic method (1) was used.	Reagent grade salts were recrystallized and dried before use. The ammonium salts were dried at 40-50°C. The sodium salt was dried at 105°C. The material designated "mixture" was prepared by mixing equimolar amounts of $NH_4H_2PO_4$ and $(NH_4)_2HPO_4$ and homogenizing them by grinding in a mortar.
	ESTIMATED ERROR:
	No information is given.
	REFERENCES:
	1. Erajzer, L.N.; Kaganskii, I.M., $Zavod.$ $Lab.$ 1967, 1, 119.

COMPONENTS:	ORIGINAL MEASUREMENTS:
(1) Sodium dihydrogenphosphate; NaH_2PO_4; [7558-80-7]	Babenko, A.M.; Vorob'eva, T.A.
(2) Ammonium dihydrogenphosphate; $NH_4H_2PO_4$; [7722-76-1]	Zh. Prikl. Khim. 1976, 49, 1502-6.
(3) Daimmonium hydrogenphosphate; $(NH_4)_2HPO_4$; [7783-28-0]	
(4) Water; H_2O; [7732-18-5]	

EXPERIMENTAL VALUES cont'd:

Part 1. Points of simultaneous crystallization of two or three solid phases in the $NH_4H_2PO_4$-$(NH_4)_2HPO_4$-NaH_2PO_4-H_2O system.

mixture[a]		$NH_4H_2PO_4$	$(NH_4)_2HPO_4$	NaH_2PO_4			solid
mass%	mol/kg[b]	mass%[b]	mass%[b]	mass%	mol/kg[b]	$t/°C$	phases[c]
24.0	2.27	11.17	12.83	33.2	6.46	56.0	C + D
20.0	2.02	9.31	10.69	40.0	8.33	55.0	C + D
20.0	2.20	9.31	10.69	43.2	9.78	67.0	C + D
0.0	0.0	0.0	0.0	56.0	10.60	40.2	B + E
1.8	0.17	0.84	0.96	55.0	10.61	43.5	B + E
4.75	0.45	2.21	2.54	52.5	10.23	46.0	B + E
8.0	0.82	3.72	4.28	52.44	11.04	42.0	D + E
8.0	0.88	3.72	4.28	55.2	12.50	50.5	D + E
8.9	1.01	4.14	4.76	55.5	12.99	49	D + E + F
0.0	0.0	0.0	0.0	60.8	12.92	57.2	E + F
1.6	0.17	0.74	0.86	60.0	13.02	59.8	E + F
4.2	0.45	1.96	2.24	58.0	12.78	63.5	E + F
11.25	1.35	5.24	6.01	55.0	13.58	68.4	E + F

Part 2. Solubility isotherms in the $NH_4H_2PO_4$-$(NH_4)_2HPO_4$-NaH_2PO_4-H_2O system.

mixture[a]		$NH_4H_2PO_4$	$(NH_4)_2HPO_4$	NaH_2PO_4		H_2O	$(N + P_2O_5)$
mass%	mol/kg[b]	mass%[b]	mass%[b]	mass%	mol/kg[b]	mass%	mass%
			temp. = -5°C				
0.0	0.0	0.0	0.0	35.4	4.56	64.6	21.0
6.7	0.45	3.12	3.58	33.0	4.56	60.3	24.5
12.0	0.79	5.59	6.41	26.4	3.57	61.6	24.5
2.392	0.17	1.11	1.28	40.2	5.83	57.408	25.5
22.0	1.43	10.24	11.76	15.6	2.08	62.4	25.6
33.4	2.25	15.55	17.85	6.66	0.92	59.94	28.78
40.2	2.72	18.71	21.48	0.0	0.0	59.8	29.9
			temp. = 0°C				
0.0	0.0	0.0	0.0	37.6	5.02	62.4	22.1
14.0	0.94	6.51	7.48	25.8	3.57	60.2	25.5
6.58	0.45	3.06	3.52	34.2	4.81	59.22	26.2
2.38	0.17	1.11	1.27	41.87	6.26	55.87	26.4
24.4	1.63	11.36	13.04	15.12	2.08	60.48	27.1
33.6	2.28	15.64	17.96	6.64	0.92	59.76	28.98
42.6	3.00	19.83	22.77	0.0	0.0	57.4	31.6
			temp. = 10°C				
0.0	0.0	0.0	0.0	42.2	6.08	57.8	25.0
6.38	0.45	2.97	3.41	36.2	5.25	57.42	26.24
19.0	1.36	8.85	10.15	24.3	3.57	56.7	28.63
29.0	2.06	13.50	15.50	14.18	2.08	56.72	29.9
2.28	0.17	1.06	1.22	44.8	7.05	52.992	28.29
40.0	3.00	18.62	21.38	6.0	0.92	54.0	33.36
47.0	3.59	21.88	25.12	0.0	0.0	53.0	35.0

[a]"Mixture" is an equimolar mixture of $NH_4H_2PO_4$ and $(NH_4)_2HPO_4$.

[b]These values were calculated by the compiler.

[c]The solid phases are: A = ice; B = $NaH_2PO_4 \cdot 2H_2O$; C = $NH_4H_2PO_4$; D = $(NH_4)_2HPO_4$; E = $NaH_2PO_4 \cdot H_2O$; F = NaH_2PO_4.

COMPONENTS:	ORIGINAL MEASUREMENTS:
(1) Sodium dihydrogenphosphate; NaH_2PO_4; [7558-80-7]	Khallieva, Sh.D.
(2) Potassium dihydrogenphosphate; KH_2PO_4; [7778-77-0]	*Izv. Akad. Nauk Turkm. SSR, Ser. Fiz.-Tekh., Khim. Geol. Nauk* 1977, 3,125-8.
(3) Sodium chloride; NaCl; [7647-14-5]	
(4) Potassium chloride; KCl; [7747-40-7]	
(5) Water; H_2O; [7732-18-5]	

VARIABLES:	PREPARED BY:
Composition at 40°C.	J. Eysseltová

EXPERIMENTAL VALUES:

Part 1.   Solubility in the Na^+, $K^+||Cl^-$, $H_2PO_4^-$-H_2O system at 40°C.

Jänecke's indices[a]

soln. no.	Na^+	K^+	Cl^-	$H_2PO_4^-$	H_2O	solid phase[b]
1	64.75	35.25	100	-----	752.69	A + B
2	64.85	35.15	96.20	3.80	725.80	"
3	64.80	35.20	93.08	6.92	697.90	"
4	65.06	34.94	90.78	9.22	696.95	"
5	65.58	34.42	89.82	10.18	694.38	"
6	66.30	33.70	88.93	11.07	680.58	"
7	66.27	33.73	88.68	11.32	660.76	"
8	68.31	31.69	87.57	12.43	657.31	"
9	68.40	31.60	87.55	12.45	633.20	A + B + C
10	-----	100	92.42	7.55	990.59	B + D
11	16.72	83.28	91.84	8.16	860.17	"
12	26.80	73.20	90.95	9.05	853.50	"
13	43.45	56.55	89.83	10.17	766.98	"
14	56.15	43.85	89.16	10.84	762.84	"
15	63.94	36.05	88.19	11.81	716.00	"
16	83.92	16.08	-----	100	449.24	C + D
17	84.43	15.57	5.93	94.07	441.34	"
18	84.01	15.99	15.52	84.48	394.02	"
19	84.38	15.62	23.46	76.54	446.46	"
20	83.87	16.13	30.41	69.59	484.40	"
21	83.83	16.17	34.66	65.34	465.75	"
22	83.62	16.38	43.12	56.88	595.07	"

(continued next page)

AUXILIARY INFORMATION

METHOD/APPARATUS/PROCEDURE:	SOURCE AND PURITY OF MATERIALS:
The standard isothermal method was used. The mixtures were stirred until the liquid phase had a constant composition. The chloride ion content was determined argentimetrically, potassium was determined as potassium tetraphenylborate, phosphorus was determined by differential colorimetry on an FEK-56 apparatus, and sodium was determined by difference.	No information is given.
	ESTIMATED ERROR:
	No information is given.
	REFERENCES:

COMPONENTS:	ORIGINAL MEASUREMENTS:
(1) Sodium dihydrogenphosphate; NaH_2PO_4; [7558-80-7]	Khallieva, Sh.D.
(2) Potassium dihydrogenphosphate; KH_2PO_4; [7778-77-0]	*Izv. Akad. Nauk Turkm. SSR, Ser. Fiz.-Tekh., Khim. Geol. Nauk* 1977, 3, 125-8.
(3) Sodium chloride; NaCl; [7647-14-5]	
(4) Potassium chloride; KCl; [7747-40-7]	
(5) Water; H_2O; [7732-18-5]	

EXPERIMENTAL VALUES cont'd:

Part 1. Solubility in the Na^+, K^+||Cl^-, $H_2PO_4^-$-H_2O system at 40°C.

Jänecke's indices[a]

soln. no.	Na^+	K^+	Cl^-	$H_2PO_4^-$	H_2O	solid phase[b]
23	83.28	16.72	47.78	52.22	578.00	C + D
24	82.42	17.58	56.32	43.68	629.80	"
25	80.03	19.97	69.54	30.48	624.76	"
26	74.85	25.15	79.28	20.72	640.42	"
27	70.02	29.98	83.72	16.28	662.04	"

[a]The units are: mol/100 mol of solute.

[b]The solid phases are: A = NaCl; B = KCl; C = $NaH_2PO_4 \cdot H_2O$; D = KH_2PO_4.

Part 2. The compiler has calculated the following results from the data in Part 1.

soln. no.	Na^+ mass%	Na^+ mol/kg	K^+ mass%	K^+ mol/kg	Cl^- mass%	Cl^- mol/kg	$H_2PO_4^-$ mass%	$H_2PO_4^-$ mol/kg	H_2O mass%
1	7.44	4.78	6.90	2.60	17.76	7.38	0.00	0.00	67.88
2	7.55	4.96	6.98	2.69	17.31	7.36	1.85	0.29	66.31
3	7.67	5.16	7.08	2.80	17.02	7.41	3.43	0.55	64.78
4	7.65	5.19	7.00	2.78	16.50	7.24	4.54	0.73	64.31
5	7.71	5.25	6.89	2.75	16.32	7.19	5.01	0.81	64.06
6	7.88	5.41	6.82	2.75	16.33	7.26	5.50	0.90	63.46
7	8.02	5.57	6.95	2.84	16.58	7.46	5.73	0.95	62.72
8	8.28	5.77	6.54	2.68	16.40	7.40	6.30	1.05	62.48
9	8.48	6.00	6.68	2.77	16.78	7.68	6.46	1.09	61.60
10	0.00	0.00	15.19	5.61	12.73	5.18	2.82	0.42	69.27
11	1.66	1.08	14.06.	5.38	14.06	5.93	3.38	0.53	66.84
12	2.68	1.74	12.48	4.76	14.06	5.92	3.79	0.59	66.99
13	4.71	3.15	10.44	4.10	15.04	6.51	4.61	0.74	65.20
14	6.16	4.09	8.19	3.19	15.10	6.49	4.97	0.79	65.59
15	7.28	4.96	7.58	3.03	15.52	6.84	5.63	0.92	63.98
16	9.52	10.38	3.11	1.99	0.00	0.00	47.43	12.37	39.96
17	9.82	10.63	3.09	1.96	1.06	0.75	45.77	11.84	40.26
18	10.53	11.85	3.42	2.25	3.01	2.19	44.30	11.91	38.74
19	10.32	10.50	3.25	1.94	4.43	2.92	39.16	9.52	42.83
20	10.06	9.56	3.32	1.85	5.67	3.49	35.12	7.98	45.84
21	10.43	10.00	3.43	1.93	6.66	4.13	34.01	7.79	45.46
22	9.47	7.81	3.16	1.53	7.55	4.02	26.95	5.31	52.87
23	9.71	8.00	3.32	1.61	8.61	4.59	25.48	5.02	52.88
24	9.41	7.27	3.42	1.55	9.93	4.97	20.86	3.85	56.38
25	9.54	7.12	4.06	1.78	12.80	6.18	15.20	2.71	58.40
26	9.03	6.49	5.17	2.18	14.77	6.88	10.45	1.80	60.58
27	8.36	5.88	6.10	2.52	15.44	7.03	8.13	1.37	61.98

COMPONENTS:	ORIGINAL MEASUREMENTS:
(1) Sodium dihydrogenphosphate; NaH_2PO_4; [7558-80-7]	Solov'ev, A.P.; Balashova, E.F.; Verendyakina, N.A.; Zjuzina, L.F.
(2) Sodium chloride; NaCl; [7647-14-5]	*Resp. Sb. Nauch. Tr.-Yaroslav. Gos. Pedagog,*
(3) Ammonium dihydrogenphosphate; $NH_4H_2PO_4$; [7722-76-1]	*In.-t.* 1978, 169, 79-84.
(4) Ammonium chloride; NH_4Cl; [12125-02-9]	
(5) Water; H_2O; [7732-18-5]	

VARIABLES:	PREPARED BY:
Composition at 25°C.	J. Eysseltová

EXPERIMENTAL VALUES:

Solubility in the Na^+, NH_4^+||Cl^-, $H_2PO_4^-$-H_2O system at 25°C.

NaH_2PO_4		NaCl		$NH_4H_2PO_4$		NH_4Cl		Na^+	$H_2PO_4^-$	solid[a]
mass%	mol/kg[b]	mass%	mol/kg[b]	mass%	mol/kg[b]	mass%	mol/kg[b]	ion%[c]	ion%[c]	phase
25.01	3.48	15.13	4.33	----	----	----	----	100.0	44.63	A + B
24.74	3.63	13.97	4.21	4.55	0.70	----	----	91.75	50.72	"
23.26	3.51	12.90	4.00	8.59	1.35	----	----	84.69	54.90	"
24.19	3.85	11.01	3.60	12.40	2.06	----	----	78.31	62.25	A + B + C
43.47	8.12	----	----	11.93	2.35	----	----	77.68	100.0	B + C
38.32	7.25	4.54	1.76	13.11	2.59	----	----	77.69	84.72	"
32.90	5.20	8.00	2.96	12.79	2.40	----	----	78.74	73.75	"
30.54	5.37	9.33	3.37	12.78	2.35	----	----	78.86	69.71	"
12.30	1.78	16.87	4.96	12.58	1.88	----	----	78.20	42.40	A + C
----	----	20.85	5.75	17.10	2.40	----	----	70.77	29.23	"
----	----	16.07	4.21	5.64	0.75	12.93	3.70	48.58	8.66	A + C + D
----	----	17.07	4.41	2.82	0.37	13.88	3.92	50.69	4.34	A + D
----	----	17.15	4.40	----	----	16.14	4.52	49.24	----	"
----	----	8.68	2.18	4.48	0.57	18.75	5.15	27.55	7.26	C + D
----	----	5.89	1.48	5.15	0.66	21.11	5.82	18.67	8.32	"
----	----	1.59	0.39	5.60	0.70	22.91	6.13	5.36	9.72	"
----	----	----	----	6.12	0.78	25.54	6.98	----	10.17	"

[a] The solid phases are: A = NaCl; B = $NaH_2PO_4 \cdot 2H_2O$; C = $NH_4H_2PO_4$; D = NH_4Cl.

[b] The mol/kg H_2O values were calculated by the compiler.

[c] The units here are: mol/100 mol of solute (compiler).

AUXILIARY INFORMATION

METHOD/APPARATUS/PROCEDURE:	SOURCE AND PURITY OF MATERIALS:
The method of invariant points was used. A third component was added to eutectic systems until a new solid phase appeared. At equilibrium both liquid and solid phases were analyzed. Chloride ion content was determined by the Volhard method. $H_2PO_4^-$ ions were precipitated as $NH_4MgPO_4 \cdot 6H_2O$, and the excess of Mg was titrated complexometrically (1). Ammonium ions were removed and the excess of base used was titrated with HCl.	No information is given.
	ESTIMATED ERROR:
	No information is given.
	REFERENCES:
	1. Shemyakin, F.M.; Zelenina, E.N., *Zavod. Lab.* 1969, 6.

COMPONENTS:	ORIGINAL MEASUREMENTS:
(1) Sodium dihydrogenphosphate; NaH_2PO_4; [7558-80-7]	Solov'ev, A.P.; Balashova, E.F.; Verendjakina, N.A.; Zjzina, L.F.
(2) Potassium dihydrogenphosphate; KH_2PO_4; [7778-77-0]	Resp. Sb. Nauchn. Tr.-Yaroslov. Gos. Pedagog. In-t. 1978, 169, 79-84.
(3) Sodium chloride; NaCl; [7647-14-6]	
(4) Potassium chloride; KCl; [7747-40-7]	
(5) Water; H_2O; [7732-18-5]	

VARIABLES:	PREPARED BY:
Composition at 25°C.	J. Eysseltová

EXPERIMENTAL VALUES:

Solubility in the K^+, $Na^+||Cl^-$, $H_2PO_4^-$-H_2O system at 25°C.

NaCl		NaH_2PO_4		KCl		KH_2PO_4		K^+	$H_2PO_4^-$	solid[c]
mass%	mol/kg[a]	mass%	mol/kg[a]	mass%	mol/kg[a]	mass%	mol/kg[a]	ion%[b]	ion%[b]	phase
15.13	4.33	25.01	3.48	----	----	----	----	----	44.63	A + B
13.74	4.31	27.72	4.23	----	----	4.01	0.54	5.95	52.58	"
12.85	4.47	30.72	5.20	----	----	7.22	1.08	10.04	58.46	A + B + C
----	----	46.65	8.71	----	----	9.14	1.51	14.79	100.0	B + C
5.93	2.21	39.75	7.21	----	----	8.37	1.34	12.45	79.48	"
17.73	5.01	11.57	1.56	----	----	10.18	1.23	15.78	36.11	A + C
21.11	5.64	----	----	4.16	0.87	10.70	1.23	27.22	15.93	"
18.62	4.75	----	----	8.69	1.74	5.62	0.62	33.15	8.68	A + C + D
19.49	4.86	----	----	8.23	1.61	3.67	0.39	29.38	5.96	A + D
20.90	5.22	----	----	10.58	2.07	----	----	28.44	----	"
14.75	3.67	----	----	11.00	2.14	5.22	0.56	42.46	8.76	C + D
7.78	1.87	----	----	16.85	3.18	4.34	0.49	65.96	8.15	"
----	----	----	----	23.50	4.28	2.90	0.29	100.0	6.33	"

[a] The mol/kg H_2O values were calculated by the compiler.

[b] The authors' ion% values are to be understood as mol/100 mol solute.

[c] The solid phases are: A = NaCl; B = $NaH_2PO_4 \cdot 2H_2O$; C = KH_2PO_4; D = KCl.

AUXILIARY INFORMATION

METHOD/APPARATUS/PROCEDURE:	SOURCE AND PURITY OF MATERIALS:
The method of invariant points was used. A third component was added to eutectic systems until a new solid appeared. The solid and liquid phases were each analyzed. Chloride ion content was determined by the Volhard method. $H_2PO_4^-$ ions were precipitated as $NH_4MgPO_4 \cdot 6H_2O$, the excess of Mg^{2+} being titrated compleximetrically (1). NH_4^+ ions were removed and the excess of base used was titrated with 0.1 M HCl.	No information is given.
	ESTIMATED ERROR:
	No information is given.
	REFERENCES:
	1. Shemjakin, F.M.; Zelenina, E.N. Zavod. Lab. 1969, 6.

COMPONENTS:	ORIGINAL MEASUREMENTS:
(1) Sodium dihydrogenphosphate; NaH_2PO_4; [7558-80-7] (2) Potassium dihydrogenphosphate; KH_2PO_4; [7778-70-0] (3) Sodium nitrate; $NaNO_3$; [7631-99-4] (4) Potassium nitrate; KNO_3; [7757-79-1] (5) Water; H_2O; [7732-18-5]	Girich, T.E.; Guljamov, Yu.M. *Vopr. Khim. Khim. Tekhnol.* <u>1979</u>, 57, 54-7.

VARIABLES:	PREPARED BY:
Composition and temperature.	J. Eysseltová

EXPERIMENTAL VALUES:

Composition of saturated solutions in the K^+, $Na^+ || NO_3^-$, $H_2PO_4^-$-H_2O system.

soln. no.	K^+ ion%	NO_3^- ion%	H_2O conc.[a]	solid phases[b]
		temp. = 323 K		
1	91.54	89.68	633.0	A + C
2	81.09	90.22	534.2	"
3	67.25	90.20	458.7	"
4	61.59	90.47	437.2	"
5	56.26	90.87	390.7	"
6	52.33	90.87	285.5	"
7	47.09	92.09	285.5	"
8	42.68	92.44	260.3	"
9	40.89	93.02	244.5	"
10	39.25	92.14	217.1	"
11	37.10	93.42	192.9	A + B + C
12	38.95	100.0	208.2	A + B
13	37.98	98.25	193.8	"
14	37.81	95.99	193.0	"
15	16.57	0.00	365.9	C + E
16	14.29	18.01	283.0	"
17	13.88	27.98	292.4	"
18	14.93	37.85	280.3	"
19	16.88	42.92	276.1	B + C + E
20	16.99	47.60	245.9	B + C
21	20.66	59.33	238.0	"
22	25.22	76.45	219.8	"
23	33.67	90.92	208.2	" (continued next page)

AUXILIARY INFORMATION

METHOD/APPARATUS/PROCEDURE:	SOURCE AND PURITY OF MATERIALS:
The isothermal method was used. Equilibrium was ascertained by repeated analysis of the solid and liquid phases. The dihydrogenphosphate ion content was determined photocolorimetrically, the sodium and potassium ion content was determined by flame photometry, and the nitrate ion content was determined by a titration with $FeSO_4$.	The solids were chemically pure and were recrystallized twice before being used.
	ESTIMATED ERROR: No details are given.
	REFERENCES:

COMPONENTS:	ORIGINAL MEASUREMENTS:
(1) Sodium dihydrogenphosphate; NaH_2PO_4; [7558-80-7]	Girich, T.E.; Guljamov, Yu.M.
(2) Potassium dihydrogenphosphate; KH_2PO_4; [7778-70-0]	*Vopr. Khim. Khim. Tekhnol.* <u>1979</u>, 57, 54-7.
(3) Sodium nitrate; $NaNO_3$; [7631-99-4]	
(4) Potassium nitrate; KNO_3; [7757-79-1]	
(5) Water; H_2O; [7732-18-5]	

EXPERIMENTAL VALUES cont'd:

Composition of saturated solutions in the K^+, $Na^+ || NO_3^-$, $H_2PO_4^- - H_2O$ system.

soln. no.	K^+ ion%	NO_3^- ion%	H_2O conc.[a]	solid phases[b]
		temp. = 323 K		
24	0.00	38.43	373.1	B + E
25	5.59	40.89	318.0	"
26	11.35	41.33	291.7	"
		temp. = 348 K		
27	100.00	95.31	373.4	A + C
28	85.49	94.68	302.1	"
29	74.33	94.36	271.4	"
30	64.73	94.37	227.0	"
31	55.08	94.30	193.5	"
32	46.39	94.17	165.3	"
33	47.01	100.0	149.0	A + B
34	46.13	96.73	148.0	"
35	44.62	94.04	147.4	A + B + C
36	42.98	92.94	156.2	B + C
37	40.83	90.53	178.0	"
38	36.91	85.19	192.5	"
39	34.44	82.00	201.4	"
40	31.00	75.19	206.2	"
41	25.11	64.30	211.7	"
42	21.83	55.68	219.8	"
43	20.91	47.29	226.3	"
44	29.06	40.92	234.0	B + C + D
45	20.05	38.47	243.5	C + D
46	21.31	29.19	258.7	"
47	21.28	19.44	266.0	"
48	21.70	11.09	276.4	"
49	20.52	0.00	289.1	"
50	0.00	51.14	267.2	B + D
51	5.31	50.00	256.0	"
52	11.15	45.96	247.3	"
53	15.31	44.96	239.5	"
		temp. = 373 K		
54	100.00	97.08	245.0	A + C
55	99.68	97.08	213.2	"
56	83.91	96.83	191.7	"
57	70.53	95.35	176.3	"
58	60.85	95.18	142.8	"
59	52.95	95.03	113.5	"
60	51.90	100.0	95.7	A + B
61	51.32	97.75	93.0	"
62	50.35	94.97	92.4	A + B + C
63	49.02	91.24	94.6	B + C
64	47.52	82.85	112.5	"
65	45.01	74.71	117.8	"
66	42.46	63.76	130.0	"
67	36.87	54.88	139.5	"
68	33.82	47.00	147.0	"
69	31.22	44.73	156.3	"
70	29.53	42.27	161.5	B + C + D
71	29.01	38.09	164.0	C + D
72	28.03	34.04	176.2	"
73	25.78	28.26	181.4	"
74	23.45	18.88	189.0	"
75	25.43	5.30	196.1	"
76	25.44	0.00	205.0	"

(continued next page)

AMO—H

COMPONENTS:	ORIGINAL MEASUREMENTS:
(1) Sodium dihydrogenphosphate; NaH_2PO_4; [7558-80-7]	Girich, T.E.; Guljamov, Yu.M.
(2) Potassium dihydrogenphosphate; KH_2PO_4; [7778-70-0]	*Vopr. Khim. Khim. Tekhnol.* <u>1979</u>, 57, 54-7.
(3) Sodium nitrate; $NaNO_3$; [7631-99-4]	
(4) Potassium nitrate; KNO_3; [7757-79-1]	
(5) Water; H_2O; [7732-18-5]	

EXPERIMENTAL VALUES cont'd:

Composition of saturated solutions in the K^+, $Na^+ || NO_3^-$, $H_2PO_4^-$-H_2O system.

soln. no.	K^+ ion%	NO_3^- ion%	H_2O conc.^{a}	solid phases^b
		temp. = 373 K		
77	0.00	55.26	229.8	B + D
78	2.66	54.97	225.6	"
79	10.00	50.22	201.3	"
80	18.36	45.95	190.0	"

^{a}The concentration units are: mol H_2O/100 g equiv of dry salts.

^{b}The solid phases are: A = KNO_3; B = $NaNO_3$; C = KH_2PO_4; D = NaH_2PO_4;

E = $NaH_2PO_4 \cdot 2H_2O$.

The compiler has calculated the following values from the data given above.

soln. no.	conc. of K^+ mass%	mol/kg	conc. of Na^+ mass%	mol/kg	conc. of NO_3^- mass%	mol/kg	conc. of $H_2PO_4^-$ mass%	mol/kg
				temp. = 323 K				
1	16.47	8.03	0.89	0.74	25.59	7.87	4.60	0.90
2	16.04	8.43	2.19	1.96	28.30	9.38	4.79	1.01
3	14.46	8.14	4.14	3.96	30.76	10.92	5.22	1.18
4	15.60	7.82	4.99	4.88	31.70	11.49	5.22	1.21
5	13.12	7.99	6.00	6.21	33.62	12.92	5.28	1.29
6	13.82	10.18	7.40	9.27	38.07	17.68	5.98	1.77
7	12.54	9.16	8.28	10.29	38.91	17.91	5.22	1.53
8	11.80	9.10	9.32	12.23	40.54	19.72	5.18	1.61
9	11.58	9.29	9.84	13.43	41.79	21.13	4.90	1.58
10	11.52	10.04	10.48	15.54	42.91	23.57	5.72	2.01
11	11.33	10.68	11.29	18.11	45.26	26.90	4.98	1.89
12	11.82	10.39	10.90	16.29	48.16	26.68	0.00	0.00
13	11.72	10.88	11.26	17.77	48.12	28.16	1.34	0.50
14	11.61	10.88	11.23	17.90	46.78	27.63	3.05	1.15
15	3.43	2.51	10.17	12.66	0.00	0.00	51.45	15.18
16	3.34	2.80	11.80	16.82	6.69	3.53	47.64	16.09
17	3.28	2.63	11.99	16.36	10.51	5.31	42.32	13.68
18	3.65	2.95	12.25	16.86	14.70	7.50	37.77	12.31
19	4.19	3.39	12.14	16.72	16.91	8.63	35.17	11.48
20	4.41	3.83	12.69	18.75	19.63	10.75	33.80	11.83
21	5.55	4.82	12.54	18.52	25.30	13.84	27.13	9.49
22	7.20	6.37	12.56	18.90	34.64	19.32	16.68	5.95
23	10.04	8.98	11.63	17.69	43.01	24.26	6.71	2.42
24	0.00	0.00	13.23	14.89	34.38	5.72	38.66	9.16
25	1.33	0.97	13.24	16.49	34.99	7.14	34.93	10.32
26	2.77	2.16	12.74	16.88	35.59	7.87	32.84	11.17
				temp. = 348 K				
27	23.00	14.87	0.00	0.00	34.77	14.18	2.67	0.69
28	21.56	15.72	2.15	2.66	37.87	17.41	3.32	0.97
29	19.66	15.21	3.99	5.25	39.59	19.31	3.70	1.15
30	18.30	15.84	5.86	8.63	42.32	23.09	3.94	1.37
31	16.47	15.81	7.90	12.89	44.74	27.07	4.23	1.63
32	14.59	15.59	9.91	18.01	46.99	31.64	4.55	1.95
33	15.39	17.52	10.20	19.74	51.94	37.28	0.00	0.00
34	15.00	17.31	10.30	20.22	49.89	36.31	2.63	1.22
35	14.44	16.81	10.53	20.87	48.27	35.44	4.78	2.24
36	13.71	15.28	10.69	20.28	47.04	33.05	5.58	2.51
37	12.57	12.74	10.71	18.46	44.23	28.25	7.23	2.95
38	11.03	10.65	11.09	18.20	40.39	24.58	10.98	4.27
39	10.11	9.50	11.32	18.08	38.20	22.61	13.11	4.96
40	8.92	8.35	11.68	18.59	34.33	20.25	17.72	6.68

(continued next page)

COMPONENTS:	ORIGINAL MEASUREMENTS:
(1) Sodium dihydrogenphosphate; NaH_2PO_4; [7558-80-7]	Girich, T.E.; Guljamov, Yu.M.
(2) Potassium dihydrogenphosphate; KH_2PO_4; [7778-70-0]	*Vopr. Khim. Khim. Tekhnol.* <u>1979</u>, 57, 54-7.
(3) Sodium nitrate; $NaNO_3$; [7631-99-4]	
(4) Potassium nitrate; KNO_3; [7757-79-1]	
(5) Water; H_2O; [7732-18-5]	

EXPERIMENTAL VALUES cont'd:

The compiler has calculated the following values from the data given above.

soln. no.	conc. of K^+ mass%	conc. of K^+ mol/kg	conc. of Na^+ mass%	conc. of Na^+ mol/kg	conc. of NO_3^- mass%	conc. of NO_3^- mol/kg	conc. of $H_2PO_4^-$ mass%	conc. of $H_2PO_4^-$ mol/kg
				temp. 348 K				
41	7.03	6.58	12.32	19.65	28.55	16.87	24.79	9.36
42	5.94	5.51	12.51	19.75	24.04	14.07	29.93	11.20
43	5.54	5.13	12.32	19.41	19.87	11.60	34.65	12.94
44	7.45	6.89	10.69	16.84	16.64	9.71	37.58	14.02
45	5.10	4.57	11.96	18.24	15.53	8.77	38.85	14.03
46	5.21	4.57	11.32	16.89	11.33	6.26	42.98	15.20
47	5.05	4.44	11.00	16.44	7.32	4.06	47.50	16.82
48	5.00	4.36	10.62	15.73	4.06	2.22	50.92	17.87
49	4.57	3.94	10.42	15.27	0.00	0.00	55.32	19.21
50	0.00	0.00	15.30	20.79	21.11	10.63	31.55	10.15
51	1.38	1.15	14.56	20.54	20.75	10.85	32.45	10.85
52	2.90	2.50	13.59	19.96	18.97	10.32	34.89	12.14
53	3.99	3.55	12.99	19.64	18.60	10.42	35.63	12.76
				temp. = 373 K				
54	26.73	22.67	0.00	0.00	41.17	22.01	1.93	0.66
55	27.74	25.97	0.05	0.08	42.86	25.29	2.01	0.76
56	24.45	24.31	2.75	4.66	44.76	28.06	2.29	0.91
57	21.25	22.22	5.22	9.28	45.58	30.04	3.47	1.46
58	19.46	23.67	7.36	15.23	48.30	37.02	3.82	1.87
59	17.89	25.91	9.34	23.02	50.93	46.51	4.16	2.43
60	18.34	30.12	9.99	27.92	56.07	58.05	0.00	0.00
61	18.10	30.65	10.10	29.08	54.71	58.39	1.96	1.34
62	17.65	30.27	10.23	29.85	52.81	57.10	4.37	3.02
63	16.96	28.78	10.37	29.93	50.07	53.58	7.51	5.14
64	15.62	23.46	10.14	25.91	43.21	40.91	13.98	8.46
65	14.38	21.22	10.33	25.93	37.88	35.23	20.05	11.92
66	12.97	18.14	10.34	24.58	30.91	27.24	27.47	15.48
67	10.93	14.68	11.00	25.14	25.81	21.85	33.19	17.96
68	9.76	12.78	11.23	25.01	21.51	17.76	37.95	20.03
69	8.87	11.09	11.49	24.44	20.17	15.89	38.98	19.64
70	8.30	10.15	11.65	24.24	18.85	14.54	40.27	19.85
71	8.05	9.82	11.58	24.04	16.77	12.90	42.63	20.97
72	7.59	8.83	11.46	22.69	14.62	10.73	44.33	20.79
73	6.86	7.89	11.61	22.73	11.93	8.65	47.36	21.97
74	6.06	6.89	11.64	22.50	7.74	5.54	52.05	23.84
75	6.31	7.20	10.88	21.12	2.08	1.50	58.31	26.82
76	6.17	6.89	10.64	20.20	0.00	0.00	60.25	27.10
77	0.00	0.00	16.18	24.17	24.13	13.35	30.55	10.81
78	0.73	0.65	15.78	23.96	24.04	13.53	30.80	11.08
79	2.78	2.75	14.75	24.83	22.20	13.85	34.42	13.73
80	5.08	5.36	13.30	23.87	20.20	13.43	37.16	15.80

COMPONENTS:	ORIGINAL MEASUREMENTS:
(1) Sodium dihydrogenphosphate; NaH_2PO_4; [7558-80-7]	Ferroni, G.; Galea, J.; Antonetti, G.
(2) 2-Propanone (acetone); C_3H_6O; [67-64-1]	*Bull. Soc. Chim. Fr.* <u>1974</u>, *12*, (Pt. 1), 273-81.
(3) Water; H_2O; [7732-18-5]	

VARIABLES:	PREPARED BY:
Composition at 25°C.	J. Eysseltová

EXPERIMENTAL VALUES:

A miscibility gap was found in the NaH_2PO_4-C_3H_6O-H_2O system.

The results for the isothermal binodal curve at 25°C are:

	u p p e r l a y e r				l o w e r l a y e r			
	NaH_2PO_4	C_3H_6O	H_2O		NaH_2PO_4	C_3H_6O	H_2O	
$\rho/g\ cm^{-3}$	mass%	mass%	mass%	$\rho/g\ cm^{-3}$	mass%	mass%	mass%	solid phase[a]
0.7940	∿0	100	∿0	----	----	----	----	A
0.8032	0.10	92.99	6.91	1.5317	54.81	0.98	44.21	
0.8343	0.23	75.74	24.03	1.4930	52.24	1.18	46.58	
0.8941	1.10	51.85	47.05	1.4684	50.51	1.25	48.24	
0.9066	1.38	41.16	57.45	1.4125	46.44	1.66	51.89	
0.9304	2.42	34.12	63.47	1.2694	34.17	2.19	63.64	
0.9807	5.75	21.35	72.90	1.1344	21.53	7.15	71.32	
1.1022	18.44	9.44	72.22		critical solution			B
1.5576	56.10	0	43.90					C

[a] The solid phases are: A = NaH_2PO_4; B = $NaH_2PO_4 \cdot H_2O$; C = $NaH_2PO_4 \cdot 2H_2O$.

AUXILIARY INFORMATION

METHOD/APPARATUS/PROCEDURE:	SOURCE AND PURITY OF MATERIALS:
The mixtures were equilibrated by stirring in a thermostat for 48 hours. This was done in the dark to prevent photodecomposition. The dihydrogenphosphate ion content was determined by an automatic potentiometric pH titration after evaporating the solution to dryness and dissolving the residue in bidistilled water. The 2-propanone content was determined iodometrically using a potentiometric titration.	Reagent grade 2-propanone was used. Reagent grade NaH_2PO_4 was dehydrated at 100°C and then stored in a vacuum over NaOH. The water was twice distilled and deaerated.

	ESTIMATED ERROR:
	The temperature was held to within ±0.1 K. The precision of the analyses was 0.5%.

	REFERENCES:

COMPONENTS:	ORIGINAL MEASUREMENTS:
(1) Sodium dihydrogenphosphate; NaH_2PO_4; [7558-80-7] (2) Sodium perchlorate; $NaClO_4$; [7601-89-0] (3) 2-Propanone (acetone); C_3H_6O; [67-64-1] (4) Water; H_2O; [7732-18-5]	Ferroni, G.; Galea, J.; Antonetti, G. *Bull. Soc. Chim. Fr.* <u>1974</u>, *12*, (Pt. 1), 273-81.

VARIABLES:	PREPARED BY:
Concentration of $NaClO_4$ at 25°C.	J. Eysseltová

EXPERIMENTAL VALUES:

Composition of the saturated solutions at 25°C.

	1 mol $NaClO_4/dm^3$		3 mols $NaClO_4/dm^3$	
H_2O c^a	NaH_2PO_4 mol/dm^3	solid phase	NaH_2PO_4 mol/dm^3	solid phase
100	4.912	$NaH_2PO_4 \cdot 2H_2O$	2.400	$NaH_2PO_4 \cdot 2H_2O$
90.9	1.089	binodal curve	0.741	binodal curve
83.3	0.241	" "	0.029	" "
66.7	0.026	" "	0.0052	" "
50.0	0.0061	" "	0.0024	$NaH_2PO_4 \cdot H_2O$
33.3	$\sim 7.5 \times 10^{-4}$	" "	4.2×10^{-4}	"
9.1	$\sim 6 \times 10^{-6}$	" "	5×10^{-6}	"
0.0	10^{-6}	NaH_2PO_4	10^{-7}	NaH_2PO_4

^{a}The concentration units are: mol/100 mols of solvent.

AUXILIARY INFORMATION

METHOD/APPARATUS/PROCEDURE:	SOURCE AND PURITY OF MATERIALS:
The mixtures were equilibrated by stirring in a thermostat for 48 hours. This was done in the dark to prevent photo-decomposition. The dihydrogenphosphate ion content was determined by an automatic potentiometric pH titration after evaporating the solution to dryness and then dissolving the residue in bidistilled water. The 2-propanone content was determined iodometrically using a potentiometric titration.	Reagent grade materials were used. The NaH_2PO_4 was dehydrated at 100°C and stored in a vacuum over NaOH. The water was bidistilled and deaerated.

	ESTIMATED ERROR:
	The temperature was constant to within ±0.1 K. The analyses had a precision of ±0.5%.

	REFERENCES:

COMPONENTS:	EVALUATOR:
(1) Disodium hydrogenphosphate; Na_2HPO_4; [7558-79-4] (2) Water; H_2O; [7732-18-5]	J. Eysseltová Charles University Prague, Czechoslovakia May 1985

CRITICAL EVALUATION:

THE BINARY SYSTEM

Solubility data have been reported for the temperature interval 273-373 K (1), for the 272-313 K interval (2), and at 273, 291 and 298 K (3). Wendrow and Kobe (4) report their own extrapolated data as well as data obtained by others (5). Older data (7-13) are cited in the article by D'Ans and Schreiner (6). But these data appear to have a systematic error and were eliminated from consideration during the first graphical examination of the material. On the other hand, some data from studies of multicomponent systems (14-19) were consistent with those reported by others (1-3) and were included in the evaluation procedure.

Several hydrates of disodium hydrogenphosphate have been reported. Wendrow and Kobe (4) stated that the transition temperatures of the dodecahydrate to the heptahydrate and of the heptahydrate to the dihydrate were 308.7 and 321.2 K, respectively. A more precise determination of these values (6) gives transition temperatures of 308.55 and 321.55 K, respectively. There is also a report of the existence of two forms of the dodecahydrate with a transition temperature of 302.8 K (2), but this has not been confirmed by any other investigators. The transition temperature of the dihydrate to the anhydrous salt was said to be 368.2 K (4).

All the experimental data that were not eliminated in the first graphical examination were evaluated by the method described in chapter 3. The data were fitted to equation [1]. The precision of the published data was estimated to be about the same as

$$\ln x/x_o = A \cdot (1/T - 1/T_o) + B \cdot \ln(T/T_o) + C \cdot (T-T_o) \tag{1}$$

that for sodium dihydrogenphosphate and hence, the criteria for the selection of relevant points were the same as those used in chapter 3. However, these criteria could be applied completely only to the data for the dodecahydrate. For the heptahydrate and the dihydrate the data in the different reports were not in sufficiently good agreement and the selection of values for x_o was based on the results of only one report (1). Table I is a summary of the solubility data.

During the iteration procedure practically all the data except those of Shiomi (1) were eliminated. Therefore, the results of this procedure are considered to be tentative.

The values for the parameters of equation [1] are given in Table II while in Table III the solubility values calculated from equation [1] are given.

MULTICOMPONENT SYSTEMS

Solubility data have been reported for several ternary and quaternary systems but in only a few instances have data for a given system been reported by more than one investigator(s). In three of the systems solid phases other than the components or their hydrates have been reported.

Two reports (3,15) give data for the $Na_2HPO_4-H_2O_2-H_2O$ system at 273 K but a comparison of the two reports cannot be made because the concentration range studied in one report (3) is too narrow. The other article (15) reports the presence of the two compounds $Na_2HPO_4 \cdot 1.5H_2O_2$ [13769-82-9] and $Na_2HPO_4 \cdot 2.5H_2O_2$ [13769-83-0] at H_2O_2 concentrations greater than 27 mass% for this system.

Data for the $Na_2HPO_4-H_3BO_3-H_2O$ system at 298 K have been reported by Beremzhanov, et al. (20). The results are analogous to those for the $NaH_2PO_4-NaBO_2-H_2O$ system reported by the same authors (21) and discussed in chapter 4. The appearance of $Na_2B_4O_7 \cdot 10H_2O$ [61028-24-8] as a solid phase suggests that the system should be treated as part of the $Na_2O-B_2O_3-P_2O_5-H_2O$ system.

A similar situation exists with respect to the $Na_2HPO_4-Na_2SiO_3-H_2O$ system. Data for this system at 293 K were reported by Manvelyan, et al. (22). The formation of $Na_3PO_4 \cdot 12H_2O$ [10101-89-0] in this system is an indication of its pseudo-ternary character.

There is no evidence for the formation of solid solutions and/or ternary compounds in the other multicomponent systems for which data are available.

There is only one report giving data for the systems $Na_2HPO_4-Na_2H_2EDTA-H_2O$ (17), $Na_2HPO_4-NaNO_3-H_2O$ (23) and $Na_2HPO_4-NaCl-H_2O$ (24). All these data were obtained at 298 K.

Makin and his co-workers have reported data for several systems containing Na_2HPO_4 as a component. There are two reports for the $Na_2HPO_4-Na_2SO_4-H_2O$ system at 298 K (18, 19). This group has also published data for two quaternary systems: the $Na_2HPO_4-NaNO_3-Na_2SO_4-H_2O$ system (25); and the $Na_2HPO_4-NaNO_3-NaCl-H_2O$ system (26), both at

(continued next page)

COMPONENTS:	EVALUATOR:
(1) Disodium hydrogenphosphate; Na_2HPO_4; [7558-79-4] (2) Water; H_2O; [7732-18-5]	J. Eysseltová Charles University Prague, Czechoslovakia May 1985

CRITICAL EVALUATION:

Table I. Solubility of Na_2HPO_4 in water.

T/K	mass%	ref.	weight init/final	T/K	mass%	ref.	weight init/final
				$Na_2HPO_4 \cdot 12H_2O$			
268.5	1.43	2	1/1	298.30	10.74	1	1/1
273.2	1.605	3	2/0	298.30	10.72	1	2/2
273.2	1.6	15	2/0	298.60	10.97	1	1/1
273.8	1.71	1	2/0	298.60	10.98	1	1/1
279.7	2.66	2	1/1	298.60	10.96	1	1/1
283.41	3.43	1	1/0	298.70	11.05	1	1/1
283.41	3.42	1	1/0	298.70	11.04	1	1/1
283.51	3.46	1	4/0	298.70	11.09	1	1/1
288.26	4.97	1	1/1	301.0	12.40	2	1/0
288.26	4.96	1	1/1	301.8	13.70	2	1/0
291.2	5.985	3	1/0	302.2	13.82	2	1/0
293.10	6.77	2	1/0	302.7	14.66	2	1/0
293.39	7.30	1	1/0	303.3	16.28	2	1/0
293.39	7.32	1	1/0	303.36	17.22	1	1/1
293.39	7.31	1	2/0	303.36	17.27	1	1/1
295.92	8.20	2	1/0	303.41	17.76	1	3/0
297.30	8.70	2	1/0	303.41	17.78	1	1/0
298.2	10.829	3	1/0	303.41	17.74	1	1/0
298.2	10.59	17	1/1	303.41	17.44	1	1/0
298.2	10.32	18, 19	2/0	303.41	17.77	1	1/0
298.2	10.80	16	1/0	303.91	18.98	1	1/0
298.2	10.4	14	1/0	303.91	18.96	1	1/0
298.2	10.60	17	1/1	303.91	18.97	1	1/0
				306.19	23.59	1	2/2
				306.29	23.89	1	1/1
				306.29	23.88	1	1/1
				$Na_2HPO_4 \cdot 7H_2O$			
309.42	31.20	1	7/7	313.44	35.42	1	2/0
309.42	31.21	1	1/1	313.44	35.41	1	1/0
309.42	31.22	1	1/1	318.29	40.68	1	1/1
309.42	31.23	1	1/1	318.29	40.69	1	1/1
310.42	32.23	1	3/3	318.29	40.71	1	1/1
310.42	32.19	1	2/2	318.29	40.72	1	1/1
310.42	32.18	1	1/1	320.5	43.37	1	1/1
310.42	32.22	1	1/1	320.5	43.36	1	1/1
313.44	35.46	1	2/0	321.3	44.45	1	2/2
313.44	35.43	1	1/0	321.3	44.48	1	1/1
				$Na_2HPO_4 \cdot 2H_2O$			
323.37	44.57	1	1/1	343.41	46.83	1	1/1
323.37	44.55	1	1/1	353.54	48.65	1	2/0
323.37	44.54	1	1/1	353.54	48.67	1	1/0
323.37	44.56	1	1/1	362.89	50.70	1	1/1
328.32	44.86	1	1/1	362.89	50.71	1	1/1
328.32	44.88	1	1/1	367.90	51.76	1	1/0
328.42	44.94	1	1/1	367.90	51.78	1	1/0
328.42	44.95	1	1/1	367.90	51.77	1	1/0
333.38	45.36	1	2/2	369.01	51.71	1	2/0
333.38	45.35	1	1/1	370.01	51.20	1	1/1
343.41	46.84	1	1/1	370.01	51.22	1	1/1
343.41	46.86	1	1/1	372.92	50.52	1	1/0
				372.92	50.53	1	2/0

(Table continued on next page)

COMPONENTS:	EVALUATOR:
(1) Disodium hydrogenphosphate; Na_2HPO_4; [7558-79-4] (2) Water; H_2O; [7732-18-5]	J. Eysseltová Charles University Prague, Czechoslovakia May 1985

CRITICAL EVALUATION:

Table II. Parameters for equation [1].

	$Na_2HPO_4 \cdot 12H_2O$		$Na_2HPO_4 \cdot 7H_2O$		$Na_2HPO_4 \cdot 2H_2O$	
Parameter	value	σ^a	value	σ^a	value	σ^a
A	-1.989×10^6	5000	-7.56×10^5	300	-1.546×10^5	500
B	-1.379×10^4	50	-4.79×10^3	20	887	5
C	24.0	0.1	7.63	0.03	-1.264	0.005
x_o	0.014714		0.064963		0.10040	
T_o	298.2		313.4		343.4	

a The standard deviation for the parameter.

Table III. Tentative values, calculated from equation [1], for the solubility of Na_2HPO_4 in water.

$Na_2HPO_4 \cdot 12H_2O$

T/K	mole fraction	mole/kg	mass%
273.2	0.0021371	0.12	1.66
278.2	0.0037242	0.21	2.87
283.2	0.0055018	0.31	4.19
288.2	0.0074979	0.42	5.63
293.2	0.010172	0.57	7.51
298.2	0.014714	0.83	10.55
303.2	0.024149	1.37	16.34
308.2	0.047504	2.77	28.25
309.45^a	0.058234	3.43	32.80

$Na_2HPO_4 \cdot 7H_2O$

309.45	0.058234	3.43	32.80
311.2	0.058862	3.47	33.06
313.2	0.064287	3.82	35.17
315.2	0.070110	4.19	37.32
317.2	0.076443	4.60	39.53
319.2	0.083431	5.06	41.82
321.2_b	0.091254	5.58	44.23
321.6^b	0.092892	5.69	44.71

a The dodecahydrate to heptahydrate transition temperature.

b The heptahydrate to dihydrate transition temperature.

Both transition temperatures were found graphically by the evaluator.

(continued next page)

COMPONENTS:	EVALUATOR:
(1) Disodium hydrogenphosphate; Na_2HPO_4; [7558-79-4] (2) Water; H_2O; [7732-18-5]	J. Eysseltová Charles University Prague, Czechoslovakia May 1985

CRITICAL EVALUATION:

Table III, contd.

$Na_2HPO_4 \cdot 2H_2O$

T/K	mole fraction	mole/kg	mass%
321.6	0.092892	5.69	44.71
323.2	0.092657	5.67	44.64
328.2	0.093014	5.70	44.75
333.2	0.094587	5.80	45.20
338.2	0.097088	5.97	45.92
343.2	0.10026	6.19	46.81
348.2	0.10386	6.44	47.78
353.2	0.10762	6.70	48.78
358.2	0.11126	6.95	49.72
363.2	0.11451	7.18	50.52
368.2	0.11707	7.36	51.15
373.2	0.11865	7.48	51.53

298 K. In the articles by Makin and his co-workers (18, 19, 23-26), data for the solubility of Na_2HPO_4 in water are also included and these form a consistent set of data. Furthermore, the data in these articles may be compared with respect to the composition of solutions saturated with respect to two salts, but in one of these reports (26) the headings "NaNO_3" and "NaCl" for two of the columns appear to have been inter-changed. It appears also that some incorrect constants have been used in calculating the mol% values in this paper. The evaluator was unable to reproduce these calculations.

No solid solutions or ternary compounds were found in the $Na_2HPO_4-MgHPO_4-H_2O$ system at 298 K (17) but potassium sodium hydrogen phosphate, $NaKHPO_4 \cdot 5H_2O$ [14518-27-5] was observed in the $Na_2HPO_4-K_2HPO_4-H_2O$ system at 273 and 298 K (16), and sodium ammonium hydrogenphosphate, $NaNH_4HPO_4$ [13011-54-6] was found to be present in the $Na_2HPO_4-(NH_4)_2HPO_4-H_2O$ system at 298 K (14). The latter compound was also observed in the quaternary system $2Na^+, 2NH_4^+ || HPO_4^{2-}, Cl^--H_2O$ at 273 and 298 K (27). However, the data in this report (27) are at variance with those of Platford (14) with respect to the composition of solutions saturated with both $NaNH_4HPO_4$ and $Na_2HPO_4 \cdot 12H_2O$ as well as with those saturated with both $NaNH_4HPO_4$ and $(NH_4)_2HPO_4$. The data in (27) also disagree with those of Makin (24) with respect to the composition of the eutonic solution of the $Na_2HPO_4-NaCl-H_2O$ system at 298 K. The values for the Na_2HPO_4 content in the work of Lauffenburger and Brodsky (27) seem to have a large negative systemtic error.

Values have been reported for three systems having an organic component. Ferroni, et al. (28) report values for the $Na_2HPO_4-CH_3COCH_3-H_2O$ system and for two sections through the $Na_2HPO_4-NaClO_4-CH_3COCH_3-H_2O$ system at 298 K. Bruder, et al. (29) report solubility data for the $Na_2HPO_4-CH_3OH-H_2O$ system at 333 K. All three systems are characterized by limited miscibility.

(continued next page)

COMPONENTS:	EVALUATOR:
(1) Disodium hydrogenphosphate; Na_2HPO_4; [7558-79-4] (2) Water; H_2O; [7732-18-5]	J. Eysseltová Charles University Prague, Czechoslovakia May 1985

CRITICAL EVALUATION:

REFERENCES

1. Shiomi, Ts. *Mem. Col. Sci. Emp. (Kyoto)* 1908, *1*, 406.
2. Hammick, D.L.; Goadby, H.K.; Booth, H. *J. Chem. Soc.* 1920, *67*, 1589.
3. Menzel, H.; Gabler, C. *Z. Anorg. Chem.* 1929, *177*, 187.
4. Wendrow, B.; Kobe, K.A. *Ind. Eng. Chem.* 1952, *44*, 1439.
5. Menzies, A.W.; Humphrey, K.C. *Orig. Com. 8th Intern. Congr. Appl. Chem.* 1912, *2*, 175. This work was quoted in ref. (4).
6. D'Ans. J.; Schreiner, O. *Z. Anorg. Chem.* 1911, *75*, 95.
7. Mulder, G.J. *Bijdragen tot de geschiedenis van het scheikundig gebonden water*, Rotterdam 1894. Quoted in ref. (6).
8. Tilden, W.A. *J. Chem. Soc.* 1884, *45*, 268. Quoted in ref. (6).
9. Ferrein, A. *Pharm. Viertelj.* 1858, *7*, 244; *Jahresber*, 1858, 117. Quoted in ref. (6).
10. Neese, N. *Russ. Zeitschr. f. Pharm.* 1863, *1*, 101; *Jahresber*, 1863, 180. Quoted in ref. (6).
11. Schiff, H. *Lieb. Ann.* 1859, *109*, 362. Quoted in ref. (6).
12. Guthrie, F. *Phil. Mag.* 1876, *5*, 212; *Phys.-Chem. Tabellen* 558. Quoted in ref. (6).
13. Muller, A. *J. f. Prakt. Chem.* 1860, *80*, 202; 1865, *95*, 52. Quoted in ref. (6).
14. Platford, R.F. *J. Chem. Eng. Data* 1974, *19*, 166.
15. Ukraintseva, E.A. *Izv. Sib. Otd. Akad. Nauk SSSR, Ser. Khim.* 1963, *3*, 14.
16. Ravich, M.I. Popova, Z.V. *Izv. Akad. Nauk SSSR, Ser. Khim.* 1942, 268.
17. Dudakov, V.G.; Shternina, E.B. *VINITI* Nr. 469-74, 1974.
18. Makin, A.V. *Uch. Zapiski Gos. Ped. In-ta* 1959, *30*, 291.
19. Druzhinin, I.G.; Makin, A.V. *Izv. Akad. Nauk Kirg. SSR, Ser. Estestv. i Tekhn. Nauk* 1960, *2*, 19.
20. Beremzhanov, B.A.; Savich, R.F.; Kunanbaeva, G.S. *Prikl. Teor. Khim.* 1978, 8.
21. Beremzhanov, B.A.; Savich, R.F.; Kunanbaeva, G.S. *Khim. Khim. Tekhnol., (Alma Ata)* 1977, *22*, 15.
22. Manvelyan, M.G.; Galstyan, V.D.; Organesyan, E.B. Sayamyan, E.A. *Arm. Khim. Zh.* 1973, *26*, 510.
23. Makin, A.V.; Karnaukhov, A.S. *Zh. Neorg. Khim.* 1957, *2*, 1420.
24. Makin, A.V.; *Zh. Neorg. Khim.* 1957, *2*, 2794.
25. Makin, A.V.; Lepeshkov, I.N. *Zh. Neorg. Khim.* 1964, *9*, 495.
26. Makin, A.V.; *Zh. Neorg. Khim.* 1958, *3*, 2764.
27. Lauffenburger, R.; Brodsky, M. *Compt. Rend.* 1938, *206*, 1383.
28. Ferroni, G.; Galea, J.; Antonetti, G. *Bull. Soc. Chim. Fr.* 1974, *12 (Pt. 1)*, 273.
29. Bruder, K.; Vohland, P.; Schuberth, H. *Z. Phys. Chem. Leipzig* 1977, *4*, 721.

COMPONENTS:	ORIGINAL MEASUREMENTS:
(1) Disodium hydrogenphosphate; Na_2HPO_4; [7558-79-4] (2) Water; H_2O; [7732-18-5]	Shiomi, Ts. *Mem. Col. Sci. Emp.(Kyoto)* 1908, *1*, 406-13.

VARIABLES:	PREPARED BY:
Composition and temperature.	J. Eysseltová

EXPERIMENTAL VALUES:

Solubility of Na_2HPO_4 in water.

$t/°C$	concn[a]	mean	mass%[b]	mol/kg[b]	$t/°C$	concn[a]	mean	mass%[b]	mol/kg[b]
0.65	1.74		1.71	0.12	30.21	20.80		17.22	1.46
0.65	1.74	1.74	1.71	0.12	30.21	20.76	20.81	17.19	1.46
0.65	1.74		1.71	0.12	30.21	20.88		17.27	1.47
10.26	3.55		3.43	0.25	30.26	21.60		17.76	1.52
10.26	3.54	3.55	3.42	0.25	30.26	21.62		17.78	1.52
10.36	3.58		3.46	0.25	30.26	21.56		17.74	1.52
10.36	3.59		3.46	0.25	30.26	21.55		17.73	1.52
10.36	3.58	3.58	3.46	0.25	30.26	21.56	21.59	17.74	1.52
10.36	3.58		3.46	0.25	30.26	21.61		17.77	1.52
15.11	5.23		4.97	0.37	30.26	21.60		17.76	1.52
15.11	5.22	5.23	4.96	0.37	30.26	21.59		17.76	1.52
20.24	7.88		7.30	0.55	30.26	21.59		17.76	1.52
20.24	7.90	7.89	7.32	0.56	30.76	23.42		18.98	1.65
20.24	7.89		7.31	0.56	30.76	23.40	23.41	18.96	1.65
20.24	7.89		7.31	0.56	30.76	23.41		18.97	1.65
25.15	12.03		10.74	0.85	33.04	30.88		23.59	2.17
25.15	12.01	12.02	10.72	0.84	33.04	30.88	30.88	23.59	2.17
25.15	12.01		10.72	0.84	33.14	31.39		23.89	2.21
25.40	12.32		10.97	0.87	33.14	31.37	31.38	23.88	2.21
25.40	12.34	12.32	10.98	0.87	36.27	45.36		31.20	3.19
25.40	12.31		10.96	0.86	36.27	45.35		31.20	3.19
25.50	12.42		11.05	0.87	36.27	45.34		31.20	3.19
25.50	12.41	12.43	11.04	0.87	36.27	45.38		31.21	3.19
25.50	12.47		11.09	0.88	36.27	45.39	45.37	31.22	3.19

(continued next page)

AUXILIARY INFORMATION

METHOD/APPARATUS/PROCEDURE:	SOURCE AND PURITY OF MATERIALS:
The isothermal method was used. Equilibrium was approached from both supersaturation and undersaturation. Samples of saturated solution were weighed, evaporated to dryness and heated strongly to form the pyrophosphate. The solubility was calculated from the weight of the pyrophosphate formed.	The Na_2HPO_4 was recrystallized twice.
	ESTIMATED ERROR: The temperature was kept constant within 0.1 K (0.6 K above 90°).
	REFERENCES:

Disodium Hydrogenphosphate

COMPONENTS:	ORIGINAL MEASUREMENTS:
(1) Disodium hydrogenphosphate; Na_2HPO_4; [7558-79-4]	Shiomi, Ts.
(2) Water; H_2O; [7732-18-5]	*Mem. Col. Sci. Emp. (Kyoto)* <u>1908</u>, *1*, 406-13.

EXPERIMENTAL VALUES cont'd:

Solubility of Na_2HPO_4 in water.

$t/°C$	concn[a]	mean	mass%[b]	mol/kg[b]	$t/°C$	concn[a]	mean	mass%[b]	mol/kg[b]
36.27	45.36		31.20	3.19	55.17	81.37		44.86	5.72
36.27	45.41		31.23	3.19	55.17	81.43	81.40	44.88	5.73
36.27	45.36		31.20	3.19	55.27	81.61		44.94	5.74
36.27	45.35		31.20	3.19	55.27	81.66	81.64	44.95	5.74
37.27	47.56		32.32	3.34	60.23	83.01		45.36	5.84
37.27	47.48		32.19	3.34	60.23	83.02	83.00	45.36	5.84
37.27	47.46		32.18	3.34	60.23	82.98		45.35	5.84
37.27	47.53	47.52	32.22	3.34	70.26	88.10		46.84	6.20
37.27	47.56		32.23	3.34	70.26	88.17	88.11	46.86	6.20
37.27	47.48		32.19	3.34	70.26	88.07		46.83	6.20
37.27	47.49		32.20	3.34	80.39	94.74		48.65	6.66
37.27	47.56		32.23	3.34	80.39	94.83	94.78	48.67	6.67
40.29	54.95		35.46	3.86	80.39	94.76		48.65	6.67
40.29	54.86		35.42	3.86	89.74	102.85		50.70	7.23
40.29	54.95		35.46	3.86	89.74	102.89	102.87	50.71	7.24
40.29	54.83	54.88	35.41	3.86	94.75	107.31		51.76	7.55
40.29	54.85		35.42	3.86	94.75	107.37	107.34	51.78	7.55
40.29	54.88		35.43	3.86	94.75	107.34		51.77	7.55
45.14	68.67		40.71	4.83	95.86	107.08		51.71	7.53
45.14	68.68	68.64	40.72	4.83	95.86	107.09	107.09	51.71	7.53
45.14	68.61		40.69	4.83	96.86	104.94		51.20	7.38
45.14	68.58		40.68	4.82	96.86	105.01	104.98	51.22	7.39
47.23	76.60		43.37	5.39	99.57	101.25		50.31	7.12
47.23	76.55	76.58	43.36	5.38	99.57	101.22	101.21	50.30	7.12
48.23	80.03	80.03	44.45	5.63	99.57	101.16		50.29	7.12
48.33	80.12		44.48	5.64	99.77	102.12		50.52	7.18
48.33	80.17	80.15	44.50	5.64	99.77	102.16	102.15	50.53	7.18
50.22	80.40		44.57	5.66	99.77	102.16		50.53	7.18
50.22	80.34	80.35	44.55	5.65					
50.22	80.32		44.54	5.65					
50.22	80.36		44.56	5.65					

[a] The concentration units are: g/100 g H_2O.

[b] These values were calculated by the compiler.

COMPONENTS:	ORIGINAL MEASUREMENTS:
(1) Disodium hydrogenphosphate; Na_2HPO_4; [7558-79-4] (2) Water; H_2O; [7732-18-5]	Hammick, D.L.; Goadby, H.K.; Booth, H. *J. Chem. Soc.* 1920, *67*, 1589-92.

VARIABLES:	PREPARED BY:
Composition and temperature.	J. Eysseltová

EXPERIMENTAL VALUES:

Solubility of Na_2HPO_4 in water.

$t/°C$	concn Na_2HPO_4			H_2O
	g/100 g H_2O	mass%[a]	mol/kg[a]	mass%[a]
-0.47	1.45	1.43	0.10	98.57
6.00	2.73	2.66	0.19	97.34
19.95	7.26	6.77	0.51	93.23
22.77	8.93	8.20	0.63	91.80
24.15	9.53	8.70	0.67	91.30
25.75	10.90	9.83	0.77	90.17
27.80	14.16	12.40	1.00	87.60
28.65	15.87	13.70	1.12	86.30
29.05	16.04	13.82	1.13	86.18
29.50	17.18	14.66	1.21	85.34
30.10	19.45	16.28	1.37	83.72
30.90	20.08	16.72	1.41	83.28
32.50	22.57	18.41	1.59	81.57
33.70	24.63	19.76	1.73	80.24
34.70	29.75	22.93	2.09	77.07
36.50	31.15	23.75	2.19	76.25
40.02	35.56	26.23	2.50	73.76

[a] These values were calculated by the compiler.

AUXILIARY INFORMATION

METHOD/APPARATUS/PROCEDURE:	SOURCE AND PURITY OF MATERIALS:
Saturated solutions were prepared by stirring the solid phase with distilled water in an electrically heated thermostat. The saturated solution was siphoned through a glass-wool filter into a weighed bottle. The composition was determined by converting the dissolved phosphate to $Mg_2P_2O_7$.	Arsenic-free $Na_2HPO_4 \cdot 12H_2O$ was recrystallized and used to prepare the other hydrates. The dihydrate was prepared by boiling finely divided dodecahydrate with ethyl alcohol. The heptahydrate was prepared by fusing together an appropriate mixture of the dihydrate and dodecahydrate and cooling.

ESTIMATED ERROR:

No information is given.

REFERENCES:

COMPONENTS:	ORIGINAL MEASUREMENTS:
(1) Disodium hydrogenphosphate; Na_2HPO_4; [7558-79-4] (2) Water; H_2O; [7732-18-5]	Menzel, G.; Gabler, C. Z. Anorg. Chem. <u>1928</u>, 177, 187-214.
VARIABLES: Temperature and composition.	PREPARED BY: J. Eysseltová

EXPERIMENTAL VALUES:

Solubility of Na_2HPO_4 in water.

concentration of Na_2HPO_4

$t/°C$	in 1000 ml soln		in 1000 g soln		in 1000 g of H_2O	
	mol	gram	mol	gram	mol	gram
0	0.1152	16.37	0.1130	16.05	0.1148	16.31
18	0.4444	63.12	0.4212	59.85	0.4482	63.67
25	0.8399	119.28	0.7625	108.29	0.8551	121.44
-0.48[a]					0.109	15.5

The solid phase was $Na_2HPO_4 \cdot 12H_2O$ [10039-32-4].

[a]This is the cryohydric point.

AUXILIARY INFORMATION

METHOD/APPARATUS/PROCEDURE:	SOURCE AND PURITY OF MATERIALS:
The apparatus is described elsewhere (1). At 0°C, the equilibrium vessel and sampling pipet were thermostated in an ice-water mixture. Equilibrium was checked by repeated analysis. The Na_2HPO_4 content was determined by titration with 0.1 M HCl using methylorange as indicator (2).	Purest Kahlbaum $Na_2HPO_4 \cdot 12H_2O$ was used.

ESTIMATED ERROR:

The temperature was controlled to ± 0.1 K. The accuracy of the cryohydric temperature is ± 0.01 K.

REFERENCES:

1. Menzel, H. Z. Anorg. Allg. Chem. <u>1927</u>, 164, 6.

2. Kolthoff, I. Massanalyse, II, p. 139, Berlin, <u>1928</u>.

COMPONENTS:	ORIGINAL MEASUREMENTS:
(1) Disodium hydrogenphosphate; Na_2HPO_4; [7558-79-4] (2) Hydrogen peroxide; H_2O_2; [7722-84-1] (3) Water; H_2O; [7732-18-5]	Menzel, H.; Gabler, C. Z. Anorg. Chem. <u>1929</u>, *177*, 187-214.

VARIABLES:	PREPARED BY:
Composition at 0°C.	J. Eysseltová

EXPERIMENTAL VALUES:

Composition of saturated solutions of Na_2HPO_4 in aqueous H_2O_2 at 0°C.

mol P:mol O_2^{2-}	H_2O_2		Na_2HPO_4	
	g/1000 g soln	mol/1000 g H_2O	g/1000 g soln	mol/1000 g H_2O
1:0	------	-------	16.05	0.1148
1:0.63	2.501	0.0750	16.52	0.1186
1:1.71	7.132	0.2149	17.42	0.1258
1:2.18	9.349	0.2825	17.89	0.1294
1:2.98	12.88	0.3910	18.60	0.1353
1:3.54	16.33	0.4977	19.23	0.1405

The equilibrium solid phase was identified as $Na_2HPO_4 \cdot 12H_2O$ [10039-32-4].

AUXILIARY INFORMATION

METHOD/APPARATUS/PROCEDURE:	SOURCE AND PURITY OF MATERIALS:
The method used was described earlier (1). The equilibrium vessel and the sampling pipet were thermostated in an ice-water mixture. The equilibrium was checked by repeated analysis. The Na_2HPO_4 content was determined by titration with 0.1 N HCl using methylorange as indicator (2). The H_2O_2 content was determined by titration with 0.1 N $KMnO_4$.	The $Na_2HPO_4 \cdot 12H_2O$ was the purest Kahlbaum grade. The H_2O_2 was the purest Merck reagent grade.

	ESTIMATED ERROR:
	The temperature was controlled to within ±0.1 K. No additional information is given.

	REFERENCES:
	1. Menzel, H. Z. Anorg. Allg. Chem. <u>1927</u>, *164*, 6. 2. Kolthoff, I. *Massanalyse*, II, p. 139, Berlin, <u>1928</u>.

COMPONENTS:	ORIGINAL MEASUREMENTS:
(1) Disodium hydrogenphosphate; Na_2HPO_4; [7558-79-4] (2) Dipotassium hydrogenphosphate; K_2HPO_4; [7758-11-4] (3) Water; H_2O; [7732-18-5]	Ravich, M.I.; Popova, Z.V. *Izv. Akad. Nauk SSSR, Ser. Khim.* 1942, 268-75.

VARIABLES:	PREPARED BY:
Composition and temperature.	J. Eysseltová

EXPERIMENTAL VALUES:

Solubility in the K_2HPO_4-Na_2HPO_4-H_2O system.

K_2HPO_4		Na_2HPO_4		H_2O	solid
mass%	mol/kg^a	mass%	mol/kg^a	mass%^{a}	phase^b
			temp. = 0°C.		
----	----	10.80	0.85	89.20	A
25.57	2.06	3.24	0.32	71.19	"
31.84	2.83	3.66	0.40	64.50	"
35.17	3.32	3.94	0.46	60.89	"
39.41	4.13	5.82	0.75	54.77	A + B
40.83	4.36	5.42	0.71	53.75	B
44.59	5.02	4.38	0.60	51.03	B + C
53.41	7.03	2.98	0.48	43.61	B
55.61	7.69	2.86	0.48	41.53	B + D^c
55.60	7.70	2.92	0.50	41.48	"
			temp. = 25°C.		
60.66	9.42	2.36	0.45	36.98	D
59.68	9.59	4.60	0.90	35.72	"
57.85	9.49	7.14	1.43	35.01	B + D
54.94	8.43	7.66	1.44	37.40	B
45.31	5.81	9.91	1.56	44.78	"
30.98	3.47	17.80	2.44	51.22	"
29.57	3.30	19.05	2.61	51.38	"
26.64	2.97	21.82	2.98	51.54	B + E
26.61	2.97	22.04	3.02	51.35	E
26.31	2.91	21.76	2.95	51.93	"

(continued next page)

AUXILIARY INFORMATION

METHOD/APPARATUS/PROCEDURE:	SOURCE AND PURITY OF MATERIALS:
The isothermal method was used but no details are given. Equilibrium was reached in one day. Phosphate was determined as $Mg_2P_2O_7$, potassium was determined as $KClO_4$, and sodium was determined as sodium zincuranylacetate after separating out H_3PO_4 with the use of zinc acetate.	The $Na_2HPO_4 \cdot 12H_2O$ and the $K_2HPO_4 \cdot 3H_2O$ were recrystallized.
	ESTIMATED ERROR: No information is given.
	REFERENCES:

COMPONENTS:	ORIGINAL MEASUREMENTS:
(1) Disodium hydrogenphosphate; Na_2HPO_4; [7558-79-4]	Ravich, M.I.; Popova, Z.V.
(2) Dipotassium hydrogenphosphate; K_2HPO_4; [7758-11-4]	*Izv. Akad. Nauk SSSR, Ser. Khim.* <u>1942</u>, 268-75.
(3) Water; H_2O; [7732-18-5]	

EXPERIMENTAL VALUES cont'd:

Solubility in the K_2HPO_4-Na_2HPO_4-H_2O system.

K_2HPO_4		Na_2HPO_4		H_2O	solid[b]
mass%	mol/kg[a]	mass%	mol/kg[a]	mass%[a]	phase

temp. = 25°C.

K_2HPO_4 mass%	mol/kg[a]	Na_2HPO_4 mass%	mol/kg[a]	H_2O mass%[a]	solid[b] phase
26.72	2.98	21.84	2.99	51.44	E
26.63	2.95	21.48	2.91	51.89	"
25.84	2.84	22.02	2.97	52.14	"
23.13	2.39	21.40	2.71	55.47	A + E
23.54	2.43	20.78	2.62	55.68	"
23.12	2.41	21.72	2.77	55.16	A
18.80	1.68	16.94	1.85	64.26	"
10.87	0.84	14.46	1.36	74.67	"
6.42	0.46	13.20	1.16	80.38	"
-----	----	10.80	0.85	89.20	"

[a] These values were calculated by the compiler.

[b] The solid phases are: A = $Na_2HPO_4 \cdot 12H_2O$; B = $KNaHPO_4 \cdot 5H_2O$; C = $K_2HPO_4 \cdot 6H_2O$;

D = $K_2HPO_4 \cdot 3H_2O$; E = $Na_2HPO_4 \cdot 7H_2O$.

[c] This is a metastable state.

COMPONENTS:	ORIGINAL MEASUREMENTS:
(1) Disodium hydrogenphosphate; Na_2HPO_4; [7558-79-4] (2) Sodium nitrate; $NaNO_3$; [7631-99-4] (3) Water; H_2O; [7732-18-5]	Makin, A.V.; Karnaukhov, A.S. *Zh. Neorg. Khim.* <u>1957</u>, *2*, 1420-3.

VARIABLES:	PREPARED BY:
Composition at 25°C.	J. Eysseltová

EXPERIMENTAL VALUES:

Solubility in the Na_2HPO_4-$NaNO_3$-H_2O system at 25°C.

NaNO$_3$		Na$_2$HPO$_4$		H$_2$O	solid
mass%	mol/kg^a	mass%	mol/kg^a	mass%	phase^b
0	0	10.32	0.81	89.68	A
3.31	0.44	8.71	0.70	87.98	"
7.80	1.08	7.50	0.62	84.70	"
12.03	1.74	6.67	0.58	81.30	"
17.06	2.63	6.51	0.60	76.43	"
21.87	3.58	6.26	0.61	71.87	"
25.67	4.45	6.26	0.66	67.91	A + B
26.01	4.53	6.46	0.67	67.53	"
26.05	4.53	6.33	0.66	67.62	"
26.46	4.62	6.19	0.65	67.35	"
26.08	4.54	6.30	0.66	67.62	"
32.06	5.86	3.57	0.39	64.37	B
36.05	6.86	2.16	0.24	61.79	"
41.72	8.57	1.00	0.12	57.28	"
47.90	10.82	0	0	52.10	"

^{a}The mol/kg H_2O values were calculated by the compiler.

^{b}The solid phases are: A = $Na_2HPO_4 \cdot 12H_2O$; B = $NaNO_3$.

AUXILIARY INFORMATION

METHOD/APPARATUS/PROCEDURE:	SOURCE AND PURITY OF MATERIALS:
The isothermal method was used. The time allowed for equilibration was 12 - 45 hours. About 1 - 2 g of liquid and solid phases were sampled simultaneously. The phases were separated from each other by filtration. The phosphate content was determined gravimetrically as $NH_4MgPO_4 \cdot 6H_2O$. The sodium ion content was determined as sodium uranylacetate after removal of the phosphate ion. Nitrate ion content was determined by difference. The water content was determined by drying at 105°C to constant weight.	The Na_2HPO_4 and the $NaNO_3$ were each recrystallized twice.
	ESTIMATED ERROR:
	No information is given. The compiler estimates the reproducibility to be about 1%.
	REFERENCES:

COMPONENTS:	ORIGINAL MEASUREMENTS:
(1) Disodium hydrogenphosphate; Na_2HPO_4; [7558-79-4] (2) Sodium chloride; NaCl; [7647-14-5] (3) Water; H_2O; [7732-18-5]	Makin, A.V. *Zh. Neorg. Khim.* <u>1957</u>, *2*, 2794-6.

VARIABLES:	PREPARED BY:
Composition at 25°C.	J. Eysseltová

EXPERIMENTAL VALUES:

Solubility in the Na_2HPO_4-NaCl-H_2O system at 25°C.

Na_2HPO_4		NaCl		H_2O	solid[b]
mass%	mol/kg[a]	mass%	mol/kg[a]	mass%	phase
10.32	0.81	----	----	89.68	A
9.26	0.76	5.09	1.02	85.65	"
9.32	0.81	9.48	2.00	81.20	"
9.51	0.89	15.72	3.60	74.77	"
8.50	0.84	19.96	4.77	71.54	A + B
9.67	0.97	20.17	5.01	69.86	"
9.50	0.96	20.99	5.17	69.51	"
9.37	0.94	20.32	4.94	70.31	"
9.12	0.91	20.54	5.00	70.34	"
8.69	0.85	19.67	4.70	71.64	"
5.31	0.52	22.24	5.25	72.45	B
3.01	0.29	23.46	5.46	73.53	"
----	----	26.42	6.14	73.58	"

[a] The mol/kg H_2O values were calculated by the compiler.

[b] The solid phases are: A = $Na_2HPO_4 \cdot 12H_2O$; B = NaCl.

In addition to the above data, the author also gives the composition of the respective eutonic solution as: 9.14 mass% Na_2HPO_4 (0.91 mol/kg--compiler) 20.33 mass% NaCl (4.93 mol/kg--compiler), and 70.53 mass% water.

AUXILIARY INFORMATION

METHOD/APPARATUS/PROCEDURE:	SOURCE AND PURITY OF MATERIALS:
The isothermal method was used. The analyses were done gravimetrically but no details are given.	No information is given.
	ESTIMATED ERROR: No details are given. The compiler estimates the reproducibility of the analyses to be about ± 3%.
	REFERENCES:

COMPONENTS:	ORIGINAL MEASUREMENTS:
(1) Disodium hydrogenphosphate; Na_2HPO_4; [7558-79-4] (2) Disodium sulfate; Na_2SO_4; [7557-82-6] (3) Water; H_2O; [7732-18-5]	1. Makin, A.V. *Uch. Zapiski Gos. Ped. In-ta* 1959, *30*, 291-6. 2. Druzhinin, I.G.; Makin, A.V. *Izv. Akad. Nauk Kirg. SSR, Ser. Estestv. i Tekhn. Nauk* 1960, *2*, 19-24.

VARIABLES:	PREPARED BY:
Composition at 25°C.	J. Eysseltová

EXPERIMENTAL VALUES:

Solubility in the Na_2HPO_4-Na_2SO_4-H_2O system at 25°C.

Na_2SO_4		Na_2HPO_4		H_2O	solid[b]
mass%	mol/kg[a]	mass%	mol/kg[a]	mass%	phase
----	----	10.32	0.81	89.68	A
4.25	0.34	8.90	0.72	86.85	"
7.79	0.65	7.99	0.67	84.22	"
9.82	0.84	7.52	0.64	82.66	"
12.77	1.12	7.38	0.65	74.85	"
15.02	1.37	7.62	0.69	77.36	A + B
15.20	1.38	7.38	0.67	77.43	"
15.03	1.37	7.62	0.69	77.35	"
15.06	1.37	7.48	0.68	77.46	"
15.06	1.37	7.44	0.68	77.50	"
15.04	1.37	7.45	0.68	77.51	"
15.04	1.37	7.54	0.68	77.42	"
16.27	1.47	5.92	0.54	77.81	B
17.83	1.61	4.14	0.37	78.03	"
19.19	1.73	2.70	0.24	78.11	"
21.98	1.98	----	----	78.02	"

[a] The mol/kg H_2O values were calculated by the compiler.

[b] The solid phases are: A = $Na_2HPO_4 \cdot 12H_2O$; B = $Na_2SO_4 \cdot 10H_2O$.

AUXILIARY INFORMATION

METHOD/APPARATUS/PROCEDURE:	SOURCE AND PURITY OF MATERIALS:
The isothermal method was used. The mixtures were placed in a water thermostat and allowed to equilibrate for 3 days. Phosphate content was determined gravimetrically as $Mg_2P_2O_7$, sulfate content was determined gravimetrically as $BaSO_4$, sodium was determined by difference, and the water content was determined by drying at 105° C to constant weight.	Both salts were purified by recrystallization.
	ESTIMATED ERROR: The temperature was controlled to within ± 0.1 K. The compiler estimates that the reproducibility of the analyses was about 0.5%.
	REFERENCES:

COMPONENTS:	ORIGINAL MEASUREMENTS:
(1) Disodium hydrogenphosphate; Na_2HPO_4; [7558-79-4] (2) Hydrogen peroxide; H_2O_2; [7722-84-1] (3) Water; H_2O; [7732-18-5]	Ukraintseva, E.A. *Izv. Sib. Otd. Akad. Nauk SSSR, Ser. Khim.* <u>1963</u>, 3,]4-24.

VARIABLES:	PREPARED BY:
Composition at 0°C.	J. Eysseltová

EXPERIMENTAL VALUES:

Solubility in the Na_2HPO_4-H_2O_2-H_2O system at 0°C.

Na_2HPO_4		H_2O_2		H_2O	solid[b]
mass%	mol/kg[a]	mass%	mol/kg[a]	mass%[a]	phase
1.6	0.11	----	----	98.4	A
4.0	0.32	9.2	3.12	86.8	"
11.0	1.21	20.0	8.52	69.0	"
12.2	1.28	20.8	9.12	67.0	"
23.6	3.29	26.0	15.16	50.4	"
24.0	3.38	26.1	15.37	49.9	"
27.3	4.15	26.4	16.76	46.3	"
42.2	9.64	27.0	25.77	30.8	A + B
40.9	9.05	27.3	25.23	31.8	B
38.7	8.95	30.9	29.88	30.4	"
37.5	8.79	32.5	31.84	30.0	"
35.4	9.47	38.3	42.81	26.3	"
35.3	11.04	42.2	55.13	22.5	"
32.9	10.06	44.1	56.36	23.0	"
26.9	9.86	53.9	82.52	19.2	C
26.7	14.23	60.1	133.8	13.2	"
29.1	46.5	66.5	444	4.4	"

[a] These values were calculated by the compiler.

[b] The solid phases are: A = $Na_2HPO_4 \cdot 12H_2O$; B = $Na_2HPO_4 \cdot 1.5H_2O_2$;

C = $Na_2HPO_4 \cdot 2.5H_2O_2$.

AUXILIARY INFORMATION

METHOD/APPARATUS/PROCEDURE:	SOURCE AND PURITY OF MATERIALS:
The only information given is that the duration of the experiments was 3 to 14 hours. The composition of the solid phases was determined by the Schreinemakers' method. Hydrogen peroxide content was determined by titration with 0.1 N $KMnO_4$ in solutions containing sulfuric acid. The phosphate was determined gravimetrically as $Mg_2P_2O_7$.	Chemically pure hydrogen peroxide without stabilizers was used. No information is given about the $Na_2HPO_4 \cdot 12H_2O$.
	ESTIMATED ERROR: No information is given.
	REFERENCES:

COMPONENTS:	ORIGINAL MEASUREMENTS:
(1) Disodium hydrogenphosphate; Na_2HPO_4; [7558-79-4]	Manvelyan, M.G.; Galstyan, V.D.; Oganesyan, E.B.; Sayamyan, E.A.
(2) Disodium silicate; Na_2SiO_3; [6834-92-0]	*Arm. Khim. Zh.* <u>1971</u>, *26*, 510-12.
(3) Water; H_2O; [7732-18-5]	

VARIABLES:	PREPARED BY:
Composition at 20°C.	J. Eysseltová

EXPERIMENTAL VALUES:

Solubility in the Na_2HPO_4-Na_2SiO_3-H_2O system at 20°C.

Na_2HPO_4		Na_2SiO_3		H_2O	solid
mass%	mol/kg[a]	mass%	mol/kg[a]	mass%[a]	phase[b]
7.2	0.55	0.5	0.04	92.3	A
9.5	0.75	1.2	0.11	89.3	"
11.8	0.96	1.8	0.17	86.4	"
12.5	1.04	2.9	0.28	84.6	"
15.3	1.36	5.3	0.55	79.4	"
15.1	1.38	8.1	0.86	76.8	B
14.9	1.42	11.5	1.28	73.6	"
9.0	0.83	14.5	1.55	76.5	"
7.9	0.74	16.5	1.79	75.6	"
4.8	0.44	18.9	2.03	76.3	"
4.9	0.46	20.5	2.25	74.7	C
2.5	0.23	21.5	2.32	76.0	"
2.5	0.23	20.0	2.11	77.5	"
1.9	0.17	20.1	2.11	78.0	"

[a] These values were calculated by the compiler.

[b] The solid phases are A = $Na_2HPO_4 \cdot 12H_2O$; B = $Na_3PO_4 \cdot 12H_2O$; C = $Na_2SiO_3 \cdot 9H_2O$.

AUXILIARY INFORMATION

METHOD/APPARATUS/PROCEDURE:	SOURCE AND PURITY OF MATERIALS:
The isothermal method was used. A month was allowed for equilibration. No details about the apparatus or the analytical methods are given.	Reagent grade $Na_2HPO_4 \cdot 12H_2O$ and $Na_2SiO_3 \cdot 9H_2O$ were used.
	ESTIMATED ERROR: No information is given.
	REFERENCES:

COMPONENTS:	ORIGINAL MEASUREMENTS:
(1) Disodium hydrogenphosphate; Na_2HPO_4; [7558-79-4] (2) Magnesium hydrogenphosphate; $MgHPO_4$; [7757-86-0] (3) Water; H_2O; [7732-18-5]	Dudakov, V.G.; Shternina, E.B. *VINITI Nr. 469-74*, <u>1974</u>.

VARIABLES:	PREPARED BY:
Composition at 25°C.	J. Eysseltová

EXPERIMENTAL VALUES:

Solubility in the $MgHPO_4$-Na_2HPO_4-H_2O system at 25°C.

$10^5 C_{Mg}$	$10^3 C_{2Na}$	$10^3 C_{HPO_4}$	$MgHPO_4{}^a$	$Na_2HPO_4{}^a$	$H_2O{}^a$		$solid_b$
g ion/1000 g H_2O			mass%	mass%	mass%	pH	phase
802	----	8.02	0.096	----	99.90	6.98	A
543	32.4	37.78	0.065	0.46	99.48	8.19	"
421	79.8	84.02	0.050	1.12	98.83	8.86	"
277	159	162.2	0.032	2.21	97.76	9.23	"
338	310	313.0	0.038	4.22	95.74	9.24	"
553	527	532.6	0.061	6.96	92.97	9.23	"
741	665	672.1	0.081	8.63	91.29	9.26	"
821	726	733.9	0.089	9.35	90.56	9.36	"
1070	817	827.9	0.115	10.39	89.49	9.35	A + B
994	811	820.9	0.107	10.33	89.57	9.35	B
849	814	822.7	0.091	10.36	89.55	9.36	"
594	820	826.4	0.063	10.43	89.50	9.33	"
321	828	830.7	0.034	10.53	89.44	9.36	"
220	825	826.9	0.023	10.49	89.48	9.35	"
---	834	833.5	-----	10.60	89.40	9.36	"

a These values were calculated by the compiler using the authors' values for the ionic concentrations.

b The solid phases are: A = $MgHPO_4 \cdot 3H_2O$; B = $Na_2HPO_4 \cdot 12H_2O$.

AUXILIARY INFORMATION

METHOD/APPARATUS/PROCEDURE:	SOURCE AND PURITY OF MATERIALS:
The isothermal method was used. Equilibrium was checked refractometrically and by repeated analysis. No details are given about the apparatus or the sampling. Magnesium was determined gravimetrically as pyrophosphate, by compleximetric titration, or colorimetrically. Sodium was determined by flame photometry. Phosphate was determined gravimetrically as magnesium pyrophosphate or as ammonium phosphomolybdate. Water was determined by drying the sample at a temperature a little above its dehydration temperature or over concentrated H_2SO_4. The solid phases were identified by Schreinemakers' method, crystallooptically, and roentgenographically.	The $MgHPO_4 \cdot 3H_2O$ was synthesized from $MgCO_3$ and H_3PO_4. Other experimental details are given in ref. (1).
	ESTIMATED ERROR: No information is given.
	REFERENCES: 1. Vorob'ev, G.I.; Rykova, G.A.; Shternina, E.B. *Zh. Neorg. Khim.* <u>1970</u>, *15*, 2644.

COMPONENTS:	ORIGINAL MEASUREMENTS:
(1) Disodium hydrogenphosphate; Na_2HPO_4; [7558-79-4] (2) Disodium ethylenediaminetetraacetate; $C_{10}H_{14}O_8Na_2$; [139-33-3] (3) Water; H_2O; [7732-18-5]	Dudakov, V.G.; Shternina, E.B. *VINITI Nr. 469-74* <u>1974</u>.
VARIABLES:	PREPARED BY:
Composition at 25°C.	J. Eysseltová

EXPERIMENTAL VALUES:

Solubility in the $NaH_2EDTA-Na_2HPO_4-H_2O$ system at 25°C.

H_2EDTA^{2-}	HPO_4^{2-}	$2\ Na^+$	Na_2H_2EDTA	Na_2HPO_4	H_2O		solid
(g ion/1000 g H_2O)			mass%^{a}	mass%^{a}	mass%^{a}	pH	phase^b
0.3083	----	0.308	10.29	----	89.71	4.35	A
0.3168	0.2793	0.596	10.19	3.43	86.38	4.58	"
0.3221	0.4642	0.786	10.11	5.56	84.32	4.82	"
0.3221	0.5799	0.902	9.97	6.86	83.17	4.87	"
0.3309	0.7872	1.12	9.97	9.06	80.97	5.02	"
0.3564	0.9237	1.28	10.50	10.39	79.11	5.25	A + B
0.3194	0.9022	1.22	9.53	10.28	80.18	7.08	B
0.2412	0.8972	1.15	7.38	10.48	82.15	7.89	"
0.1773	0.8793	1.06	5.54	10.50	83.96	8.68	"
0.1232	0.8828	1.01	3.92	10.71	85.37	8.94	"
0.0501	0.8498	0.851	1.64	10.60	87.76	9.27	"
----	0.8335	0.834	----	10.59	89.41	9.33	"

^{a}These values were calculated by the compiler and were based on the concentration values given by the authors.

^{b}The solid phases are: A = $Na_2H_2EDTA\cdot 2H_2O$; B = $Na_2HPO_4\cdot 12H_2O$.

AUXILIARY INFORMATION

METHOD/APPARATUS/PROCEDURE:	SOURCE AND PURITY OF MATERIALS:
The isothermal method was used. Equilibrium was checked refractometrically and by repeated analysis. Sodium was determined by flame photometry, EDTA was determined by retitration using $Bi(NO_3)_3$, phosphate was determined gravimetrically as magnesium diphosphate or as ammonium phosphomolybdate, depending on its expected concentration. Water content was determined by drying the sample at a temperature slightly above that for the dehydration of the respective hydrate, or over H_2SO_4. The composition of the solid phases was determined by the Schreinemakers' method and crystallo-optically and roentgenographically.	This has been described elsewhere (1).
	ESTIMATED ERROR:
	No information is given.
	REFERENCES:
	1. Vorob'ev, G.I.; Rykova, G.A.; Shternina, E.B. *Zh. Neorg. Khim.* <u>1970</u>, *15*, 2644.

COMPONENTS:	ORIGINAL MEASUREMENTS:
(1) Disodium hydrogenphosphate; Na_2HPO_4; [7558-79-4] (2) Diammonium hydrogenphosphate; $(NH_4)_2HPO_4$; [7783-28-0] (3) Water; H_2O; [7732-18-5]	Platford, R.F. J. Chem. Eng. Data 1974, 19, 166-8.

VARIABLES:	PREPARED BY:
Composition at 25°C.	J. Eysseltová

EXPERIMENTAL VALUES:

Solubility in the Na_2HPO_4-$(NH_4)_2HPO_4$-H_2O system at 25°C.

Na_2HPO_4		$(NH_4)_2HPO_4$		solid[b]
mass%	mol/kg[a]	mass%	mol/kg[a]	phase
10.4	0.82	0.00	0.00	A
12.3	1.01	2.0	0.18	"
16.0	1.39	3.3	0.31	A + B
15.6	1.38	4.6	0.44	B
13.6	1.19	5.9	0.55	"
11.3	0.97	7.0	0.65	"
10.5	0.92	8.9	0.84	"
9.7	0.88	12.8	1.25	"
9.1	0.84	14.4	1.42	"
8.2	0.78	18.0	1.85	"
8.4	0.83	20.4	2.17	"
8.0	0.85	25.5	2.90	"
7.8	0.90	31.2	3.87	"
8.5	1.16	40.1	5.91	B + C
7.1	0.97	41.3	6.06	C
4.8	0.62	40.8	5.68	"
1.8	0.22	41.3	5.50	"
0.0	0.00	41.5	5.37	"

[a] The mol/kg H_2O values were calculated by the compiler.

[b] The solid phases are: A = $Na_2HPO_4 \cdot 12H_2O$; B = $NaNH_4HPO_4 \cdot 4H_2O$; C = $(NH_4)_2HPO_4$.

AUXILIARY INFORMATION

METHOD/APPARATUS/PROCEDURE:	SOURCE AND PURITY OF MATERIALS:
Conventional measurements were made on aliquots of saturated solutions. The ammonium salt was determined gravimetrically as ammoniumtetraphenylborate (1) and the total salt content was determined by evaporation to constant weight in vacuum over H_2SO_4. The sodium salt was then estimated by difference. The composition of the eutonics was checked by an isopiestic method (2).	The AR grade phosphates were recrystallized once from water. The Na_2HPO_4 was dried at 105°C. The $(NH_4)_2HPO_4$ was dried in vacuum over sulfuric acid at room temperature.

ESTIMATED ERROR:

Nothing is given.

REFERENCES:

1. Vogel, A.I. Quantitative Inorganic Analysis, Wiley. New York, 1961, p. 566.
2. Platford, R.F. Amer. J. Sci. 1972, 272, 959.

COMPONENTS:	ORIGINAL MEASUREMENTS:
(1) Disodium hydrogenphosphate; Na_2HPO_4; [7558-79-4] (2) Boric acid; H_3BO_3; [11113-50-1] (3) Water; H_2O; [7732-18-5]	Beremzhanov, B.A.; Savich, R.F.; Kunanbaeva, G.S. *Prikl. Teor. Khim.* <u>1978</u>, 8-14.

VARIABLES:	PREPARED BY:
Composition at 25°C.	J. Eysseltová

EXPERIMENTAL VALUES:

Solubility in the Na_2HPO_4–H_3BO_3–H_2O system at 25°C.

Na_2HPO_4			H_3BO_3				refr.	solid[b]
mass%	mol%	mol/kg[a]	mass%	mol%	mol/kg[a]	pH	index	phase
12.00	1.70	0.96	----	----	----	9.93	1.520	A
15.46	2.31	1.33	2.68	0.91	0.52	9.60	1.495	"
16.60	2.32	1.45	3.11	1.00	0.63	----	1.491	"
18.11	2.83	1.63	3.77	1.32	0.78	9.85	-----	"
18.95	2.92	1.74	4.46	1.59	0.94	9.47	1.487	"
20.43	3.10	1.91	4.38	1.59	0.94	----	-----	"
20.82	3.24	1.97	4.86	1.79	1.06	9.56	1.480	"
23.82	2.83	2.38	5.84	2.24	1.34	9.44	1.477	A + B
19.81	3.01	1.90	6.68	2.50	1.47	9.19	1.464	B
19.44	2.99	1.85	6.76	2.51	1.48	----	1.468	"
17.51	2.76	1.67	8.94	3.31	1.96	8.65	1.453	"
16.28	2.64	1.58	11.30	4.21	2.52	----	-----	"
15.89	2.59	1.57	12.82	4.83	2.91	7.77	1.442	B + C
12.80	1.96	1.15	8.96	3.10	1.85	7.47	1.429	C
11.00	1.61	0.94	6.62	2.24	1.30	----	-----	"
10.08	1.44	0.84	5.75	1.92	1.10	6.18	1.395	"
8.34	1.18	0.68	5.51	1.80	1.03	----	-----	"
5.16	0.69	0.41	5.54	1.75	1.00	5.69	1.382	"
4.24	0.58	0.33	5.37	1.60	1.15	5.32	1.374	"
0	0	0	5.00	1.50	0.85	4.10	1.340	"

[a] The mol/kg H_2O values were calculated by the compiler.

[b] The solid phases are: A = Na_2HPO_4; B = $Na_2B_4O_7$; C = H_3BO_3.

AUXILIARY INFORMATION

METHOD/APPARATUS/PROCEDURE:	SOURCE AND PURITY OF MATERIALS:
The standard isothermal method was used. Two series of experiments were performed. In one series, one component was added to saturated solutions of the other. In the other series, solutions of different concentrations of one component were prepared and the other component was then added to these solutions until saturation. Sodium content was determined by flame photometry, phosphate content was determined gravimetrically and the boric acid was determined by titration. No other details are given.	No information is given.
	ESTIMATED ERROR:
	No information is given.
	REFERENCES:

COMPONENTS:	ORIGINAL MEASUREMENTS:
(1) Disodium hydrogenphosphate; Na_2HPO_4; [7558-79-4] (2) Diammonium hydrogenphosphate; $(NH_4)_2HPO_4$; [7783-28-0] (3) Sodium chloride; NaCl; [7647-14-5] (4) Ammonium chloride; NH_4Cl; [12125-02-9] (5) Water; H_2O; [7732-18-5]	Lauffenburger, R.; Brodsky, N. *Compt. Rend.* <u>1938</u>, *206*, 1383-5.
VARIABLES: Composition and temperature.	**PREPARED BY:** J. Eysseltová

EXPERIMENTAL VALUES:

Part 1. Composition of the solutions saturated simultaneously by two solids in the $2 Na^+$, $2 NH_4^+ || HPO_4^{2-}$, $2 Cl^- - H_2O$ system.

	Na_2HPO_4		$(NH_4)_2HPO_4$		NaCl		NH_4Cl		
$t/°C.$	mass%[a]	mol/kg	mass%[a]	mol/kg	mass%[a]	mol/kg	mass%[a]	mol/kg	solid[b] phase
0	----	----	----	----	9.96	4.89	10.20	2.73	C + D
25	----	----	----	----	17.61	4.41	14.07	3.85	"
0	----	----	25.58	3.47	-----	----	18.58	6.22	B + D
25	----	----	12.97	1.53	-----	----	22.84	6.65	"
0	0.80	0.08	29.08	3.14	-----	----	-----	----	B + E
25	1.17	0.14	39.96	5.14	-----	----	-----	----	"
0	3.23	0.25	5.77	0.48	-----	----	-----	----	A + E
25	11.90	1.00	3.68	0.33	-----	----	-----	----	"
0	9.54	1.00	-----	----	23.39	5.95	-----	----	A + C
25	5.06	0.50	-----	----	23.72	5.70	-----	----	"

(continued next page)

AUXILIARY INFORMATION

METHOD/APPARATUS/PROCEDURE:	SOURCE AND PURITY OF MATERIALS:
The isothermal method was used. Four days were allowed for equilibration. Phosphate, chloride and ammonia were analyzed. Sodium and water were determined by difference.	All materials were "pur." grade.
	ESTIMATED ERROR: The temperature was constant to within ± 0.05 K.
	REFERENCES:

COMPONENTS:	ORIGINAL MEASUREMENTS:
(1) Disodium hydrogenphosphate; Na_2HPO_4; [7558-79-4]	Lauffenburger, R.; Brodsky, N.
(2) Diammonium hydrogenphosphate; $(NH_4)_2HPO_4$; [7783-28-0]	*Compt. Rend.* <u>1938</u>, *206*, 1383-5.
(3) Sodium chloride; NaCl; [7647-14-5]	
(4) Ammonium chloride; NH_4Cl; [12125-02-9]	
(5) Water; H_2O; [7732-18-5]	

EXPERIMENTAL VALUES cont'd:

Part 2. Composition of solutions saturated simultaneously by three solid phases in the 2 Na^+, 2 NH_4^+ || HPO_4^{2-}, 2 Cl^--H_2O system.

	Na^+		NH_4^+		HPO_4^{2-}		Cl^-		
$t/°C.$	mass%[a]	mol/kg	mass%[a]	mol/kg	mass%[a]	mol/kg	mass%[a]	mol/kg	solid phases[b]
0	7.83	4.90	3.32	2.80	0.53	0.08	18.65	7.55	C + D + E
25	6.62	4.34	5.28	4.67	1.47	0.23	20.15	8.55	"
25	10.57	6.51	0.48	0.40	3.12	0.46	15.03	5.99	A + C + E
0	0.43	0.30	10.64	10.1	15.76	2.65	11.20	5.1	B + C + E
25	0.67	0.45	9.97	9.04	10.21	1.64	14.28	6.21	"

[a] These values were calculated by the compiler.

[b] The solid phases are: A = Na_2HPO_4; B = $(NH_4)_2HPO_4$; C = NaCl; D = NH_4Cl;

E = $NH_4NaHPO_4·4H_2O$.

COMPONENTS:	ORIGINAL MEASUREMENTS:
(1) Disodium hydrogenphosphate; Na_2HPO_4; [7558-79-4] (2) Sodium chloride; NaCl; [7647-14-5] (3) Sodium nitrate; $NaNO_3$; [7631-99-4] (4) Water; H_2O; [7732-18-5]	Makin, A.V. Zh. Neorg. Khim. 1958, 3, 2764-6.

VARIABLES:	PREPARED BY:
Composition at 25°C.	J. Eysseltová

EXPERIMENTAL VALUES:

Points of simultaneous crystallization of two or three solid phases in the $NaNO_3$-Na_2HPO_4-NaCl-H_2O system at 25°C.

$NaNO_3$[a]		Na_2HPO_4		NaCl[a]		solid phases[c]
mass%	conc[b]	mass%	conc[b]	mass%	conc[b]	
20.32	68.00	9.17	31.10	----	----	A + B
18.17	54.72	7.93	23.80	7.10	21.48	"
17.21	48.64	7.70	21.90	10.47	29.46	"
14.85	40.36	7.18	19.48	14.80	40.16	"
13.20	35.44	6.63	17.82	17.42	46.74	"
11.43	29.52	6.26	16.20	21.04	54.28	"
9.61	23.97	5.73	14.28	24.78	61.75	"
8.05	19.30	5.66	13.59	27.97	67.11	A + B + C
-----	-----	6.34	19.60	26.05	80.40	B + C
1.95	5.32	5.95	17.18	27.30	77.50	"
3.70	8.28	5.81	16.42	27.91	75.30	"
5.81	13.80	5.71	14.78	28.10	71.42	"
8.05	19.30	5.64	13.61	27.99	67.07	A + B + C
12.31	28.75	----	----	30.51	71.25	A + C
10.50	24.61	1.78	4.18	30.38	71.21	"
9.33	22.10	2.90	6.89	29.98	71.01	"
8.47	19.80	4.71	10.30	28.70	69.90	"
8.05	19.30	5.63	13.57	27.94	67.13	A + B + C

[a] These are probably incorrect headings--see the critical evaluation.

[b] The concentration units are: mol/100 mol of solute.

[c] The solid phases are: A = NaCl; B = $Na_2HPO_4 \cdot 12H_2O$; C = $NaNO_3$.

AUXILIARY INFORMATION

METHOD/APPARATUS/PROCEDURE:	SOURCE AND PURITY OF MATERIALS:
The method of invariant points was used. To a solution saturated with two salts, a third salt was added until ternary eutonic equilibrium was reached. Analyses were made gravimetrically: the hydrogenphosphate ion was precipitated as NH_4MgPO_4; sodium was precipitated as sodium zinc uranylacetate after removing the phosphate. The nitrate ion content was determined by difference.	No details are given.
	ESTIMATED ERROR:
	No details are given. The compiler considers the reproducibility to be about ± 0.1%.
	REFERENCES:

COMPONENTS:	ORIGINAL MEASUREMENTS:
(1) Disodium hydrogenphosphate; Na_2HPO_4; [7558-79-4]	Makin, A.V.; Lepeshkov, I.N.
(2) Disodium sulfate; Na_2SO_4; [7757-82-6]	*Zh. Neorg. Khim.* 1964, 9, 495-8.
(3) Sodium nitrate; $NaNO_3$; [7631-99-4]	
(4) Water; H_2O; [7732-18-5]	

VARIABLES:	PREPARED BY:
Composition at 25°C.	J. Eysseltová

EXPERIMENTAL VALUES:

Part 1. Points of simultaneous crystallization of two or three solid phases in the $NaNO_3$-Na_2HPO_4-Na_2SO_4-H_2O system at 25°C.

soln. no.	Na_2SO_4 mass%	Na_2SO_4 mol%[a]	Na_2HPO_4 mass%	Na_2HPO_4 mol%[a]	$NaNO_3$ mass%	$NaNO_3$ mol%[a]	H_2O mol%[a]	solid phase[b]
1	15.07	66.72	7.52	33.28	----	----	2767.3	A + B
2	14.03	54.40	6.92	26.81	2.85	18.79	2325.8	"
3	13.54	48.19	6.90	24.54	4.56	27.27	2104.8	"
4	13.39	45.12	6.86	23.16	5.65	31.72	1969.9	"
5	12.32	36.13	6.83	19.99	8.95	43.88	1662.5	"
6	12.35	32.48	6.63	17.44	11.52	50.08	1234.5	"
7	12.02	28.47	6.49	15.29	14.29	56.24	1248.9	"
8	10.96	23.49	6.07	13.07	17.67	63.44	1106.7	A + B + C
9	10.06	20.67	6.04	10.91	20.50	68.42	1027.3	"
10	8.84	16.53	5.50	10.26	23.46	73.21	901.9	"[c]
11	6.53	12.05	4.55	8.41	27.52	79.54	905.1	"
12	5.25	8.95	4.10	6.97	29.65	84.08	817.2	"
13	4.99	7.96	3.76	6.01	32.35	86.03	742.0	A + C + D
14	14.94	28.67	----	----	22.24	71.33	952.6	B + C
15	13.98	27.45	1.25	2.45	21.39	70.10	980.5	"
16	13.99	27.57	2.72	5.36	20.37	67.07	950.4	"
17	11.81	24.9	4.00	8.16	19.88	67.75	1035.0	"
18	11.28	23.8	5.43	11.46	18.37	64.74	1080.5	"
19	11.22	23.87	5.65	12.03	18.09	64.10	1093.0	"
20	10.99	23.61	6.04	12.97	17.67	63.42	1106.7	A + B + C
21	5.02	6.42	----	-----	43.75	93.58	517.5	C + D
22	5.01	7.12	1.83	2.60	37.96	90.28	619.9	"
23	5.04	7.58	3.06	4.59	34.96	87.83	677.7	"

(continued next page)

AUXILIARY INFORMATION

METHOD/APPARATUS/PROCEDURE:	SOURCE AND PURITY OF MATERIALS:
The method of invariant points was used. At least 6 days were allowed for equilibration. All analyses were done gravimetrically; phosphate was determined as $NH_4MgPO_4 \cdot 6H_2O$; sulfate as $BaSO_4$; sodium as zinc uranylacetate. Water and nitrate contents were determined by difference.	No information is given.
	ESTIMATED ERROR:
	No information is given.
	REFERENCES:

COMPONENTS:	ORIGINAL MEASUREMENTS:
(1) Disodium hydrogenphosphate; Na$_2$HPO$_4$; [7558-79-4] (2) Disodium sulfate; Na$_2$SO$_4$; [7757-82-6] (3) Sodium nitrate; NaNO$_3$; [7631-99-4] (4) Water; H$_2$O; [7732-18-5]	Makin, A.V.; Lepeshkov, I.N. *Zh. Neorg. Khim.* <u>1964</u>, *9*, 495-8.

EXPERIMENTAL VALUES cont'd:

Part 1. Points of simultaneous crystallization of two or three solid phases in the NaNO$_3$-Na$_2$HPO$_4$-Na$_2$SO$_4$-H$_2$O system at 25°C.

soln. no.	Na$_2$SO$_4$ mass%	Na$_2$SO$_4$ mol%[a]	Na$_2$HPO$_4$ mass%	Na$_2$HPO$_4$ mol%[a]	NaNO$_3$ mass%	NaNO$_3$ mol%[a]	H$_2$O mol%[a]	solid phase[b]
24	4.99	7.96	3.76	6.01	32.35	86.03	742.0	A + C + D
25	----	----	6.34	12.76	26.05	87.24	1089.1	A + D
26	1.39	2.44	5.74	10.06	29.87	87.50	871.5	"
27	3.19	5.25	5.05	8.52	30.88	86.23	803.0	"
28	4.22	6.94	4.46	7.31	31.20	85.72	780.0	"
29	4.99	7.96	3.76	6.01	32.35	86.03	742.0	A + C + D

[a] This should be: mol/100 mol of solute--compiler.

[b] The solid phases are: A = Na$_2$HPO$_4$·12H$_2$O; B = Na$_2$SO$_4$·10H$_2$O;

C = NaNO$_3$·Na$_2$SO$_4$·H$_2$O; D = NaNO$_3$.

[c] This is an obvious error: the compiler suggests that the correct phases are A + C.

Part 2. The compiler has calculated the following values from the data in Part 1.

soln. no.	Na$_2$SO$_4$ mol/kg	Na$_2$HPO$_4$ mol/kg	NaNO$_3$ mol/kg	H$_2$O mass%
1	1.37	0.68	----	77.41
2	1.30	0.64	0.44	76.20
3	1.25	0.64	0.71	75.15
4	1.27	0.65	0.90	74.10
5	1.21	0.67	1.46	71.90
6	1.25	0.67	1.95	69.50
7	1.26	0.68	2.50	67.20
8	1.18	0.65	3.18	65.30
9	1.12	0.67	3.80	63.40
10	1.00	0.62	4.44	62.20
11	0.75	0.52	5.27	61.40
12	0.60	0.47	5.72	61.00
13	0.60	0.45	6.46	58.90
14	1.67	----	4.16	62.82
15	1.55	0.14	3.97	63.38
16	1.56	0.30	3.81	62.92
17	1.29	0.44	3.64	64.31
18	1.22	0.59	3.33	64.92
19	1.21	0.61	3.27	65.04
20	1.18	0.65	3.18	65.30
21	0.69	----	10.05	51.23
22	0.64	0.23	8.09	55.20
23	0.62	0.38	7.22	56.94
24	0.60	0.45	6.46	58.90
25	----	0.66	4.53	67.61
26	0.16	0.64	5.58	63.00
27	0.37	0.58	5.97	60.88
28	0.49	0.52	6.10	60.12
29	0.60	0.45	6.46	58.90

COMPONENTS:	ORIGINAL MEASUREMENTS:
(1) Disodium hydrogenphosphate; Na_2HPO_4; [7558-79-4] (2) 2-Propanone (acetone); C_3H_6O; [67-64-1] (3) Water; H_2O; [7732-18-5]	Ferroni, G.; Galea, J.; Antonetti, G. *Bull. Soc. Chim. Fr.* <u>1974</u>, *12*, Pt. 1, 273-81.

VARIABLES:	PREPARED BY:
Composition at 25°C.	J. Eysseltová

EXPERIMENTAL VALUES:

A miscibility gap was found in the system. The following data are for the isothermal binodal curve at 25°C.

u p p e r   l a y e r   l o w e r   l a y e r

density g/cm^3	Na_2HPO_4 mass%	C_3H_6O mass%	H_2O mass%	density g/cm^3	Na_2HPO_4 mass%	C_3H_6O mass%	H_2O mass%	solid$_\alpha$ phase
0.7922	~0	100	0	----	----	----	----	A
0.793	~10^{-7}	95.60	4.40	----	----	----	----	B
0.821	~10^{-6}	84.30	15.70	----	----	----	----	"
0.8331	~10^{-6}	75.79	24.21	1.328	40.02	1.24	58.74	C
0.8612	0.043	61.29	38.47	1.311	38.75	2.04	59.21	"
0.9097	0.602	38.97	60.93	1.321	38.52	1.98	59.50	"
0.9524	2.170	23.84	73.99	1.307	38.21	1.58	60.21	"
				1.304	37.21	----	62.79	"

a The solid phases are: A = Na_2HPO_4; B = $Na_2HPO_4 \cdot 2H_2O$; C are supersaturated solutions which solidified when seeded.

AUXILIARY INFORMATION

METHOD/APPARATUS/PROCEDURE:	SOURCE AND PURITY OF MATERIALS:
The samples were stirred in a thermostat for 48 hours. The equilibration was done in the dark. The HPO_4^{2-} ion content was determined by a pH titration after evaporating the sample to dryness and dissolving the residue in bidistilled water. The 2-propanone content was determined iodometrically.	Merck reagent grade Na_2HPO_4 was dehydrated at 100°C and stored in a vacuum over NaOH. BLB reagent grade 2-propanone was used. The water was distilled twice and deaerated.
	ESTIMATED ERROR: The temperature was held within ± 0.1 K. The analyses had a precision of 0.5%.
	REFERENCES:

COMPONENTS:	ORIGINAL MEASUREMENTS:
(1) Disodium hydrogenphosphate; Na_2HPO_4; [7558-79-4] (2) Sodium perchlorate; $NaClO_4$; [7601-89-0] (3) 2-Propanone (acetone); C_3H_6O; [67-64-1] (4) Water; H_2O; [7732-18-5]	Ferroni, G.; Galea, J.; Antonetti, G. *Bull. Soc. Chim. Fr.* <u>1974</u>, *12*, Pt. 1, 273-81.

VARIABLES:	PREPARED BY:
Two concentrations of $NaClO_4$ at 25°C.	J. Eysseltová

EXPERIMENTAL VALUES:

Composition of saturated solutions of Na_2HPO_4 in aqueous $NaClO_4$ at 25°C.

C_{NaClO_4}/mol dm^{-3}	concn H_2O^b	$C_{Na_2HPO_4}$/mol dm^{-3}	solid phase^a
1	100	0.805	A
"	90.9	0.050	binodal curve
"	83.3	0.023	" "
"	66.7	0.0011	" "
"	50.0	0.33×10^{-3}	A
"	33.3	$\sim6.9 \times 10^{-5}$	B
"	9.1	$\sim1.5 \times 10^{-6}$	"
"	0.0	$<10^{-7}$	C
3	100	0.201	A
"	90.9	0.013	"
"	83.3	0.0048	"
"	66.7	1.17×10^{-3}	"
"	50.0	0.84×10^{-3}	"
"	33.3	$\sim1.9 \times 10^{-4}$	B
"	9.1	$\sim3.3 \times 10^{-6}$	"
"	0.0	$<10^{-7}$	C

^{a}The solid phases are A = $Na_2HPO_4 \cdot 12H_2O$; B is probably $Na_2HPO_4 \cdot 2H_2O$; C = Na_2HPO_4.

^{b}The concentration unit is: mol/100 mol of solvent.

AUXILIARY INFORMATION

METHOD/APPARATUS/PROCEDURE:	SOURCE AND PURITY OF MATERIALS:
The samples were stirred in a thermostat for 48 hours. Equilibration was done in the dark. The HPO_4^{2-} ion concentration was determined by an automatic potentiometric titration after evaporating the sample to dryness and dissolving the residue in bidistilled water. The 2-propanone content was determined iodometrically.	Merck reagent grade Na_2HPO_4 was dehydrated at 100°C and stored over NaOH in a vacuum. The $NaClO_4$ was reagent grade. The water was bidistilled and deaerated. BLB reagent grade 2-propanone was used.
	ESTIMATED ERROR:
	The temperature was controlled to within ± 0.1 K. The analyses had a precision of ± 0.5%.
	REFERENCES:

COMPONENTS:	ORIGINAL MEASUREMENTS:
(1) Disodium hydrogenphosphate; Na_2HPO_4; [7558-79-4] (2) Methanol; CH_4O; [67-56-1] (3) Water; H_2O; [7732-18-5]	Bruder, K.; Vohland, P.; Schuberth, H. Z. Phys. Chem. (Leipzig) 1977, 4, 721-9.

VARIABLES:	PREPARED BY:
Composition at 60°C.	J. Eysseltová

EXPERIMENTAL VALUES:

Part 1. Solubility in the $CH_4O-H_2O-Na_2HPO_4$ system at 60°C.

CH_3OH conc^a	Na_2HPO_4 mol%	mass%^{b}	CH_3OH conc^a	Na_2HPO_4 mol%	mass%^{b}
0.000	0.0918	44.39	0.313^c	0.0045	2.79
0.040	0.0710	36.92	0.400	0.00204	1.22
0.050	0.0672	35.38	0.500	0.00080	0.453
0.100	0.0488	27.32	0.600	0.00021	0.113
0.150	0.0295	17.69	0.700	0.00011	0.0562
0.191	0.0130	8.30	0.800	0.000076	0.0369
0.200	0.0114	7.30	0.900	0.000060	0.0278
0.300	0.0047	2.93	0.950	0.000055	0.0249
			1.000	0.000054	0.0240

Part 2. The miscibility gap (conjugated solutions) in the $CH_4O-H_2O-Na_2HPO_4$ system.

CH_3OH conc^a	Na_2HPO_4 mol%	mass%^{b}	CH_3OH conc^a	Na_2HPO_4 mol%	mass%^{b}
0.040	0.071	36.92	0.191	0.013	8.30
0.073$_d$	0.051	28.65	0.124	0.032	19.23
0.097^d	0.041	23.89			

^{a}The concentration unit is: mol/100 mol of solvent.

^{b}These values were calculated by the compiler.

^{c}The point of transition of crystalline dihydrate into unsolvated salt.

^{d}Critical solution.

AUXILIARY INFORMATION

METHOD/APPARATUS/PROCEDURE:	SOURCE AND PURITY OF MATERIALS:
Some mixtures were prepared by allowing the solid salt to equilibrate in the complex solvent. Other mixtures were prepared by adding methanol to aqueous solutions of Na_2HPO_4 until precipitation occurred. In all cases the liquid phase was evaporated to dryness and the residue was weighed. All data are designated as smoothed. No details are given about the computational procedure.	The water was distilled twice. The methanol was obtained by rectification (n_D^{20} = 1.3286, d^{20} = 0.7913, b.p. = 64.6°C). The $Na_2HPO_4 \cdot 10H_2O$, a product of VEB Laborchemie Apolda, was dried by the method suggested by Menzies (1). The analysis by a Karl Fischer titration gave the water content as 0.04%.
	ESTIMATED ERROR: No details are given.
	REFERENCES: 1. Menzies, A.W.; Humphrey, E.C. C. R. Congr. Int. Appl. Electrocat. Electrochim. 1912, 2, 175.

COMPONENTS:	EVALUATOR:
(1) Trisodium phosphate; Na_3PO_4; [7601-54-9] (2) Water; H_2O; [7732-18-5]	J. Eysseltová Charles University Prague, Czechoslovakia May 1985

CRITICAL EVALUATION:

THE BINARY SYSTEM

There is a good deal of uncertainty about this system. There is disagreement about the solubility and about the composition of the solid phase. The data published by Apfel (1) for the solubility of trisodium phosphate over the temperature range of 273-356 K disagree with the data of Mulder (2) as quoted by others (3). Kobe and Leipper (4) reported the solubility of a substance having the composition $Na_3PO_4 \cdot 1/7NaOH \cdot 12H_2O$ and their results agree to some extent with those of Apfel (1). Ravich and Shcherbakova (5) reported the existence of solid solutions $mNa_3PO_4 \cdot nNaH_2PO_4$ in equilibrium with saturated solutions having Na/P ratios = 1/3 above 523 K.

The matter of the hydrates of Na_3PO_4 has also been the subject of disagreement. Most authors consider $Na_3PO_4 \cdot 12H_2O$ to be the solid phase at temperatures below 323 K, but Wendrow and Kobe (3) suggest that it is $Na_3PO_4 \cdot 1/4NaOH \cdot 12H_2O$. The solid phase in equilibrium with saturated solutions is reported to be $Na_3PO_4 \cdot 10H_2O$ over the temperature range 323-333 K and $Na_3PO_4 \cdot 8H_2O$ at temperatures from 343 to 348 K (1). But others (3), on the basis of extrapolated data, suggest that the equilibrium solid phases are $Na_3PO_4 \cdot 1/4NaOH \cdot 12H_2O$ at temperatures up to 328 K, $Na_3PO_4 \cdot 8H_2O$ from 328-338 K, and $Na_3PO_4 \cdot 6H_2O$ from 338-373 K. More work is needed to clarify the nature of the solid phases before the solubility data can be evaluated.

Schroeder, et al. (6) made solubility measurements over the temperature interval 348-523 K. They reported the equilibrium solid phase to be $Na_3PO_4 \cdot H_2O$ in the temperature interval 393-488 K and the anhydrous Na_3PO_4 to be the solid phase above 488 K. Attempts to fit these data to the general solubility equations described and discussed in the section on NaH_2PO_4 (chap. 3) were unsuccessful. The number of experimental points remaining after the iteration was too small to consider the results to be reasonable. Perhaps a different model is needed to treat these data.

MULTICOMPONENT SYSTEMS

The phase diagrams for systems in which Na_3PO_4 is a component differ substantially from those in which NaH_2PO_4 or Na_2HPO_4 are components. The latter usually form simple eutonic systems, while with systems containing Na_3PO_4 the formation of solid solutions or complex compounds has often been reported. Solubility values reported for these systems often disagree with each other. This is probably due to the chemical complexity of the systems and the fact that the analyses are complicated by the high pH values of these systems.

No solid solutions or complex compounds have been reported as equilibrium solid phases for the following systems:

Na_3PO_4-$NaNO_2$-H_2O at 298 K (7);

Na_3PO_4-$NaCl$-H_2O at 298 and 378 K (8);

Na_3PO_4-Na_2WO_4-H_2O at 503 K (9);

Na_3PO_4-Na_2CO_3-H_2O at 273 and 293 K (10), and 298 K (11).

The presence of a small amount of NaOH in one study of the last system above (4) resulted in only a small increase in the concentration of the other salt components.

Solid solutions have been reported as solid phases for the following systems:

Na_3PO_4-$Na_2B_4O_7$-H_2O at 293 K (12);

Na_3PO_4-$NaClO_4$-H_2O at 293 K (12);

Na_3PO_4-NaF-H_2O at 298 K (13).

Complex compounds that have been reported as solid phases are:
$Na_3PO_4 \cdot 4.5H_2O_2$ in the Na_3PO_4-H_2O_2-H_2O system at 253 K (14);
$Na_3PO_4 \cdot 4.5H_2O_2$ and $Na_3PO_4 \cdot H_2O_2$ in the above system at 273 K (14);
$Na_3PO_4 \cdot Cu_3(PO_4)_2$ in the Na_3PO_4-$Cu_3(PO_4)_2$-H_2O system at 298 and 323 K (15).

(continued next page)

COMPONENTS:

(1) Trisodium phosphate; Na_3PO_4; [7601-54-9]

(2) Water; H_2O; [7732-18-5]

EVALUATOR:

J. Eysseltová
Charles University
Prague, Czechoslovakia

May, 1985

CRITICAL EVALUATION:

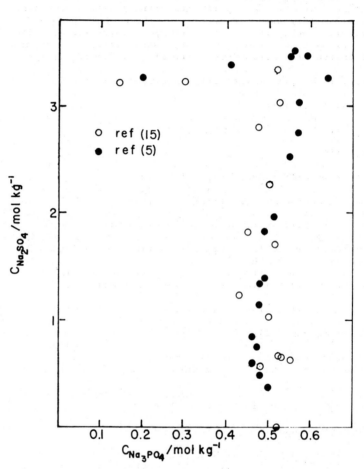

Figure 1.   Solubility in the Na_3PO_4-Na_2SO_4-H_2O system at 523 K.

COMPONENTS:	EVALUATOR:
(1) Trisodium phosphate; Na_3PO_4; [7601-54-9] (2) Water; H_2O; [7732-18-5]	J. Eysseltová Charles University Prague, Czechoslovakia May 1985

CRITICAL EVALUATION: (cont'd)

Three groups of systems have been studied in more detail.

1. The Na_3PO_4-Na_2SO_4-H_2O system. This system has been studied at 523 K (16) and at 423, 473, 523 and 573 K (6). (In the latter paper some data were also reported for the Na_3PO_4-Na_2SO_4-$NaOH$-H_2O system.) The results obtained at 523 K, Figure 1, agree reasonably well with each other. However, one group reports a solid phase of $Na_2SO_4 \cdot 2Na_3PO_4$ (6) while the other group (16) reports instead two types of phases having varying compositions. Additional work is needed to settle this matter.

2. Abduragimova, et al. have studied the following systems at 298 K:

Na_3PO_4-$NaVO_3$-H_2O (17)

Na_3PO_4-Na_2SO_4-H_2O (17)

Na_3PO_4-$NaVO_3$-Na_2SO_4-H_2O (18)

$(NaAlO_2 + NaOH)$-Na_3PO_4-$NaVO_3$-H_2O (19)

$(NaAlO_2 + NaOH)$-Na_3PO_4-Na_2SO_4-H_2O (19).

The following solid phases were reported as being present:

$4Na_2O \cdot P_2O_5 \cdot V_2O_5 \cdot 30H_2O$; $4Na_2O \cdot P_2O_5 \cdot V_2O_5 \cdot 18H_2O$; and $3Al_2O_3 \cdot 4Na_2O \cdot P_2O_5 \cdot 15H_2O$.

However, these data are not considered to be reliable because of many obvious errors in the tabular data. The errors make it difficult to interpret the data.

3. A group of Armenian authors has studied the following systems:

Na_3PO_4-Na_2SiO_3-H_2O (20,21);

Na_3PO_4-K_3PO_4-H_2O (22);

Na_3PO_4-Na_2SiO_3-$NaOH$-H_2O (23);

Na_3PO_4-Na_2SiO_3-Na_2CO_3-H_2O (21);

$3K^+,3Na^+ || PO_4^{3-},3/2SiO_3^{2-}$-$H_2O$ (24).

Ref. (23) contains only graphical data and in the other papers the data consist of limits within which the individual phases exist, rather than precise solubility data. The most recent report (21) maintains that no solid solutions of Na_2SiO_3 and Na_3PO_4 are formed, but does not substantiate this statement. Therefore, more work is needed before this set of papers can be evaluated.

The system Na_3PO_4-CH_3COCH_3-H_2O has also been studied (25). In contrast to the NaH_2PO_4-CH_3COCH_3-H_2O and Na_2HPO_4-CH_3COCH_3-H_2O systems (26), no limited miscibility has been observed.

(continued next page)

COMPONENTS:	EVALUATOR:
(1) Trisodium phosphate; Na_3PO_4; [7601-54-9] (2) Water; H_2O; [7732-18-5]	J. Eysseltová Charles University Prague, Czechoslovakia May 1985

CRITICAL EVALUATION: (cont'd)

References

1. Apfel, O. Dissertation, Technical University, Darmstadt 1911.
2. Mulder, G.J. *Bijdragen tot de geschiedenis van het scheikundig gebonden water*, Rotterdam 1894. Quoted in Landolt-Bornstein, p. 558.
3. Wendrow, B.; Kobe, K.A. *Ind. Eng. Chem.* 1952, *44*, 1439.
4. Kobe, K.A.; Leipper, A. *Ind. Eng. Chem.* 1940, *32*, 198.
5. Ravich, M.I.; Shcherbakova, L.G. *Izv. Sektora Fiz.-Khim. Analiza, Inst. Obshch. Neorgan., Khim. Akad. Nauk SSSR* 1955, *26*, 248.
6. Schroeder, W.C.; Berk, A.A.; Gabriel, A. *J. Am. Chem. Soc.* 1937, *59*, 1783.
7. Protsenko, P.I.; Ivleva, T.I.; Rubleva, V.V.; Berdyukova, V.A.; Edush, T.V. *Zh. Prikl. Khim.* 1975, *48*, 1055.
8. Obukhov, A.P.; Mikhailova, M.N. *Zh. Prikl. Khim.* 1935, *8*, 1148.
9. Urosova, M.A.; Balyashko, V.M.; Rakova, N.N.; Zelikman, A.N.; Yevdokimova, G.V. *Zh. Neorg. Khim.* 1975, *20*, 2585.
10. Gyunashyan, A.P. *Arm. Khim. Zh.* 1979, *32*, 868.
11. Korf, D.M.; Balyasnaya, A.M. *Zh. Prikl. Khim.* 1941, *14*, 475.
12. Babayan, G.G.; Darbinyan, G.M. *Arm. Khim. Zh.* 1972, *25*, 482.
13. Roslyakova, O.N.; Petrov, M.R.; Zhikharev, M.I. *Zh. Neorg. Khim.* 1979, *24*, 206.
14. Ukraintseva, E.A. *Izv. Sib. Otd. Akad. Nauk SSSR, Ser. Khim.* 1963, *3*, 14.
15. Druzhinin, I.G.; Tusheva, L.A. *Izv. Vusov, Khim. Khim. Tekhnol.* 1974, *17*, 1513.
16. Ravich, M.I.; Yastrebova, L.F. *Zh. Neorg. Khim.* 1958, *3*, 2771.
17. Abduragimova, R.A.; Rza-Zade, P.F. *Issled. Obl. Neorg. Fiz. Khim.* 1971, 179.
18. Abduragimova, R.A.; Rza-Zade, P.F. *Issled. Obl. Neorg. Fiz. Khim.* 1971, 191.
19. Abduragimova, R.A.; Rza-Zade, P.F.; Abduragimov, A.A. *Dokl. Akad. Nauk Azerb. SSR* 1971, *27*, 41.
20. Babayan, G.G.; Sayamyan, E.A.; Darbinyan, G.M. *Arm. Khim. Zh.* 1970, *23*, 986.
21. Gyunashyan, A.P. *Arm. Khim. Zh.* 1979, *32*, 868.
22. Manvelyan, M.G.; Galstyan, V.D.; Voskanyan, S.S. *Arm. Khim. Zh.* 1974, *27*, 810.
23. Manvelyan, M.G.; Galstyan, V.D.; Gyunashyan, A.P.; Sayamyan, E.A.; Oganesyan, E.B.; Grigoryan, K.G. *Arm. Khim. Zh.* 1977, *30*, 219.
24. Manvelyan, M.G.; Galstyan, V.D.; Sayamyan, E.A.; Gyunashyan, A.P.; Oganesyan, E.B. *Arm. Khim. Zh.* 1973, *26*, 632.
25. Nirenberg, Z.; Solenchyk, B.; Yaron, I. *J. Chem. Eng. Data* 1977, *22*, 47.
26. Ferroni, G.; Galea, J.; Antonetti. G. *Bull. Soc. Chim. Fr.* 1974, *12 (Pt.1)*, 273.

COMPONENTS:	ORIGINAL MEASUREMENTS:
(1) Trisodium phosphate; Na_3PO_4; [7601-54-9] (2) Water; H_2O; [7732-18-5]	Apfel, O. Dissertation, Technical University, Darmstadt, <u>1911</u>.

VARIABLES:
 	
  Temperature and Composition
	PREPARED BY:
	
	  J. Eysseltová

EXPERIMENTAL VALUES:

Composition of saturated solutions of Na_3PO_4 in water.

$t/°C$	PO_4^{3-} mol/kg sln	Na_3PO_4 [a] mass%	Na_3PO_4 [a] mol/kg	solid phase [b]
0	0.26	4.27	0.27	A
25	0.75	12.31	0.86	"
37	0.98	16.08	1.17	B
40	1.02	16.74	1.23	"
44	1.09	17.89	1.33	"
50	1.38	22.65	1.78	B + C [c]
55	1.595	26.18	2.16	C
65	1.84	30.20	2.64	"
70	1.99	32.66	2.96	"
75	2.14	35.12	3.30	"

[a] These values were calculated by the compiler.

[b] The solid phases are: A = $Na_3PO_4 \cdot 12H_2O$; B = $Na_3PO_4 \cdot 10H_2O$; C = $Na_3PO_4 \cdot 8H_2O$.

[c] The octahydrate is said to exist in the region 50 to 75°C "with great probability".

AUXILIARY INFORMATION

METHOD/APPARATUS/PROCEDURE:	SOURCE AND PURITY OF MATERIALS:
All the experiments were performed in a water thermostat. The attainment of equilibrium was checked by repeated analysis of the liquid phase. The liquid phase was separated from the solid phase by filtration through a mat of platinum wires. Phosphate was precipitated as $NH_4MgPO_4 \cdot 6H_2O$ and weighed as $Mg_2P_2O_7$. Sodium was determined as Na_2SO_4 after phosphoric acid had been removed as lead phosphate.	Nothing given.
	ESTIMATED ERROR: Nothing given.
	REFERENCES:

COMPONENTS:	ORIGINAL MEASUREMENTS:
(1) Trisodium phosphate; Na_3PO_4; [7601-54-9] (2) Water; H_2O; [7732-18-5]	Schroeder, W.C.; Berk, A.A.; Gabriel, A. *J. Am. Chem. Soc.* <u>1937</u>, 59, 1783-90.
VARIABLES: Temperature and Composition	PREPARED BY: J. Eysseltová

EXPERIMENTAL VALUES:

Solubility of Na_3PO_4 in water at 83 to 350°C.

concn of Na_3PO_4

$t/°C$	g(1)/100 g(2)	mass% ^{a}	mol/kg ^{a}	time/h ^{b}
83	61.1	37.93	3.72	39
83	62.2	38.35	3.79	39
101	78.4	43.95	4.78	43
101	76.8	43.44	4.68	43
115	88.6	46.98	5.40	48
115	90.3	47.45	5.50	48
115	89.8	47.31	5.47	48
121	93.2	48.24	5.68	86
129	91.1	47.67	5.55	45
129	89.3	47.17	5.44	45
139	88.2	46.85	5.37	39
139	88.7	47.00	5.40	39
139	88.8	47.03	5.41	39
150	83.9	45.62	5.11	16
150	79.8	44.38	4.86	16
150	83.1	45.38	5.06	44
150	78.9	44.10	4.81	44
150	82.2	45.12	5.01	44
150	84.1	45.68	5.12	18
150	78.6	44.01	4.79	18
159	76.0	43.18	4.63	66

(continued next page)

AUXILIARY INFORMATION

METHOD/APPARATUS/PROCEDURE:	SOURCE AND PURITY OF MATERIALS:
Self-constructed high temperature solubility bomb with sampler ensuring the sampling at the operating temperature. The time of equilibration varied from case to case, because of the difficulty in attaining true equilibrium. Phosphate determinations were made by a colorimetric method using aminonaphtholsulfonic acid (1).	Merck CP $Na_3PO_4 \cdot 12H_2O$ was used. The actual phosphate content of this material was determined by analysis but the results are not given. In some cases the dodecahydrate was dried at 120°C to give approximately the monohydrate or it was recrystallized at 250°C to give the anhydrous salt.
	ESTIMATED ERROR: Phosphate determination: the error not greater than 1%.
	REFERENCES: 1. Fiske, C.H.; Subbarow, J.T. *J. Biol. Chem.* <u>1925</u>, *66*, 375.

COMPONENTS:	ORIGINAL MEASUREMENTS:
(1) Trisodium phosphate; Na_3PO_4; [7601-54-9]	Schroeder, W.C.; Berk, A.A.; Gabriel, A.
(2) Water; H_2O; [7732-18-5]	J. Am. Chem. Soc. 1937, 59, 1783-90.

EXPERIMENTAL VALUES cont'd:

Solubility of Na_3PO_4 in water at 83 to 350°C.

concn of Na_3PO_4

t/°C	g(1)/100 g(2)	mass%[a]	mol/kg[a]	time/h[b]
169	71.9	41.83	4.38	47
169	70.2	41.24	4.28	47
185	66.2	39.83	4.03	48
185	65.0	39.39	3.96	48
187	63.1	38.69	3.84	67
187	62.0	38.27	3.78	67
204	62.0	38.27	3.78	71
204	60.8	37.81	3.70	71
214	50.0	33.33	3.05	90
214	50.8	33.69	3.09	90
216	48.8	32.80	2.97	65
216	47.6	32.25	2.90	65
225	25.2	20.13	1.54	15
225	33.7	25.20	2.05	15
225	27.3	21.44	1.66	18
225	27.8	21.75	1.69	18
235	17.9	15.18	1.09	17
250	8.6	7.92	0.52	17
250	8.6	7.92	0.52	17
250	8.5	7.83	0.52	17
300	2.4	2.34	0.15	18
350	0.15	0.15	0.01	19

[a] These values were calculated by the compiler.

[b] This is the time allowed for equilibration.

COMPONENTS:	ORIGINAL MEASUREMENTS:
(1) Trisodium phosphate; Na_3PO_4; [7601-54-9]	Obukhov, A.P; Mikhailova, M.N.
(2) Sodium chloride; NaCl; [7647-14-5]	*Zh. Prikl. Khim.* 1935, *8*, 1148-51.
(3) Water; H_2O; [7732-18-5]	

VARIABLES:	PREPARED BY:
Two temperatures: 25 and 105°C	J. Eysseltová
Composition	

EXPERIMENTAL VALUES:

Composition of saturated solutions in the
Na_3PO_4-NaCl-H_2O system.

Na_3PO_4			NaCl			solid phase [b]
mass %	mol %	mol/kg[a]	mass %	mol %	mol/kg[a]	
			temp. = 25°C.			
13.40	1.67	0.94	----	----	----	A
9.51	1.15	0.65	2.00	0.68	0.39	"
7.24	0.87	0.49	3.48	1.17	0.67	"
5.14	0.61	0.35	6.12	2.06	1.18	"
4.70	0.57	0.32	7.45	2.53	1.45	"
4.28	0.53	0.30	10.44	3.61	2.09	"
3.07	0.38	0.22	12.50	4.34	2.53	"
2.32	0.29	0.18	17.05	6.09	3.62	"
1.96	0.26	0.16	21.92	8.12	4.93	"
1.90	0.25	0.16	25.26	9.62	5.93	A + C
0.95	0.12	0.08	25.76	9.75	6.01	C
----	----	----	26.40	9.94	6.14	"

(continued on next page)

AUXILIARY INFORMATION

METHOD/APPARATUS/PROCEDURE:	SOURCE AND PURITY OF MATERIALS:
Isothermal method using an oil thermostat. The mixtures were equilibrated for 2-3 hours at 25° and for 30 min at 105°C. At 105°C the samples of saturated solution were drawn into glass tubes and allowed to solidify. The tubes were then weighed and the samples were washed out and analyzed. P_2O_5 was determined gravimetrically using the method of Schmitz (no reference is given). Chloride was determined by the Volhard method. The composition of the solid phases was determined by Schreinemakers' method.	Nothing given.
	ESTIMATED ERROR:
	Nothing given.
	REFERENCES:

COMPONENTS:	ORIGINAL MEASUREMENTS:
(1) Trisodium phosphate; Na_3PO_4, [76-54-9]	Obukhov, A.P., Mikhailova M.N.
(2) Sodium chloride; NaCl, [7647-14-5]	*Zh. Prikl. Khim.* <u>1935</u>, *8*, 1148-51.
(3) Water, H_2O, [7732-18-5]	

EXPERIMENTAL VALUES cont'd:

Composition of saturated solutions in the
Na_3PO_4-$NaCl$-H_2O system.

Na_3PO_4			NaCl			solid [b]
mass %	mol %	mol/kg[a]	mass %	mol %	mol/kg[a]	phase

temp. = 105°C.

mass %	mol %	mol/kg	mass %	mol %	mol/kg	phase
49.7.	9.78	6.01	----	----	----	B
44.50	8.33	5.11	2.56	1.34	0.83	"
43.35	8.03	4.93	3.15	1.63	1.01	"
39.70	7.26	4.50	6.67	3.42	2.13	"
37.33	6.67	4.13	7.75	3.89	2.41	"
35.75	6.29	3.89	8.40	4.15	2.57	"
34.62	6.09	3.80	9.95	4.91	3.07	B + C
33.89	5.93	3.69	10.30	5.06	3.16	C
21.72	3.43	2.13	16.27	7.21	4.49	"
14.60	2.19	1.36	20.08	8.46	5.26	"
10.78	1.57	0.98	22.23	9.12	5.68	"
3.30	0.46	0.28	26.51	10.37	6.46	"
1.31	0.18	0.11	27.65	10.68	6.66	"
-----	----	----	28.25	10.81	6.74	"

[a] These values were calculated by the compiler.

[b] The solid phases are: A = $Na_3PO_4 \cdot 12H_2O$; B = Na_3PO_4; C = NaCl.

COMMENT: According to the authors, the most concentrated solutions of
Na_3PO_4 did attack the glass of the vessels containing the mixtures.

COMPONENTS:	ORIGINAL MEASUREMENTS:
(1) Trisodium phosphate; Na_3PO_4; [7601-54-9]	Schroeder, W.C.; Berk, A.A.; Gabriel, A.
(2) Disodium sulfate; Na_2SO_4; [7757-82-6]	J. Am. Chem. Soc. 1937, 59, 1783-90.
(3) Water; H_2O; [7732-18-5]	

VARIABLES:	PREPARED BY:
Composition at temperatures of 150° to 350°C.	J. Eysseltová

EXPERIMENTAL VALUES:

Composition of saturated solutions in the Na_3PO_4-Na_2SO_4-H_2O system.

Na_3PO_4			Na_2SO_4			H_2O
g/100g H_2O	mass%[a]	mol/kg[a]	g/100g H_2O	mass%[a]	mol/kg[a]	mass%[a]

temp. = 150°C.

3.1	2.35	0.21	41.5	28.70	2.93	68.95
6.1	4.56	0.41	40.6	27.67	2.88	67.76
9.0	6.63	0.61	39.8	26.75	2.83	66.62
11.9	8.64	0.80	39.1	25.89	2.78	65.47
15.4	10.97	1.04	38.4	24.97	2.74	64.06
21.4	14.77	1.46	37.2	23.46	2.67	61.77
30.5	20.14	2.08	34.6	20.96	2.50	58.90
40.6	25.57	2.77	31.2	18.16	2.27	56.27
51.7	30.97	3.51	27.3	15.25	2.00	53.78
57.4	33.56	3.87	24.9	13.66	1.82	52.78
60.8	35.20	4.06	21.8	11.94	1.59	52.86
65.9	37.28	4.38	20.2	10.85	1.47	51.86
70.9	39.33	4.67	17.7	9.38	1.29	51.29
69.7	39.36	4.50	13.5	7.37	0.97	53.27
67.4	38.65	4.33	12.6	7.00	0.90	54.35
69.4	40.08	4.35	6.6	3.75	0.47	56.17
70.7	41.02	4.36	2.9	1.67	0.20	57.31
74.9	42.58	4.60	1.8	1.02	0.13	56.40
76.5	42.90	4.73	3.3	1.84	0.23	55.27
80.0	44.04	4.94	3.0	1.64	0.21	54.32
79.8	44.26	4.88	0.9	0.50	0.06	55.24

(continued next page)

AUXILIARY INFORMATION

METHOD/APPARATUS/PROCEDURE:	SOURCE AND PURITY OF MATERIALS:
A self constructed high temperature solubility bomb was used. Samples were taken at the operating temperature. The time allowed for equilibration is not specified. Phosphate determinations were made by a colorimetric method using aminonaphtholsulfonic acid (1). Sulfate was determined gravimetrically as $BaSO_4$. Care was taken to avoid adsorption of phosphate.	Merck chemically pure $Na_3PO_4 \cdot 12H_2O$ was used. If necessary, the dodecahydrate was dried at 120°C to give approximately the monohydrate or it was recrystallized at 250°C to give the anhydrous salt. No other details are given.

ESTIMATED ERROR:

The error in the phosphate determination is less than 1%. No other details are given.

REFERENCES:

1. Fiske, C.H.; Subbarow, J.T. J. Biol. Chem. 1925, 66, 375.

COMPONENTS:	ORIGINAL MEASUREMENTS:
(1) Trisodium phosphate; Na_3PO_4; [7601-54-9]	Schroeder, W.C.; Berk, A.A.; Gabriel, A.
(2) Disodium sulfate; Na_2SO_4; [7757-82-6]	J. Am. Chem. Soc. 1937, 59, 1783-90.
(3) Water; H_2O; [7732-18-5]	

EXPERIMENTAL VALUES cont'd:

Composition of saturated solutions in the Na_3PO_4-Na_2SO_4-H_2O system.

Na_3PO_4			Na_2SO_4			H_2O
g/100g H_2O	mass%[a]	mol/kg[a]	g/100g H_2O	mass%[a]	mol/kg[a]	mass%[a]
			temp. = 200°C.			
2.9	2.18	0.20	44.6	30.24	3.15	67.58
5.8	4.28	0.40	44.9	29.79	3.18	65.93
10.0	7.18	0.69	45.6	29.30	3.25	63.52
15.1	10.50	1.05	46.2	28.64	3.31	60.85
22.2	14.79	1.57	47.2	27.86	3.42	57.34
17.9	12.27	1.25	45.8	27.98	3.30	59.75
17.8	12.24	1.24	44.9	27.60	3.23	60.16
20.1	13.70	1.40	43.6	26.63	3.14	59.67
20.5	13.93	1.43	43.8	26.66	3.16	59.41
19.3	13.46	1.31	37.8	24.06	2.71	62.48
20.3	14.10	1.38	37.3	23.67	2.68	62.23
22.5	15.87	1.49	29.2	19.25	2.09	64.88
23.7	16.74	1.56	27.0	17.92	1.93	65.35
23.9	16.96	1.56	25.4	17.01	1.81	66.03
26.4	18.75	1.71	21.2	14.36	1.51	66.88
28.4	20.06	1.83	19.5	13.18	1.39	66.76
29.3	20.88	1.87	16.0	11.01	1.14	68.11
29.5	20.99	1.88	16.1	11.06	1.14	67.95
30.9	21.96	1.96	14.2	9.79	1.01	68.25
35.4	25.16	2.20	7.6	5.31	0.54	69.53
34.6	24.72	2.15	7.6	5.34	0.54	69.93
48.9	32.41	3.01	3.0	1.97	0.21	65.61
56.0	35.82	3.42	0.5	0.32	0.04	63.86
			temp. = 250°C.			
2.9	2.16	0.20	46.6	31.17	3.29	66.67
6.0	4.38	0.41	47.4	30.90	3.36	64.72
8.1	5.82	0.56	48.9	31.15	3.48	63.04
8.0	5.76	0.55	48.5	30.99	3.45	63.25
8.5	6.10	0.59	48.5	30.89	3.45	63.01
9.3	6.70	0.64	45.9	29.57	3.27	63.73
8.4	6.15	0.57	42.7	28.26	3.03	65.59
8.5	6.30	0.57	38.8	26.34	2.75	67.36
8.3	6.24	0.55	35.6	24.74	2.52	69.02
8.0	6.24	0.52	27.3	20.18	1.93	73.58
7.6	5.99	0.49	25.6	19.22	1.81	74.79
7.8	6.32	0.49	19.9	15.58	1.40	78.09
7.6	6.29	0.48	16.3	13.16	1.15	80.55
7.4	6.30	0.46	11.9	9.97	0.84	83.72
7.6	6.52	0.47	10.7	9.04	0.75	84.44
7.4	6.45	0.46	8.5	7.33	0.60	86.22
7.8	6.86	0.48	6.7	5.85	0.47	87.28
8.8	7.83	0.54	4.1	3.63	0.29	88.54
8.1	7.33	0.49	2.7	2.44	0.19	90.24
8.5	7.69	0.52	2.2	1.99	0.15	90.32
9.5	8.60	0.58	1.1	0.99	0.08	90.41
8.3	7.63	0.50	0.5	0.46	0.04	91.91
8.6	7.92	0.52	trace	----	----	92.08

(continued next page)

COMPONENTS:	ORIGINAL MEASUREMENTS:
(1) Trisodium phosphate; Na_3PO_4; [7601-54-9] (2) Disodium sulfate; Na_2SO_4; [7757-82-6] (3) Water; H_2O; [7732-18-5]	Schroeder, W.C.; Berk, A.A.; Gabriel, A. *J. Am. Chem. Soc.* <u>1937</u>, *59*, 1783-90.

EXPERIMENTAL VALUES cont'd:

Composition of saturated solutions in the Na_3PO_4-Na_2SO_4-H_2O system.

Na_3PO_4			Na_2SO_4			H_2O
g/100g H_2O	mass%[a]	mol/kg[a]	g/100g H_2O	mass%[a]	mol/kg[a]	mass%[a]
temp. = 300°C.						
2.0	1.60	0.13	30.7	23.13	2.16	75.27
4.9	3.74	0.32	36.8	25.97	2.60	70.28
3.5	2.84	0.22	25.2	19.58	1.78	77.58
2.7	2.30	0.17	17.8	14.77	1.25	82.93
1.8	1.66	0.11	6.9	6.34	0.48	91.99
1.7	1.63	0.10	2.8	2.68	0.20	95.96
3.7	3.49	0.22	2.4	2.26	0.17	94.25
1.9	1.84	0.12	1.4	1.36	0.10	96.80
temp. = 350°C.						
0.14	0.14	0.01	2.02	1.98	0.01	97.88
0.11	0.11	0.01	1.01	1.00	0.01	98.89
0.21	0.21	0.01	0.10	0.10	0.01	99.69

[a] These values were calculated by the compiler.

COMMENTS: At 150°C the analyses of the solutions containing 51.7 g Na_3PO_4 per 100 g H_2O and greater are said to be not consistent enough to be plotted with the other data. At 200°C $Na_2SO_4 \cdot 2Na_3PO_4$ and $Na_2SO_4 \cdot 5Na_3PO_4$ are supposed to exist as the equilibrium solid phases. At 250°C only the first compound is mentioned.

COMPONENTS:	ORIGINAL MEASUREMENTS:
(1) Trisodium phosphate; Na_3PO_4; [7601-54-9] (2) Disodium carbonate; Na_2CO_3; [497-19-8] (3) Water; H_2O; [7732-18-5]	Korf, D.M.; Balyasnaya, A.M. *Zh. Prikl. Khim.* <u>1941</u>, *14*, 475-7.

VARIABLES:	PREPARED BY:
Composition at 25°C.	J. Eysseltová

EXPERIMENTAL VALUES:

Composition of saturated solutions in the Na_3PO_4-Na_2CO_3-H_2O system at 25°C.

concn of Na_3PO_4			concn of Na_2CO_3			solid
mass %	mol %	mol/kg ^{a}	mass %	mol %	mol/kg ^{a}	phase ^{b}
13.4	16.71	0.94	----	----	----	A
8.95	11.20	0.63	4.91	9.52	0.54	"
7.48	9.55	0.54	8.65	17.10	0.97	"
7.12	9.25	0.53	11.25	22.68	1.30	"
6.78	9.14	0.53	14.91	31.00	1.80	"
6.40	9.05	0.52	19.30	42.10	2.45	A + B
6.34	8.93	0.52	19.32	42.00	2.45	A + B
5.82	8.10	0.47	19.00	40.90	2.38	B
4.73	5.90	0.38	20.03	42.80	2.51	"
3.93	5.41	0.32	20.50	44.60	2.56	"
1.87	2.55	0.15	21.00	44.3	2.57	"
----	----	----	22.75	47.5	2.78	"

^{a} The mol/kg H_2O values were calculated by the compiler.

^{b} The solid phases are: A = $Na_3PO_4 \cdot 12H_2O$; B = $Na_2CO_3 \cdot 10H_2O$.

AUXILIARY INFORMATION

METHOD/APPARATUS/PROCEDURE:	SOURCE AND PURITY OF MATERIALS:
The isothermal method with the use of a thermostat. Na_3PO_4, Na_2CO_3 and NaOH were determined simultaneously by an acido-basic titration (1). The P_2O_5 content was also checked gravimetrically. The composition of the solid phases was determined microscopically and also by Schreinemakers' method.	The $Na_3PO_4 \cdot 12H_2O$ containing 1.80 % NaOH was crystallized three times before use. No details are given about the Na_2CO_3.

	ESTIMATED ERROR:
	The temperature deviation was less than ± 0.2 K.

	REFERENCES:
	1. Smith, J.H. *J. Soc. Chem. Ind.* <u>1917</u>, *36*, 415.

COMPONENTS:	ORIGINAL MEASUREMENTS:
(1) Trisodium phosphate; Na_3PO_4; [7601-54-9] (2) Disodium sulfate; Na_2SO_4; [7757-82-6] (3) Water; H_2O; [7732-18-5]	Ravich, M.I.; Yastrebova, L.F. *Zh. Neorg. Khim.* 1958, 3, 2771-80

VARIABLES:	PREPARED BY:
Composition at 250°C.	J. Eysseltová

EXPERIMENTAL VALUES:

Composition of saturated solutions in the Na_3PO_4-Na_2SO_4-H_2O system at 250°C.

Na_3PO_4		Na_2SO_4		$NaOH$ [b]	solid phase		Na_2HPO_4 [b]
					Na_3PO_4	Na_2SO_4	
mass%	mol/kg [a]	mass%	mol/kg [a]	mass%	mass%	mass%	mass%
7.8	0.52	----	----	1.1	89.7	----	10.3
7.6	0.52	3.0	0.24	0.06	86.9	12.1	1.0
7.5	0.52	5.3	0.43	----	85.0	15.0	----
7.3	0.52	8.0	0.66	----	82.8	17.2	----
7.6	0.55	7.8	0.65	----	82.7	17.3	----
6.8	0.48	7.0	0.57	----	79.2	20.8	----
6.8	0.48	6.9	0.56	----	77.6	22.4	----
7.3	0.52	7.9	0.66	----	76.2	23.8	----
6.7	0.50	12.1	1.05	----	75.2	24.8	----
5.7	0.43	14.1	1.24	----	74.9	25.1	----
6.3	0.51	18.2	1.70	----	73.1	26.9	----
5.6	0.45	19.4	1.82	----	71.3	28.7	----
5.9	0.51	23.6	2.36	----	68.9	31.1	----
5.2	0.47	26.9	2.79	----	58.8	41.2	----
5.6	0.52	28.4	3.03	----	55.4	44.6	----
5.5	0.52	30.4	3.34	----	51.6	48.4	----
5.4	0.51	30.4	3.33	----	20.1	79.9	----
3.3	0.30	30.5	3.24	----	1.0	99.0	----
1.6	0.14	30.9	3.22	----	0	100	----

[a] The mol/kg H_2O values were calculated by the compiler.

[b] These products were produced by hydrolysis.

AUXILIARY INFORMATION

METHOD/APPARATUS/PROCEDURE:	SOURCE AND PURITY OF MATERIALS:
The phosphates and the NaOH produced by hydrolysis reactions were determined by titration with acid. The determination of phosphates was done gravimetrically as $Mg_2P_2O_7$. Sulfates were determined gravimetrically as $BaSO_4$ after removing the phosphates as iron phosphate. The composition of the solid phases was determined from the composition of the phase complex and that of the corresponding saturated solution. Corrections were made for water of evaporation.	The Na_2SO_4 was chemically pure. Solutions of Na_3PO_4 were prepared from twice recrystallized Na_2HPO_4 and approximately 50% solution of chemically pure NaOH.
	ESTIMATED ERROR:
	Temperature was constant to within ± 0.5 K. The error in the solid phase composition was less than ± 0.6%. No other details are given.
	REFERENCES:

COMPONENTS:	ORIGINAL MEASUREMENTS:
(1) Trisodium phosphate; Na_3PO_4; [7601-54-9] (2) Hydrogen peroxide; H_2O_2; [7722-84-1] (3) Water; H_2O; [7732-18-5]	Ukraintseva, E.A. *Izv. Sib. Otd. Akad. Nauk SSSR, Ser. Khim.* <u>1961</u>, *3*, 14-24.

VARIABLES:	PREPARED BY:
Composition at -20° and 0°C.	J. Eysseltová

EXPERIMENTAL VALUES:

Composition of saturated solutions in the Na_3PO_4-H_2O_2-H_2O system.

Na_3PO_4		H_2O_2		H_2O	solid[b]
mass%	mol/kg[a]	mass%	mol/kg[a]	mass%[a]	phase

temp. = -20°C.

Na_3PO_4 mass%	mol/kg	H_2O_2 mass%	mol/kg	H_2O mass%	solid phase
13.1	1.32	26.6	12.97	60.3	A
13.1	1.34	27.5	13.61	59.4	"
10.8	1.10	29.7	14.67	59.5	"
10.4	1.07	30.4	15.09	59.2	"
7.5	0.86	39.3	21.71	53.2	"
8.1	0.85	34.0	17.26	57.9	"
6.7	0.83	44.0	26.23	49.3	"
9.3	1.33	48.2	33.34	42.5	"
9.3	1.67	56.9	49.48	33.8	B
9.3	1.93	60.1	58.30	30.3	"
12.3	3.45	66.0	89.40	21.7	"

temp. = 0°C.

Na_3PO_4 mass%	mol/kg	H_2O_2 mass%	mol/kg	H_2O mass%	solid phase
5.1	0.33	0	0	94.9	C
3.7	0.24	1.5	0.46	94.8	"
4.3	0.28	3.6	1.15	92.1	"
5.3	0.36	5.3	1.74	89.4	"
4.9	0.33	5.4	1.77	89.7	C + D
3.6	0.24	6.9	2.27	89.5	D

(continued next page)

AUXILIARY INFORMATION

METHOD/APPARATUS/PROCEDURE:	SOURCE AND PURITY OF MATERIALS:
The duration of the experiments is given as 3 to 14 hours and more. The compiler supposes that this is the time of equilibration under isothermal conditions. The hydrogen peroxide content was determined by titration with permanganate. The phosphate content was determined gravimetrically as $Mg_2P_2O_7$. The composition of the solid phases was determined by the Schreinemakers' method.	The hydrogen peroxide was chemically pure and free of stabilizers. No other information is given.
	ESTIMATED ERROR: No details are given.
	REFERENCES:

COMPONENTS:	ORIGINAL MEASUREMENTS:
(1) Trisodium phosphate; Na_3PO_4; [7601-54-9]	Ukraintseva, E.A.
(2) Hydrogen peroxide; H_2O_2; [7722-84-1]	*Izv. Sib. Otd. Akad. Nauk SSSR, Ser. Khim.* <u>1961</u>, 3, 14-24.
(3) Water; H_2O; [7732-18-5]	

EXPERIMENTAL VALUES cont'd:

Composition of saturated solutions in the Na_3PO_4-H_2O_2-H_2O system.

Na_3PO_4		H_2O_2		H_2O	solid
mass%	mol/kg[a]	mass%	mol/kg[a]	mass%[a]	phase[b]

temp = 0°C.

mass%	mol/kg	mass%	mol/kg	mass%	phase
4.9	0.35	10.7	3.73	84.4	D
7.3	0.58	15.4	5.86	77.3	"
10.9	0.91	16.5	6.68	72.6	"
11.5	0.99	17.9	7.45	70.6	"
9.4	0.80	19.2	7.90	71.4	"
15.4	1.66	28.0	14.54	56.6	A + D
17.6	1.92	26.7	14.09	55.7	A + D
16.7	1.80	26.7	13.87	56.6	A + D
15.5	1.66	27.5	14.18	57.0	A
16.0	1.72	27.3	14.15	56.7	"
16.1	1.73	27.3	14.18	56.6	"
16.8	1.85	27.8	14.75	55.4	"
16.4	1.80	28.1	14.88	55.5	"
14.9	1.63	29.4	15.52	55.7	"
11.6	1.22	30.6	15.56	57.8	"
11.7	1.26	31.6	16.38	56.7	"
11.8	1.29	32.4	17.07	55.8	"
10.2	1.16	36.3	19.94	53.5	"
10.2	1.17	36.7	20.32	53.1	"
10.3	1.26	39.9	23.55	49.8	"
9.7	1.16	39.2	22.55	51.1	"
11.9	1.51	40.2	24.67	47.9	"
11.1	1.55	45.2	30.40	43.7	B
9.7	1.32	45.7	30.12	44.6	"
9.7	1.48	50.5	37.30	39.8	"
9.6	1.50	51.4	38.74	39.0	"
9.6	1.50	51.5	38.92	38.9	"
12.1	3.15	64.5	81.02	23.4	"

[a] These values were calculated by the compiler.

[b] The solid phases are: A = $Na_3PO_4 \cdot 4.5H_2O_2$; B = the solid decomposes; C = $Na_3PO_4 \cdot 12H_2O$; D = $Na_3PO_4 \cdot H_2O_2$.

COMPONENTS:	ORIGINAL MEASUREMENTS:
(1) Trisodium phosphate; Na_3PO_4; [7601-54-9]	Babayan, G.G.; Sayamyan, E.A.; Darbinyan, G.M.
(2) Disodium silicate; Na_2SiO_3; [6834-92-0]	*Arm. Khim. Zh.* 1970, *23*, 986-9.
(3) Water; H_2O; [7732-18-5]	

VARIABLES:	PREPARED BY:
Composition at 0 and 25°C.	J. Eysseltová

EXPERIMENTAL VALUES:

Solubility isotherms at 0 and 25°C are presented only in graphical form.

The following data are reported in the text:

t/°C	Na_3PO_4 mass%	Na_2SiO_3 mass%	
0	0-0.78	9.83-10.6	$Na_2SiO_3 \cdot 9H_2O$
0	"from 1.04"	10.7-11.13	solid solutions
0	2.39-9.2	8.75-1.04	$Na_3PO_4 \cdot 12H_2O$
25	0-1.02	22.2-22.3	$Na_2SiO_3 \cdot 9H_2O$
25	1.1-3.68	23.1-13.45	solid solutions
25	4.67-11.8	1.44-10.02	$Na_3PO_4 \cdot 12H_2O$

AUXILIARY INFORMATION

METHOD/APPARATUS/PROCEDURE:	SOURCE AND PURITY OF MATERIALS:
The isothermal method was used with equilibrium being attained in 3 days. SiO_2 was determined gravimetrically by precipitation as H_2SiO_3 and decomposition at 1000°C. Phosphate was precipitated with ammonium molybdate. The phosphomolybdate was dissolved in alkali hydroxide and the excess hydroxide was titrated acidimetrically. The composition of the solid phases was determined by Schreinemakers' method but no data are reported.	The sodium phosphate was chemically pure and the sodium metasilicate was reagent grade. The extent of hydration is not specified.
	ESTIMATED ERROR:
	No information is given.
	REFERENCES:

COMPONENTS:	ORIGINAL MEASUREMENTS:
(1) Trisodium phosphate; Na_3PO_4; [7601-54-9] (2) Sodium vanadate; $NaVO_3$; [13718-23-8] (3) Water; H_2O; [7732-18-5]	Abduragimova, R.A.; Rza-Zade, P.F. *Issled. Obl. Neorg. Fiz. Khim.* <u>1971</u>, 179-85.

VARIABLES:	PREPARED BY:
Composition at 25°C.	J. Eysseltová

EXPERIMENTAL VALUES:

Composition of saturated solutions in the Na_3PO_4-$NaVO_3$-H_2O system at 25°C.

Na_3PO_4		$NaVO_3$		
mass%	mol/kg[a]	mass%	mol/kg[a]	solid phase
12.30	0.85	----	----	Na_3PO_4 [b]
----	----	17.40	1.73	$NaVO_3$ [b]
11.0	0.76	0.40	0.04	$Na_3PO_4 \cdot 12H_2O$
8.15	0.55	1.98	0.18	"
7.38	0.50	2.81	0.26	"
5.41	0.36	4.43	0.40	"
4.39	0.30	5.93	0.54	"
3.85	0.26	7.60	0.70	"
3.80	0.26	9.12	0.86	"
4.18	0.30	10.45	1.00	"
5.21	0.38	12.05	1.19	$Na_3PO_4 \cdot 12H_2O + 4Na_2O \cdot P_2O_5 \cdot V_2O_5 \cdot 30H_2O$
4.48	0.33	12.26	1.21	$4Na_2O \cdot P_2O_5 \cdot V_2O_5 \cdot 30H_2O$
3.23	0.24	13.68	1.35	"
2.40	0.17	13.81	1.35	"
1.26	0.09	15.40	1.52	"
1.25	0.09	16.82	1.68	"
1.03	0.08	17.01	1.70	$4Na_2O \cdot P_2O_5 \cdot V_2O_5 \cdot 30H_2O + NaVO_3 \cdot 2H_2O$

[a] The mol/kg H_2O values were calculated by the compiler.

[b] The extent of hydration is ignored.

AUXILIARY INFORMATION

METHOD/APPARATUS/PROCEDURE:	SOURCE AND PURITY OF MATERIALS:
The isothermal method was used but no experimental details are given. The vessels were molybdenum vessels, probably made from molybdenum glass. The determination of alkali metals was made by using 0.5 mol dm^{-3} HCl. The V and P content were determined photocolorimetrically. No details are given.	Chemically pure $Na_3PO_4 \cdot 12H_2O$ and $NaVO_3 \cdot 2H_2O$ were recrystallized before use.
	ESTIMATED ERROR: No information is given.
	REFERENCES:

COMPONENTS:	ORIGINAL MEASUREMENTS:
(1) Trisodium phosphate; Na_3PO_4; [7601-54-9] (2) Disodium sulfate; Na_2SO_4; [7757-82-6] (3) Water; H_2O; [7732-18-5]	Abduragimova, R.A.; Rza-Zade, P.F. *Issled. Obl. Neorg. Fiz. Khim.* <u>1971</u>, 179-85.

VARIABLES:	PREPARED BY:
Composition at 25°C.	J. Eysseltová

EXPERIMENTAL VALUES:

Composition of saturated solutions in the Na_2SO_4-Na_3PO_4-H_2O system at 25°C.

Na_2SO_4		Na_3PO_4		
mass%	mol/kg[a]	mass%	mol/kg[a]	solid phase
21.88	1.97	----	----	Na_2SO_4[b]
----	----	12.2780	0.85	Na_3PO_4[b]
20.7500	1.89	2.1023	0.17	$Na_2SO_4 \cdot 10H_2O$ + $Na_3PO_4 \cdot 12H_2O$
19.4120	1.73	1.7860	0.14	$Na_3PO_4 \cdot 12H_2O$ "
19.3920	1.73	1.7970	0.14	"
15.6640	1.34	2.1920	0.16	"
15.6600	1.34	2.2730	0.17	"
13.5800	1.14	2.6500	0.19	"
13.5679	1.14	2.6190	0.19	"
10.2530	0.84	3.8140	0.27	"
10.2960	0.84	3.7990	0.27	"
7.4310	0.60	5.1920	0.36	"
7.3830	0.60	5.2710	0.37	"
4.4920	0.36	6.9630	0.48	"
4.4318	0.35	7.0250	0.48	"
1.5130	0.12	11.2200	0.78	"
1.4570	0.12	10.672	0.74	"
----	----	10.613	0.72	"

[a] The mol/kg H_2O values were calculated by the compiler.

[b] The extent of hydration was apparently ignored (compiler).

AUXILIARY INFORMATION

METHOD/APPARATUS/PROCEDURE:	SOURCE AND PURITY OF MATERIALS:
The isothermal method was used but no details are given. Molybdenum vessels were used. The alkali metal content was determined using 0.5 mol dm^{-3} HCl. Sulfate was determined gravimetrically as $BaSO_4$. Phosphate was determined colorimetrically.	Both salts were of chemically pure grade and were recrystallized before being used.

	ESTIMATED ERROR:
	No details are given.

	REFERENCES:

COMPONENTS:	ORIGINAL MEASUREMENTS:
(1) Trisodium phosphate; Na_3PO_4; [7601-54-9]	Babayan, G.G.; Darbinyan, G.M.
(2) Sodium perchlorate; $NaClO_4$; [7601-89-0]	*Arm. Khim. Zh.* <u>1972</u>, 25, 482-7.
(3) Water; H_2O, [7732-18-5]	

VARIABLES:	PREPARED BY:
Composition at 20°C.	J. Eysseltová

EXPERIMENTAL VALUES:

Composition of saturated solutions in the
Na_3PO_4-$NaClO_4$-H_2O system at 20°C.

Na_3PO_4		$NaClO_4$		H_2O	
mass %	mol/kg ^{a}	mass %	mol/kg ^{a}	mass % ^{a}	solid phase
0.39	0.06	58.40	11.58	41.21	$NaClO_4 \cdot H_2O$
0.40	0.05	52.80	9.21	46.80	solid solution
0.47	0.06	47.70	7.52	51.83	" "
2.14	0.23	41.60	6.04	56.26	" "
1.80	0.19	41.35	5.94	56.85	" "
2.54	0.25	35.87	4.76	61.59	" "
2.90	0.28	34.70	4.54	62.40	" "
3.20	0.27	26.00	3.00	70.80	$Na_3PO_4 \cdot 12H_2O$
4.80	0.40	21.80	2.42	73.40	"
12.46	0.97	9.80	1.03	77.74	"
18.42	1.45	4.48	0.47	77.10	"
14.46	1.14	8.20	0.86	77.34	"
23.90	1.95	1.60	0.18	74.50	"

^{a} The mass % of H_2O and the mol/kg H_2O values were calculated by the compiler.

AUXILIARY INFORMATION

METHOD/APPARATUS/PROCEDURE:	SOURCE AND PURITY OF MATERIALS:
The isothermal method was used with 5 to 7 days being allowed for equilibration. The phosphate content was determined with the use of ammonium molybdate. The perchlorate content was determined gravimetrically but no further details are given. The composition of the solid phases was determined by the Schreinemakers' method.	The Na_3PO_4 was reagent grade and the $NaClO_4$ is described as pure.
	ESTIMATED ERROR:
	No information is given.
	REFERENCES:

COMPONENTS:	ORIGINAL MEASUREMENTS:
(1) Trisodium phosphate; Na_3PO_4; [7601-54-9] (2) Disodium tetraborate; $Na_2B_4O_7$; [1330-43-4] (3) Water; H_2O; [7732-18-5]	Babayan, G.G.; Darbinyan, G.M. *Arm. Khim. Zh.* <u>1972</u>, *25*, 482-7.

VARIABLES:	PREPARED BY:
Composition at 20°C.	J. Eysseltová

EXPERIMENTAL VALUES:

Composition of saturated solutions in the
Na_3PO_4-$Na_2B_4O_7$-H_2O system at 20°C.

Na_3PO_4		$Na_2B_4O_7$		H_2O	
mass %	mol/kg [a]	mass %	mol/kg [a]	mass % [a]	solid phase
2.46	0.16	4.80	0.26	92.74	$Na_2B_4O_7 \cdot 10H_2O$
3.10	0.20	4.30	0.23	92.60	"
3.09	0.20	4.49	0.24	92.42	"
4.27	0.28	4.50	0.24	91.23	"
5.09	0.34	3.90	0.21	91.01	solid solution
5.34	0.36	3.77	0.21	90.89	" "
5.83	0.39	3.57	0.20	90.60	" "
6.28	0.42	3.37	0.18	90.35	" "
7.39	0.50	3.40	0.19	89.21	" "
7.82	0.53	3.39	0.19	88.79	" "
8.18	0.56	3.21	0.18	88.61	" "
8.81	0.61	3.30	0.19	87.89	" "
9.34	0.65	3.15	0.18	87.51	" "
10.00	0.70	3.17	0.18	86.83	" "
11.75	0.84	3.24	0.19	85.01	" "
10.95	0.77	2.92	0.17	86.13	" "
11.68	0.83	2.60	0.15	85.72	" "
15.20	1.12	2.30	0.14	82.50	$Na_3PO_4 \cdot 12H_2O$
16.75	1.26	2.70	0.17	80.55	"
16.80	1.26	1.84	0.11	81.36	"

[a] These values were calculated by the compiler.

AUXILIARY INFORMATION

METHOD/APPARATUS/PROCEDURE:	SOURCE AND PURITY OF MATERIALS:
The isothermal method was used with 5-7 days being allowed for equilibration. Phosphate content was determined by the use of ammonium molybdate. The tetraborate content was determined titrimetrically (no other details are given). The composition of the solid phases was determined by the Schreinemakers' method.	The sodium phosphate and sodium tetraborate were reagent grade materials.
	ESTIMATED ERROR: No information is given.
	REFERENCES:

COMPONENTS:	ORIGINAL MEASUREMENTS:
(1) Trisodium phosphate, Na_3PO_4; [7601-54-9]	Druzhinin, I.G.; Tusheva, L.A.
(2) Copper phosphate; $Cu_3(PO_4)_2$; [30981-48-7]	*Izv. Vuzov, Khim. Khim. Tekhnol*, <u>1974</u>, *17*,
(3) Water; H_2O; [7732-18-5]	1513-6.

VARIABLES:	PREPARED BY:
Composition at 25 and 50°C.	J. Eysseltová

EXPERIMENTAL VALUES:

Composition of saturated solutions in the
Na_3PO_4-$Cu_3(PO_4)_2$-H_2O system.

$Cu_3(PO_4)_2$		Na_3PO_4		
mass%	mol/kg[a]	mass%	mol/kg[a]	solid phase[b]
		temp. = 25°C.		
0.067	0.0018	----	----	A
0.062	0.0017	3.93	0.25	A + B
0.054	0.0015	6.05	0.39	B
0.038	0.0011	9.01	0.60	"
0.029	0.0009	13.41	0.94	B + C
----	----	13.42	0.94	C
		temp. = 50°C.		
0.21	0.0055	----	----	A
0.20	0.0054	4.02	0.25	"
0.19	0.0054	7.48	0.49	"
0.19	0.0054	7.49	0.49	B
0.07	0.0024	22.71	1.79	"
0.07	0.0024	22.71	1.79	B + D
0.07	0.0024	22.71	1.79	D
----	----	22.71	1.79	"

[a] The mol/kg H_2O values were calculated by the compiler.

[b] The solid phases are: A = $Cu_3(PO_4)_2$; B = $Na_3PO_4 \cdot Cu_3(PO_4)_2$; C = $Na_3PO_4 \cdot 12H_2O$;

D = $Na_3PO_4 \cdot 10H_2O$.

COMMENT: At 50°C solid phase B hydrolyzes to CuO in the region where it crystallizes out.

AUXILIARY INFORMATION

METHOD/APPARATUS/PROCEDURE:	SOURCE AND PURITY OF MATERIALS:
The isothermal method was used. Solid sodium phosphate was added to saturated solutions of copper phosphate and the mixtures were equilibrated for 8-10 hours. The copper content was determined as CuO. The phosphate content was determined colorimetrically and the sodium content was determined as sodium zinc uranyl acetate after separation with the aid of an ion exchanger.	Both the phosphates were of reagent grade quality.

	ESTIMATED ERROR:
	The temperature was held constant to within ± 0.1 K.

	REFERENCES:

COMPONENTS:	ORIGINAL MEASUREMENTS:
(1) Trisodium phosphate; Na_3PO_4; [7601-54-9] (2) Tripotassium phosphate; K_3PO_4; [7778-53-2] (3) Water; H_2O; [7732-18-5]	Manvelyan, M.G.; Galstyan, V.D.; Voskanyan, S.S. *Arm. Khim. Zh.* <u>1974</u>, *26*, 810.

VARIABLES:	PREPARED BY:
Composition at 20°C.	J. Eysseltová

EXPERIMENTAL VALUES:

The only information given is the limiting concentrations for the existence of individual solid phases:

0-26.85 mass% K_3PO_4 in liquid phase--$Na_3PO_4 \cdot 12H_2O$ crystallizes

31.53-47-63 mass% K_3PO_4 in liquid phase--solid solutions

from 49.40 mass% K_3PO_4 in liquid phase--$K_3PO_4 \cdot 7H_2O$ crystallizes

AUXILIARY INFORMATION

METHOD/APPARATUS/PROCEDURE:	SOURCE AND PURITY OF MATERIALS:
The isothermal method was used. The liquid phases were analyzed for Na_2O, K_2O and P_2O_5 (1).	No information is given.
	ESTIMATED ERROR: No information is given.
	REFERENCES: 1. Manvelyan, M.G.; Galstyan, V.D.; Voskanyan, S.S. *VINITI* 2323-74 Dep., <u>1974</u>.

COMPONENTS:	ORIGINAL MEASUREMENTS:
(1) Trisodium phosphate; Na_3PO_4; [7601-54-9] (2) Sodium nitrite; $NaNO_2$; [7632-00-0] (3) Water; H_2O; [7732-18-5]	Procenko, P.I.; Ivleva, T.I.; Rubleva, V.V.; Berdyukova, V.A.; Edush, T.V. Zh. Prikl. Khim. 1975, 48, 1055-9.

VARIABLES:	PREPARED BY:
Composition at 25°C.	J. Eysseltová

EXPERIMENTAL VALUES:

Composition of saturated solutions in the $NaNO_2$-Na_3PO_4-H_2O system at 25°C.

concn of $NaNO_2$			concn of Na_3PO_4			solid [b]
mass %	mol/1000 mol H_2O	mol/kg [a]	mass %	mol/1000 mol H_2O	mol/kg [a]	phase
----	----------	----	12.24	15.3000	0.85	A
4.10	12.24	0.68	8.53	10.7130	0.60	"
11.90	37.43	2.08	5.08	6.7500	0.38	"
25.49	91.58	5.09	1.90	2.8757	0.16	"
36.92	156.31	8.68	1.46	2.5991	0.14	
44.15	206.20	11.46	0.71	1.3955	0.08	A + B
44.30	207.49	11.53	0.75	1.4867	0.08	A + B
44.55	210.65	11.70	0.28	0.5570	0.03	B
45.76	220.33	12.24	----	------	----	"

[a] These values were calculated by the compiler.

[b] The solid phases are: A = $Na_3PO_4 \cdot 12H_2O$; B = $NaNO_2$.

AUXILIARY INFORMATION

METHOD/APPARATUS/PROCEDURE:	SOURCE AND PURITY OF MATERIALS:
Isothermal method with the use of a water thermostat. About 10-12 hours were allowed for equilibration. Nitrite content was determined iodometrically. Phosphate content was determined gravimetrically by weighing as $Mg_2P_2O_7$. The composition of the solid phases was determined by Schreinemakers' method.	$NaNO_2$ was recrystallized and had a purity of 99.92%. $Na_3PO_4 \cdot 12H_2O$ was recrystallized to give a purity of 99.54% (calculated as anhydrous salt).
	ESTIMATED ERROR: The temperature was controlled to within 0.1K.
	REFERENCES:

COMPONENTS:	ORIGINAL MEASUREMENTS:
(1) Trisodium phosphate; Na_3PO_4; [7601-54-9] (2) Disodium tungstate; Na_2WO_4; [13472-45-2] (3) Water; H_2O; [7732-18-5]	Urusova, M.A.; Balyashko, V.M.; Rakova, N.N.; Zelikman, A.N.; Yevdokimova, G.V. *Zh. Neorg. Khim.* 1975, *20*, 2585-6.

VARIABLES:	PREPARED BY:
One temperature: 230°C Composition	J. Eysseltová

EXPERIMENTAL VALUES:

Composition of saturated solutions of the Na_3PO_4-Na_2WO_4-H_2O system at 230°C.

concn of Na_2WO_4		concn of Na_3PO_4		concn of H_2O	
mass %	mol/kg [a]	mass %	mol/kg [a]	mass % [a]	solid phase
----	----	18.9	1.42	81.1	Na_3PO_4
----	----	19.3	1.46	80.7	"
5.6	0.24	15.6	1.21	78.8	"
8.2	0.36	15.0	1.20	76.8	"
10.6	0.47	13.3	1.06	76.1	"
16.5	0.79	12.1	1.03	71.4	"
22.4	1.14	10.6	0.96	67.0	"
30.2	1.70	9.4	0.95	60.4	"
34.9	2.05	7.1	0.74	58.0	"
37.5	2.32	7.6	0.84	54.9	"
44.3	3.05	6.3	0.78	49.4	Na_3PO_4 + Na_2WO_4
44.7	3.14	6.8	0.85	48.5	"
44.3	3.04	6.2	0.76	49.5	"
45.1	3.07	4.9	0.60	50.0	Na_2WO_4
45.7	3.07	3.6	0.43	50.7	"
46.9	3.01	----	----	53.1	"

[a] These values were calculated by the compiler.

AUXILIARY INFORMATION

METHOD/APPARATUS/PROCEDURE:	SOURCE AND PURITY OF MATERIALS:
Isothermal method, self-constructed apparatus and sampling device. Analyses: phosphorus was determined as $Mg_2P_2O_7$ in presence of 3 g limonic acid per 1 g WO_3; tungstate was determined gravimetrically as WO_3 after removing of the phosphates.	Na_2WO_4 and Na_3PO_4 were prepared by drying of unspecified hydrates at 150-200°C and 250-300°C respectively.
	ESTIMATED ERROR: Nothing given.
	REFERENCES:

COMPONENTS:	ORIGINAL MEASUREMENTS:
(1) Trisodium phosphate; Na_3PO_4; [7601-54-9]	Gyunashyan, A.P.
(2) Disodium carbonate; Na_2CO_3; [497-19-8]	Arm. Khim. Zh. 1979, 32, 868-73.
(3) Water; H_2O; [7732-18-5]	

VARIABLES:	PREPARED BY:
Composition at 0° and 20°C.	J. Eysseltová

EXPERIMENTAL VALUES:

Solubility isotherms at 0° and 20°C are given only in graphical form.

The following numerical values are given only for solutions saturated with $Na_3PO_4 \cdot 12H_2O$ and $Na_2CO_3 \cdot 10H_2O$ simultaneously.

	concn of Na_3PO_4		concn of Na_2CO_3		concn of H_2O
$t/°C$	mass %	mol/kg [a]	mass %	mol/kg [a]	mass %
0	2.60	0.17	5.70	0.59	91.70
20	6.35	0.49	14.30	1.70	79.35

[a] The mol/kg H_2O values were calculated by the compiler.

AUXILIARY INFORMATION

METHOD/APPARATUS/PROCEDURE:	SOURCE AND PURITY OF MATERIALS:
Nothing is specified.	Nothing is specified.
	ESTIMATED ERROR:
	Nothing is given.
	REFERENCES:

COMPONENTS:	ORIGINAL MEASUREMENTS:
(1) Trisodium phosphate; Na_3PO_4; [7601-54-9] (2) Disodium silicate; Na_2SiO_3; [6834-92-0] (3) Water; H_2O; [7732-18-5]	Gyunashyan, A.P. *Arm. Khim. Zh.* <u>1979</u>, *32*, 868-73.

VARIABLES:	PREPARED BY:
Composition at 20°C.	J. Eysseltová

EXPERIMENTAL VALUES:

The solubility isotherm for the Na_2SiO_3-Na_3PO_4-H_2O system at 20°C is given in graphical form. The composition of the solution in equilibrium with $Na_2SiO_3 \cdot 9H_2O$ and $Na_3PO_4 \cdot 12H_2O$ simultaneously is specified as 16.80 mass% Na_2SiO_3 (1.75 mol/kg--compiler) and 4.75 mass% Na_3PO_4 (0.37 mol/kg--compiler).

AUXILIARY INFORMATION

METHOD/APPARATUS/PROCEDURE:	SOURCE AND PURITY OF MATERIALS:
No details are given.	No information is given.
	ESTIMATED ERROR: No information is given.
	REFERENCES:

COMPONENTS:	ORIGINAL MEASUREMENTS:
(1) Trisodium phosphate; Na_3PO_4; [7601-54-9] (2) Sodium fluoride; NaF; [7681-49-4] (3) Water; H_2O; [7732-18-5]	Roslyakova, O.N.; Petrov, M.R.; Zhikharev, M.I. Zh. Neorg. Khim. <u>1979</u>, 24, 206-8.

VARIABLES:	PREPARED BY:
Composition at 25°C.	J. Eysseltová

EXPERIMENTAL VALUES:

Composition of saturated solutions in the Na_3PO_4-NaF-H_2O system at 25°C.

NaF		Na_3PO_4		H_2O			solid$_b$
mass%	mol/kg^a	mass%	mol/kg^a	mass%	d/kg m^{-3}	ν/m^2 s^{-1}	phase^b
3.77	0.93	----	----	96.23	1.035	1.108	A
2.45	0.60	0.25	0.02	97.30	1.037	1.109	"
2.04	0.50	0.49	0.03	97.47	1.039	1.122	"
1.77	0.43	0.82	0.05	97.41	1.041	1.130	"
1.54	0.38	1.43	0.09	97.03	1.045	1.138	A + B
1.41	0.35	1.55	0.10	97.04	1.044	1.135	B
1.18	0.29	1.84	0.12	96.98	1.045	1.138	"
0.96	0.24	2.29	0.14	96.75	1.045	1.138	"
0.80	0.20	2.79	0.18	96.41	1.046	1.140	"
0.67	0.16	2.96	0.19	96.37	1.048	1.152	"
0.52	0.13	3.65	0.23	95.83	1.052	1.168	"
0.41	0.10	4.49	0.29	95.10	1.057	1.175	"
0.34	0.09	5.21	0.34	94.25	1.063	1.190	"
0.26	0.06	5.91	0.38	93.83	1.070	1.231	"
0.21	0.05	7.00	0.46	92.79	1.078	1.316	"
0.20	0.05	9.05	0.61	90.75	1.103	1.458	"
0.18	0.05	9.32	0.63	90.50	1.106	1.519	B + C
0.10	0.03	10.00	0.68	89.90	1.116	1.581	C
0.05	0.01	10.62	0.72	89.33	1.122	1.617	"
----	----	12.38	0.86	87.62	1.152	1.911	"

a These values were calculated by the compiler.

b The solid phases are: A = NaF; B = solid solution; C = Na_3PO_4 (probably $Na_3PO_4 \cdot 12H_2O$-compiler).

AUXILIARY INFORMATION

METHOD/APPARATUS/PROCEDURE:	SOURCE AND PURITY OF MATERIALS:
The isothermal method was used. Equilibration required 8 to 10 hours. Fluoride content was determined colorimetrically (1), after distillation as H_2SiF_6. The phosphate content was determined colorimetrically with ammonium molybdate and reduction of the complex formed with ascorbic acid. The composition of the solid phases was determined microscopically and with the use of the Schreinemakers' method.	No information is given.

	ESTIMATED ERROR: Nothing is stated.
	REFERENCES: 1. Kukisheva, T.N.; Sinicyna, E.S.; Efimova, N.S. Zh. Anal. Khim. <u>1971</u>, 26, 954.

COMPONENTS:	ORIGINAL MEASUREMENTS:
(1) Trisodium phosphate; Na_3PO_4; [7601-54-9]	Schroeder, W.C.; Berk, A.A.; Gabriel, A.
(2) Sodium hydroxide; NaOH; [1310-73-2]	J. Am. Chem. Soc. 1937, 59, 1783-90.
(3) Disodium sulfate; Na_2SO_4; [7757-82-6]	
(4) Water; H_2O, [7732-18-5]	

VARIABLES:	PREPARED BY:
Composition and temperature.	J. Eysseltová

EXPERIMENTAL VALUES:

Composition of saturated solutions in the Na_2SO_4-Na_3PO_4-NaOH-H_2O system.

NaOH			Na_3PO_4			Na_2SO_4			H_2O
w^a	mass%^{b}	mol/kg^b	w^a	mass%^{b}	mol/kg^b	w^a	mass%^{b}	mol/kg^b	mass%^{b}
				temp. = 150°C					
7.8	5.59	1.95	3.0	2.15	0.18	28.7	20.57	2.02	71.68
7.9	5.55	1.98	6.0	4.22	0.36	28.4	19.96	2.00	70.27
8.0	5.53	2.00	8.6	5.94	0.52	28.1	19.42	1.98	69.11
7.6	5.07	1.90	14.7	9.80	0.90	27.7	18.45	1.95	66.67
19.1	13.84	4.78	2.9	2.10	0.18	16.0	11.59	1.13	72.46
19.8	13.98	4.95	5.9	4.17	0.36	15.9	11.29	1.12	70.62
19.5	13.53	4.88	8.7	6.04	0.53	15.9	11.03	1.12	69.40
19.8	13.39	4.95	12.2	8.25	0.74	15.9	10.75	1.12	67.61
				temp. = 250°C					
8.1	5.50	2.02	1.5	1.02	0.09	37.7	25.59	2.65	67.89
8.0	5.33	2.00	3.0	2.00	0.18	39.1	26.05	2.75	66.62
8.1	5.36	2.02	3.7	2.45	0.22	39.3	26.01	2.77	66.18
7.9	5.22	1.98	3.8	2.51	0.23	39.5	26.12	2.78	66.14
8.1	5.35	2.02	3.8	2.51	0.23	39.4	26.04	2.77	66.09
8.2	5.60	2.05	4.0	2.73	0.24	34.1	23.31	2.40	68.35
8.4	6.14	2.10	3.5	2.56	0.21	25.0	18.26	1.76	73.05
8.4	6.47	2.10	3.4	2.62	0.21	18.0	13.87	1.27	77.04
8.4	6.84	2.10	3.3	2.69	0.20	11.0	8.96	0.77	81.50

(continued next page)

AUXILIARY INFORMATION

METHOD/APPARATUS/PROCEDURE:

A high temperature bomb was used. The samples were withdrawn at the operating temperature. The time allowed for equilibration is not given. Phosphate determinations were made colorimetrically using aminonaphtholsulfonic acid (1). Hydroxide content was determined by titration to the methyl red end point (2 equivalents/mol of phosphate present were deducted). Sulfate was determined gravimetrically as $BaSO_4$.

SOURCE AND PURITY OF MATERIALS:

Merck chemically pure $Na_3PO_4 \cdot 12H_2O$ was used. The actual phosphate content of this material was determined by analysis but the results are not reported. If necessary, the dodecahydrate was dried at 120°C to give approximately the monohydrate, or it was recrystallized at 250°C to give the anhydrous salt. No other information is given.

ESTIMATED ERROR:

The error in the phosphate determination is less than 1%. No other details are given.

REFERENCES:

1. Fiske, C.H.; Subbarow, J.T. J. Biol. Chem. 1925, 66, 375.

COMPONENTS:	ORIGINAL MEASUREMENTS:
(1) Trisodium phosphate; Na_3PO_4; [7601-54-9]	Schroeder, W.C.; Berk, A.A.; Gabriel, A.
(2) Sodium hydroxide; NaOH; [1310-73-2]	J. Am. Chem. Soc. 1937, 59, 1783-90.
(3) Disodium sulfate; Na_2SO_4; [7757-82-6]	
(4) Water; H_2O; [7732-18-5]	

EXPERIMENTAL VALUES cont'd:

Composition of saturated solutions in the Na_2SO_4-Na_3PO_4-NaOH-H_2O system.

NaOH			Na_3PO_4			Na_2SO_4			H_2O
w^a	mass%^{b}	mol/kg^b	w^a	mass%^{b}	mol/kg^b	w^a	mass%^{b}	mol/kg^b	mass%^{b}
				temp. = 250°C					
8.3	7.31	2.08	2.7	2.38	0.16	2.6	2.29	0.18	88.03
8.4	7.40	2.10	4.9	4.32	0.30	0.2	0.18	0.01	88.10
20.0	12.86	5.00	2.7	1.74	0.16	32.8	21.09	2.31	64.31
19.8	12.74	4.95	2.8	1.80	0.17	32.8	21.11	2.31	64.35
20.0	12.93	5.00	2.9	1.87	0.18	31.8$_c$	20.56	2.24	64.64
20.6	13.79	5.15	2.7	1.81	0.16	26.1^c	17.47	1.84	66.93
20.5	14.52	5.12	2.7	1.91	0.16	18.0	12.75	1.27	70.82
20.7	15.40	5.17	2.7	2.01	0.16	11.0$_c$	8.17	0.78	74.40
20.6	15.98	5.15	2.7	2.09	0.16	5.6^c	4.34	0.39	77.58
				temp. = 350°C					
8.0	6.17	2.00	0.6	0.46	0.04	21.0^c	16.20	1.48	77.16
8.0	6.17	2.00	0.6	0.46	0.04	21.0^c	16.20	1.48	77.16
8.0	6.23	2.00	0.4	0.31	0.02	20.0^c	15.58	1.41	77.88
8.0	6.55	2.00	0.4	0.33	0.02	13.8^c	11.29	0.97	81.83
8.0	7.02	2.00	0.3	0.26	0.02	5.7^c	5.00	0.40	87.72
8.4	7.46	2.10	0.3	0.27	0.02	3.9$_c$	3.46	0.27	88.81
8.0	7.26	2.00	0.3	0.27	0.02	1.9^c	1.72	0.13	90.74
21.0	11.85	5.25	1.9	1.07	0.12	54.3$_c$	30.64	3.82	56.43
21.0	11.73	5.25	2.6	1.45	0.16	55.4^c	30.95	3.90	55.86
21.0	11.71	5.25	2.7	1.50	0.16	56.6^c	31.01	3.91	55.77
21.0	11.84	5.25	3.1	1.75	0.19	53.3^c	30.04	3.75	56.37
21.3	12.52	5.33	2.2	1.29	0.13	46.6^c	27.40	3.28	58.79
21.0	13.60	5.25	1.8	1.16	0.11	31.6^c	20.47	2.22	64.77
21.0	14.72	5.25	1.4	0.98	0.08	20.3^c	14.22	1.43	70.08
21.0	15.61	5.25	1.2	0.89	0.07	12.3^c	9.14	0.86	74.35
21.0	16.63	5.25	1.0	0.79	0.06	4.3^c	3.40	0.30	79.18
21.0	17.06	5.25	1.4	1.14	0.08	0.7	0.57	0.05	81.23

a This concentration is expressed as g/100g H_2O.

b These values were calculated by the compiler.

c These values were calculated by the authors from the initial concentrations.

COMPONENTS:	ORIGINAL MEASUREMENTS:
(1) Trisodium phosphate; Na_3PO_4; [7601-54-9] (2) Disodium carbonate; Na_2CO_3; [497-19-8] (3) Sodium hydroxide; NaOH; [1310-73-2] (4) Water; H_2O, [7732-18-5]	Kobe, K.A.; Leipper, A. *Ind. Eng. Chem.* 1940, *32*, 198-203.

VARIABLES:	PREPARED BY:
Temperature, ratio of Na_3PO_4/Na_2CO_3 at a fixed ratio of $Na_3PO_4/NaOH \overset{\circ}{=} 7/1$.	J. Eysseltová

EXPERIMENTAL VALUES:

Composition of saturated solutions in the[a]
$Na_3PO_4 \cdot 1/7NaOH - Na_2CO_3 - H_2O$ system.

	concn of $Na_3PO_4 \cdot 1/7NaOH$			concn of Na_2CO_3			solid[d]
$t/°C$	g/kg[b]	mass %[c]	mol/kg[c]	g/kg[b]	mass %[c]	mol/kg[c]	phase
-2.48	1.8	1.68	0.10	5.5	5.12	0.52	A + B
-2.10	0.0	0.0	0.0	6.1	5.75	0.58	B
-1.21	4.2	4.03	0.27	0.0	0.0	0.0	A
0	4.58	4.38	0.27	0.00	0.00	0.00	A
	2.58	2.37	0.15	6.43	5.90	0.61	A + B
	0.00	0.00	0.00	6.93	6.48	6.65	B
25	11.9	10.63	0.70	0.00	0.00	0.00	A
	10.7	9.36	0.63	3.60	3.15	0.34	"
	9.30	8.00	0.55	6.96	5.99	0.66	"
	8.05	6.65	0.47	13.0	10.74	1.23	"
	7.01	5.54	0.41	19.4	15.35	1.83	"
	5.79	4.33	0.34	28.0	20.93	2.64	A + B
	3.44	2.61	0.20	28.5	21.60	2.69	B
	0.00	0.00	0.00	29.4	22.72	2.77	"
40	20.8	17.22	1.22	0.00	0.00	0.00	A
	15.1	11.60	0.89	15.1	11.60	1.42	"
	11.6	7.91	0.68	35.0	23.87	3.30	"
	11.1	7.20	0.65	43.1	27.95	4.07	A + C
	0.0	0.0	0.0	49.2	32.98	4.64	C

(continued next page)

AUXILIARY INFORMATION

METHOD/APPARATUS/PROCEDURE:	SOURCE AND PURITY OF MATERIALS:
The isothermal method was used. Samples were withdrawn through a coarse filter paper into a weighed 10 ml pipet, weighed, diluted and analyzed acidimetrically (1). The cryohydric points of the system were found by adding the solid salts to ice and measuring the temperature with a Beckmann thermometer. When a constant minimum value was reached, samples were withdrawn and analyzed.	Baker's sodium phosphate was used. Analysis showed it had a constant composition of $Na_3PO_4 \cdot 1/7NaOH$. The Na_2CO_3 was Baker's anhydrous. For determinations at 0° and 25°C the decahydrate was prepared.

	ESTIMATED ERROR:
	The temperature regulation was: 0 ± 0.05°C; 25 ± 0.05°C; 40 ± 0.1°C; 60 ± 0.1°C; 80 ± 0.3°C; 100 ± 0.3°C. No other details are given.

	REFERENCES:
	1. Smith, J.H. *J. Soc. Chem. Ind.* 1917, *36*, 415.

COMPONENTS:	ORIGINAL MEASUREMENTS:
(1) Trisodium phosphate; Na_3PO_4; [7601-54-9]	Kobe, K.A.; Leipper, A.
(2) Disodium carbonate; Na_2CO_3; [497-19-8]	*Ind. Eng. Chem.* <u>1940</u>, *32*, 198-203.
(3) Sodium hydroxide; NaOH; [1310-73-2]	
(4) Water, H_2O; [7732-18-5]	

EXPERIMENTAL VALUES cont'd.

Composition of saturated solutions in the [a]
$Na_3PO_4 \cdot 1/7NaOH-Na_2CO_3-H_2O$ system.

	concn of $Na_3PO_4 \cdot 1/7NaOH$			concn of Na_2CO_3			
$t/°C$	g/kg[b]	mass %[c]	mol/kg[c]	g/kg[b]	mass %[c]	mol/kg[c]	solid phase[d]
60	41.8	29.48	2.46	0.0	0.0	0.0	A
	36.6	24.78	2.15	11.1	7.52	1.05	"
	31.0	20.06	1.82	23.5	15.21	2.22	"
	28.0	17.59	1.65	31.2	19.60	2.94	A + C
	11.4	7.53	0.67	40.0	26.42	3.77	C
	0.0	0.0	0.0	46.3	31.65	4.37	"
80	63.8	38.95	3.76	0.0	0.0	0.0	A
	52.3	30.35	3.08	20.0	11.61	1.89	A + C
	0.0	0.0	0.0	45.1	31.08	4.25	C
100	90.0	47.36	5.30	0.0	0.0	0.0	A
	88.0	44.24	5.18	10.9	5.48	1.03	A + C
	67.1	36.37	3.95	17.4	9.43	1.64	C
	62.1	34.35	3.66	18.7	10.34	1.76	"
	39.0	23.40	2.30	27.7	16.62	2.61	"
	23.0	14.54	1.35	35.2	22.25	3.32	"
	0.0	0.0	0.0	44.8	30.94	4.23	"

[a] For more information on the phosphate component see the Critical Evaluation.

[b] This is an obvious error. According to the compiler it should be g/100 g H_2O.

[c] These values were calculated by the compiler on the assumption given in footnote [b].

[d] The solid phases are: A = Na_3PO_4 (the NaOH and H_2O content are not specified);
B = $Na_2CO_3 \cdot 10H_2O$; C = $Na_2CO_3 \cdot H_2O$.

COMPONENTS:	ORIGINAL MEASUREMENTS:
(1) Trisodium phosphate; Na_3PO_4; [7601-54-9]	Abduragimova, R.A.; Rza-Zade, P.F.;
(2) Sodium aluminate; $NaAlO_2$; [1302-42-7]	Abduragimov, A.A.
(3) Sodium vanadate; $NaVO_3$; [13718-23-8]	Dokl. Akad. Nauk Azerb. SSR 1971, 27, 41-5.
(4) Sodium hydroxide; NaOH; [1310-73-2]	
(5) Water; H_2O; [7732-18-5]	

VARIABLES:	PREPARED BY:
Composition at 25°C and one ratio of $NaAlO_2/NaOH = 1$.	J. Eysseltová

EXPERIMENTAL VALUES:

Composition of saturated solutions in the
$(NaAlO_2 + NaOH)-NaVO_3-Na_3PO_4-H_2O$ system at 25°C.

$NaAlO_2$ + NaOH		$NaVO_3$		Na_3PO_4		H_2O	solid[a]
mass%	mol/kg[b]	mass%	mol/kg[b]	mass%	mol/kg[b]	mass%[b]	phase
----	----	17.42	1.73	----	----	82.58	A
----	----	----	----	12.30	0.85	87.70	B
----	----	12.05	1.19	5.21	0.38	82.74	B + C
----	----	17.00	1.72	1.92	0.14	81.08	A + C
32.64	4.09	1.99	0.25	----	----	65.37	A + D
29.60	3.45	----	----	0.66	0.06	69.74	B + E
42.17	6.00	0.22	0.03	----	----	57.61	D
40.77	5.69	0.49	0.07	----	----	58.74	"
46.48	7.16	----	----	0.26	0.03	53.26	D + E
42.11	6.00	----	----	0.36	0.04	57.53	"
40.09	5.54	0.29	0.04	0.26	0.03	59.36	"
34.27	4.32	0.44	0.06	0.18	0.02	65.11	"
30.13	3.58	0.59	0.07	0.21	0.02	69.07	"
----	----	12.49	1.23	4.35	0.32	83.16	C
----	----	16.01	1.59	1.31	0.10	82.68	"
----	----	1.00	0.09	9.03	0.61	89.97	B
----	----	4.43	0.40	5.41	0.36	90.16	"
----	----	10.82	1.05	4.40	0.32	84.78	"
1.98	0.20	13.61	1.35	1.49	0.11	82.92	A + C + F
21.36	2.27	1.18	0.12	0.19	0.01	77.27	A + F
15.44	1.55	2.40	0.24	0.26	0.02	81.90	"

(continued next page)

AUXILIARY INFORMATION

METHOD/APPARATUS/PROCEDURE:	SOURCE AND PURITY OF MATERIALS:
The isothermal method was used with metallic vessels having mechanical stirrers. The time for equilibration was 155 hours. Saturated solutions were sampled by filtration and analyzed for Na_2O, Al_2O_3, V_2O_5 and P_2O_5 by volumetric, gravimetric, photocolorimetric and nephelometric methods. The composition of the solid phase was determined by Schreinemakers' method.	No information is given.
	ESTIMATED ERROR:
	Nothing is stated.
	REFERENCES:

COMPONENTS:	ORIGINAL MEASUREMENTS:
(1) Trisodium phosphate; Na_3PO_4; [7601-54-9]	Abduragimova, R.A.; Rza-Zade, P.F.; Abduragimov, A.A.
(2) Sodium aluminate; $NaAlO_2$; [1302-42-7]	*Dokl. Akad. Nauk Azerb. SSR* 1971, 27, 41-5.
(3) Sodium vanadate; $NaVO_3$; [13718-23-8]	
(4) Sodium hydroxide; NaOH; [1310-73-2]	
(5) Water; H_2O; [7732-18-5]	

EXPERIMENTAL VALUES cont'd:

Composition of saturated solutions in the
$(NaAlO_2 + NaOH)-NaVO_3-Na_3PO_4-H_2O$ system at 25°C.

$NaAlO_2 + NaOH$		$NaVO_3$		Na_3PO_4		H_2O	solid[a] phase
mass%	mol/kg[b]	mass%	mol/kg[b]	mass%	mol/kg[b]	mass%[b]	
0.84	0.08	9.20	0.88	4.16	0.29	85.80	B + F
0.29	0.02	0.19	0.02	7.34	0.48	92.18	"
23.18	2.51	0.83	0.09	0.27	0.02	75.72	E + F + B
19.93	2.06	0.61	0.06	0.22	0.02	79.24	B + F
13.04	1.24	0.14	0.01	0.41	0.03	86.41	"
1.81	0.17	10.61	1.01	1.09	0.08	86.49	C + F
1.01	0.09	8.25	0.77	2.60	0.18	88.14	"
0.93	0.09	7.00	0.65	4.09	0.28	87.98	B + C
0.63	0.06	1.51	0.09	5.49	0.36	92.37	"

[a] The solid phases are: A = $NaVO_3 \cdot 2H_2O$; B = $Na_3PO_4 \cdot 12H_2O$; C = $4Na_2O \cdot P_2O_5 \cdot V_2O_5 \cdot 30H_2O$; D = $Al_2O_3 \cdot 2Na_2O \cdot 10H_2O$; E = $Al_2O_3 \cdot 2.5Na_2O \cdot 14H_2O$; F = $3Al_2O_3 \cdot 4Na_2O \cdot P_2O_5 \cdot 15H_2O$.

[b] These values were calculated by the compiler.

COMPONENTS:	ORIGINAL MEASUREMENTS:
(1) Trisodium phosphate; Na_3PO_4; [7601-54-9] (2) Sodium aluminate; $NaAlO_2$; [1302-42-7] (3) Sodium hydroxide; NaOH; [1310-73-2] (4) Disodium sulfate; Na_2SO_4; [7757-82-6] (5) Water; H_2O; [7732-18-5]	Abduragimova, R.A.; Rza-Zade, P.F.; Abduragimov, A.A. *Dokl. Akad. Nauk Azerb. SSR* <u>1971</u>, *27*, 41-5.
VARIABLES:	PREPARED BY:
Composition at 25°C and one ratio of $NaAlO_2$/NaOH = 1.	J. Eysseltová

EXPERIMENTAL VALUES:

Composition of saturated solutions in the
$NaAlO_2$-NaOH-Na_2SO_4-Na_3PO_4-H_2O system at 25°C.

$NaAlO_2$ + NaOH		Na_2SO_4		Na_3PO_4		H_2O	solid[a] phase
mass%	mol/kg[b]	mass%	mol/kg[b]	mass%	mol/kg[b]	mass%[b]	
----	----	21.90	1.97	----	----	78.10	A
----	----	----	----	12.30	0.85	87.70	B
21.15	2.43	7.46	0.74	----	----	71.39	A + C
----	----	20.75	1.89	2.10	0.16	77.15	B + D
50.13	8.29	0.29	0.04	----	----	49.58	C
47.72	7.54	0.39	0.05	----	----	51.89	"
46.48	7.15	----	----	0.25	0.03	53.27	E
42.11	6.00	----	----	0.36	0.04	57.53	"
----	----	19.41	1.73	1.78	0.14	78.81	B
----	----	16.01	1.37	1.79	0.13	82.20	"
----	----	8.29	0.67	4.42	0.31	87.29	"
----	----	1.51	0.12	11.22	0.78	87.27	"
1.85	0.18	12.34	1.02	0.81	0.06	85.00	A + B + F
32.16	3.96	0.92	0.10	0.38	0.03	66.54	C + E + F
0.74	0.07	0.39	0.03	7.86	0.52	91.01	B + F
28.34	3.33	1.65	0.17	0.28	0.02	69.73	A + C + F
26.23	2.98	1.28	0.12	0.30	0.02	72.19	B + E + F
46.00	7.08	0.44	0.06	0.26	0.03	53.30	E + C
42.62	6.20	0.63	0.08	0.43	0.05	56.32	"
37.92	5.14	1.01	0.12	0.55	0.06	60.52	"

(continued next page)

AUXILIARY INFORMATION

METHOD/APPARATUS/PROCEDURE:	SOURCE AND PURITY OF MATERIALS:
The isothermal method was used with metallic vessels having a mechanical stirrer. The time for equilibration was 155 hours. Saturated solutions were sampled by filtration and analyzed for Na_2O, Al_2O_3, and P_2O_5 by volumetric, gravimetric, photocolorimetric and nephelometric methods. The composition of the solid phases was determined by Schreinemakers' method.	No details are given.
	ESTIMATED ERROR:
	No information is given.
	REFERENCES:

COMPONENTS:	ORIGINAL MEASUREMENTS:
(1) Trisodium phosphate; Na_3PO_4; [7601-54-9]	Abduragimova, R.A.; Rza-Zade, P.F.;
(2) Sodium aluminate; $NaAlO_2$; [1302-42-7]	Abduragimov, A.A.
(3) Sodium hydroxide; NaOH; [1310-73-2]	*Dokl. Akad. Nauk Azerb. SSR* <u>1971</u>, 27, 41-5.
(4) Disodium sulfate; Na_2SO_4; [7757-82-6]	
(5) Water, H_2O; [7732-18-5]	

EXPERIMENTAL VALUES cont'd:

Composition of saturated solutions in the
$NaAlO_2$-NaOH-Na_2SO_4-Na_3PO_4-H_2O system at 25°C.

$NaAlO_2$ + NaOH		Na_2SO_4		Na_3PO_4		H_2O	solid
mass%	mol/kg[b]	mass%	mol/kg[b]	mass%	mol/kg[b]	mass%[b]	phase[a]
30.23	3.64	1.31	0.14	0.46	0.04	68.00	E + C
28.42	3.35	1.61	0.16	0.39	0.03	69.58	"
28.00	3.27	1.45	0.14	0.28	0.02	70.27	A + F
22.94	2.54	2.81	0.27	0.29	0.02	73.96	"
26.07	2.97	1.36	0.13	0.54	0.04	72.03	B + E + F
26.10	2.95	1.21	0.12	0.19	0.02	72.50	"
22.00	2.38	1.61	0.15	0.64	0.05	75.75	B + F
18.06	1.84	1.09	0.10	0.58	0.04	80.27	"
2.42	0.20	0.01	0.00	0.81	0.05	96.76	"
0.65	0.06	2.46	0.18	2.06	0.13	94.83	"

[a]The solid phases are: A = $Na_2SO_4 \cdot 10H_2O$; B = $Na_3PO_4 \cdot 12H_2O$; C = $Al_2O_3 \cdot 3Na_2O \cdot 7.5H_2O$;
D = $Al_2(SO_4)_3 \cdot 10H_2O$; E = $Al_2O_3 \cdot 2.5Na_2O \cdot 14H_2O$; F = $3Al_2O_3 \cdot 4Na_2O \cdot P_2O_5 \cdot 15H_2O$.
It should be noted that the "solid phase" column in the source paper contains a
great number of typographic errors.

[b]The mol/kg H_2O values were calculated by the compiler.

COMPONENTS:	ORIGINAL MEASUREMENTS:
(1) Trisodium phosphate; Na_3PO_4; [7601-54-9]	Abduragimova, R.A.; Rza-Zade, P.F.
(2) Sodium vanadate; $NaVO_3$; [13718-23-8]	*Issled. Obl. Neorg. Fiz. Khim.* 1971, 191-5.
(3) Disodium sulfate; Na_2SO_4; [7757-82-6]	
(4) Water; H_2O; [7732-18-5]	

VARIABLES:	PREPARED BY:
Composition at 25°C.	J. Eysseltová

EXPERIMENTAL VALUES:

Composition of saturated solutions in the Na_3PO_4-$NaVO_3$-Na_2SO_4-H_2O system at 25°C.

Na_3PO_4		$NaVO_3$		Na_2SO_4		solid[c] phase
mass%[a]	mol/kg[b]	mass%[a]	mol/kg[b]	mass%[a]	mol/kg[b]	
12.30	0.85	----	----	----	----	A
----	----	17.40	1.73	----	----	B
----	----	----	----	21.90	1.97	C
0.92	0.07	14.40	1.41	0.88	0.07	B + D
1.98	0.15	11.80	1.17	3.70	0.32	"
2.6	0.19	9.80	0.99	6.15	0.53	"
2.20	0.16	6.42	0.64	8.82	0.75	B + D + E
1.60	0.12	3.65	0.36	12.60	1.08	B + E
1.06	0.08	1.98	0.20	15.09	1.30	"
0.95	0.07	1.52	0.16	18.39	1.64	"
0.83	0.06	1.31	0.14	20.42	1.86	B + C + E
5.21	0.39	12.65	1.26	----	----	A + E
4.35	0.31	8.20	0.78	1.20	0.10	A + D
4.02	0.28	6.42	0.60	2.60	0.21	"
3.45	0.24	4.25	0.39	3.45	0.27	"
3.01	0.20	2.68	0.24	3.13	0.24	"
2.43	0.17	1.46	0.14	8.80	0.71	A + D + E
2.03	0.14	0.87	0.08	9.60	0.77	A + E
1.79	0.12	0.80	0.08	10.40	0.84	"
1.62	0.12	0.63	0.06	13.15	1.09	"
1.59	0.12	0.45	0.04	15.80	1.35	"
1.12	0.09	0.21	0.02	20.60	1.86	A + C + E

(continued next page)

AUXILIARY INFORMATION

METHOD/APPARATUS/PROCEDURE:	SOURCE AND PURITY OF MATERIALS:
The method of the third component was used. Equilibrium was checked by analysis. The alkali metal content was determined by using 0.5 mol dm^{-3} HCl. Sulfate was determined gravimetrically as $BaSO_4$. P and V were determined photocolorimetrically.	All the salts were of chemically pure grade and were recrystallized before being used.
	ESTIMATED ERROR:
	No information is given.
	REFERENCES:

COMPONENTS:	ORIGINAL MEASUREMENTS:
(1) Trisodium phosphate; Na_3PO_4; [7601-54-9]	Abduragimova, R.A.; Rza-Zade, P.F.
(2) Sodium vanadate; $NaVO_3$; [13718-23-8]	*Issled. Obl. Neorg. Fiz. Khim.* 1971, 191-5.
(3) Disodium sulfate; Na_2SO_4; [7757-82-6]	
(4) Water; H_2O; [7732-18-5]	

EXPERIMENTAL VALUES Cont'd:

Composition of saturated solutions in the $Na_3PO_4-NaVO_3-Na_2SO_4-H_2O$ system at 25°C.

Na_3PO_4		$NaVO_3$		Na_2SO_4		solid[c]
mass%[a]	mol/kg[b]	mass%[a]	mol/kg[b]	mass%[a]	mol/kg[b]	phase
1.25	0.09	16.82	1.68	----	----	B + D
2.18	0.16	5.12	0.50	8.16	0.68	D + E
1.87	0.13	3.81	0.36	6.82	0.55	"
1.63	0.11	3.16	0.29	6.23	0.49	"
----	----	1.58	0.17	20.80	1.89	B + C
2.10	0.16	----	----	20.78	1.90	A + C
1.40	0.09	4.01	0.36	2.29	0.17	B
1.80	0.13	11.98	1.19	3.89	0.33	"
2.03	0.15	10.18	1.02	5.76	0.49	"
2.97	0.23	9.73	1.01	8.03	0.71	"
1.93	0.14	7.22	0.72	8.65	0.74	"
1.73	0.13	5.67	0.58	12.63	1.11	"
1.40	0.11	3.40	0.35	16.12	1.44	D
1.98	0.14	13.94	1.38	1.33	0.11	"
2.93	0.22	12.88	1.28	1.74	0.15	"
3.55	0.25	8.65	0.83	2.45	0.20	"
3.82	0.27	8.13	0.78	2.03	0.17	"
3.09	0.22	7.33	0.69	2.78	0.22	"
2.13	0.15	6.12	0.59	6.83	0.57	"
2.67	0.19	4.68	0.45	6.91	0.58	"
1.2	0.08	4.13	0.39	8.2	0.67	E
1.94	0.14	3.92	0.37	7.99	0.65	"
2.06	0.15	3.11	0.30	9.01	0.74	"
2.09	0.15	2.89	0.28	10.14	0.84	"
1.92	0.16	2.43	0.27	12.38	0.94	"
1.88	0.14	1.63	0.17	16.47	1.45	"
1.21	0.09	1.13	0.12	18.32	1.62	"
2.63	0.17	2.48	0.22	1.34	0.10	A
3.36	0.22	2.13	0.19	1.43	0.11	"
4.48	0.30	1.77	0.16	1.60	0.12	"
2.88	0.19	1.18	0.10	1.80	0.14	"
4.69	0.31	0.96	0.08	2.05	0.16	"
5.13	0.34	0.76	0.07	2.28	0.17	"
5.49	0.37	0.63	0.06	2.90	0.22	"
6.35	0.43	0.43	0.04	3.19	0.25	"
8.42	0.58	0.21	0.02	3.28	0.26	"
9.80	0.69	0.22	0.02	3.32	0.27	"

[a] The compiler supposes this column to have this meaning. In the source paper nothing is specified.

[b] These values were calculated by the compiler on the assumption stated in footnote [a].

[c] The solid phases are: A = $Na_3PO_4 \cdot 12H_2O$; B = $NaVO_3 \cdot 2H_2O$; C = $Na_2SO_4 \cdot 10H_2O$; D = $4Na_2O \cdot P_2O_5 \cdot V_2O_5 \cdot 30H_2O$; E = $4Na_2O \cdot P_2O_5 \cdot V_2O_5 \cdot 18H_2O$.

COMPONENTS:	ORIGINAL MEASUREMENTS:
(1) Trisodium phosphate; Na_3PO_4; [7601-54-9]	Manvelyan, M.G.; Galstyan, V.D.; Sayamyan, E.A.; Gyunashyan, A.P.; Oganesyan, E.B.
(2) Disodium silicate; Na_2SiO_3; [6834-92-0]	
(3) Dipotassium silicate, K_2SiO_3; [10006-28-7]	*Arm. Khim. Zh.* 1973, 26, 632-7.
(4) Tripotassium phosphate; K_3PO_4; [7778-53-2]	
(5) Water; H_2O; [7732-18-5]	

VARIABLES:	PREPARED BY:
Composition at 20°C.	J. Eysseltová

EXPERIMENTAL VALUES:

Jänecke coordinates of the solutions coexisting with two or more solid phases in the system: Na^+, $K^+ \parallel SiO_3^{2-}$, PO_4^{3-} $-H_2O$ at 20°C.

filtrate				solid phases [a]
Na^+	K^+	SiO_3^{2-}	PO_4^{3-}	
100.0		75.45	24.55	A + F
100.0		89.57	10.43	B + F
74.25	25.75	93.56	6.44	B + C + F
62.28	37.72	73.30	27.70	A + C + F
36.46	63.54	22.38	77.62	A + C + E
15.00	85.00	40.20	59.20	C + D + E
41.40	58.60	–	100.0	A + E
13.08	86.92	–	100.0	D + E
93.80	6.05	95.02	4.82	A + B
89.42	10.50	95.42	4.59	A + B
37.88	62.12	31.08	68.90	A + C
6.55	93.45	55.00	45.00	D

[a] The solid phases are: A = $Na_3PO_4 \cdot 12H_2O$; B = $Na_2SiO_3 \cdot 9H_2O$; C = $Na_3PO_4 \cdot 8H_2O$; D = $K_3PO_4 \cdot 7H_2O$; E = solid solutions formed by A and D; F = simultaneous crystallization of A and B.

AUXILIARY INFORMATION

METHOD/APPARATUS/PROCEDURE:	SOURCE AND PURITY OF MATERIALS:
The only information given is that the method of invariant points was used.	No information is given.
	ESTIMATED ERROR:
	No information is given.
	REFERENCES:

COMPONENTS:	ORIGINAL MEASUREMENTS:
(1) Trisodium phosphate; Na_3PO_4; [7601-54-9] (2) 2-Propanone (acetone); C_3H_6O; [67-64-1] (3) Water; H_2O, [7732-18-5]	Nirenberg, Z.; Solenchyk, B.; Yaron, I. *J. Chem. Eng. Data* <u>1977</u>, *22*, 47-8.

VARIABLES:	PREPARED BY:
Composition and temperature.	J. Eysseltová

EXPERIMENTAL VALUES:

I. Ternary solid-liquid equilibrium in the Na_3PO_4-C_3H_6O-H_2O system.

t/°C	Na_3PO_4 mass%	C_3H_6O mass%	H_2O mass%	t/°C	Na_3PO_4 mass%	C_3H_6O mass%	H_2O mass%
10	4.65	----	95.35	20	7.73	----	92.27
10	2.43	8.10	89.47	20	3.20	7.98	88.82
10	1.06	15.75	83.19	20	1.30	16.75	81.95
10	0.50	24.10	75.40	20	0.75	26.40	72.85
10	0.20	33.30	66.50	20	0.17	32.46	67.37
31.5	10.74	----	89.26	40	15.10	----	84.90
31.5	6.00	6.57	87.43	40	8.80	5.95	85.25
31.5	3.00	13.70	83.30	40	4.90	14.10	81.00
31.5	0.62	24.87	74.50	40	2.25	23.50	74.25
31.5	0.34	34.40	65.06	40	0.90	34.10	65.00
31.5	0.19	42.60	57.20	40	0.27	44.50	55.23
31.5	0.07	52.00	47.93	40	0.09	52.00	47.91

X-ray diffraction showed that the solid phase was $Na_3PO_4 \cdot 12H_2O$ and remained unaltered.

(continued next page)

AUXILIARY INFORMATION

METHOD/APPARATUS/PROCEDURE:

Acetone-water mixtures were placed in flasks in a thermostat. In experiments at 31.5°C the phosphate was progressively added to the solvent until saturation was reached. When acetone/water ratios exceeded 2/3, a 9% water solution of $Na_3PO_4 \cdot 12H_2O$ was added in aliquots of 0.5 ml. At other temperatures an excess of phosphate was always added. Mixtures were equilibrated for 24 hours. Samples were taken, diluted immediately and analyzed for Na, P_2O_5 and acetone. Phosphorus was determined spectrophotometrically at 426 nm using the vanadamolybdate method. Acetone was determined iodometrically. Sodium was determined using a Pye Unicam SP90 spectrometer.

SOURCE AND PURITY OF MATERIALS:

Analytical reagent grade materials were used and the purity was rechecked. Doubly distilled water was used in all experiments.

ESTIMATED ERROR:

Temperature was controlled to within ± 0.1 K. P_2O_5: accuracy ±0.25 ppm at 50 ppm; acetone: accuracy ±0.048 mg; Na: accuracy ± 2 ppm at 200 ppm.

REFERENCES:

COMPONENTS:	ORIGINAL MEASUREMENTS:
(1) Trisodium phosphate; Na_3PO_4; [7601-54-9] (2) 2-Propanone (acetone); C_3H_6O; [67-64-1] (3) Water; H_2O; [7732-18-5]	Nirenberg, Z.; Solenchyk, B.; Yaron, I. J. Chem. Eng. Data 1977, 22, 47-8.

EXPERIMENTAL VALUES cont'd:

II. Solubility isotherms in the Na_3PO_4-C_3H_6O-H_2O system in
various ratios of acetone to 100 ml of solution.

	10°C		20°C		31.5°C	
ml/100ml	Na_3PO_4 g/100 ml	sp.gr.	Na_3PO_4 g/100 ml	sp. gr.	Na_3PO_4 g/.100 ml	sp. gr.
0	4.86	1.04	8.35	1.08	11.50	1.07
10	2.37	0.98	3.30	1.02	6.39	1.05
20	1.05	0.98	1.25	0.98	3.02	1.00
30	0.47	0.97	0.72	0.95	0.61	0.96
40	0.19	0.95	0.16	0.94	0.32	0.93
50	trace	----	trace	----	0.18	0.90
60	0	----	0	----	0.07	0.90
100	0	----	0	----	0	----

	40°C	
0	17.05	1.13
10	9.67	1.10
20	5.11	1.04
30	2.25	1.00
40	0.87	0.97
50	0.26	0.94
60	0.08	0.91
100	0	----

COMPONENTS:	EVALUATOR:
(1) Tripotassium phosphate; K$_3$PO$_4$; [7778-53-2] (2) Phosphoric acid; H$_3$PO$_4$; [7664-38-2] (3) Potassium hydroxide; KOH; [1310-58-3] (4) Water; H$_2$O; [7732-18-5]	J. Eysseltová Charles University Prague, Czechoslovakia July, 1986

CRITICAL EVALUATION:

The K$_2$O-P$_2$O$_5$-H$_2$O system

The K$_2$O-P$_2$O$_5$-H$_2$O system has been the subject of study in fifteen papers (1-15). Some of these (1-8) report the solubility over a wide range of K/P ratios; others (9-15) limit the study to a narrow range of K/P ratios. In the latter papers the study is often limited to the solubility of compounds such as KH$_2$PO$_4$ or KH$_5$(PO$_4$)$_2$.

Figures 1-3 illustrate the difficulties encountered in attempting to determine solubilities in this system. One of the characteristics of this system is its tendency to form supersaturated solutions. This fact has been stressed by those who have made the most comprehensive studies (3-7) and is especially characteristic at K/P ratios ≥ 2. In such solutions the reported solubilities vary widely.

Only two conclusions can be deduced from a camparison of solubility studies made on this system.
1. The determination of the solubility of all potassium phosphates except KH$_2$PO$_4$ and KH$_5$(PO$_4$)$_2$ is extremely difficult and considerable care must be given to the conditions under which the determinations are made.
2. Most of the results reported by Jänecke (2) have a systematic error and should be rejected.

Because of the complexity of the system, a large number of equilibrium solid phases have been reported. They are listed below.

K$_3$PO$_4$·7H$_2$O	[22763-02-6]	3K$_2$HPO$_4$·KH$_2$PO$_4$·2H$_2$O	[101056-48-8]
K$_3$PO$_4$·3H$_2$O	[22763-03-7]	K$_5$H$_4$(PO$_4$)$_3$·H$_2$O	[101056-49-9]
K$_3$PO$_4$	[7778-53-2]	K$_2$HPO$_4$·KH$_2$PO$_4$·3H$_2$O	[101056-50-2]
K$_2$HPO$_4$·6H$_2$O	[101056-47-7]	K$_2$HPO$_4$·KH$_2$PO$_4$·2H$_2$O	[66922-99-4]
K$_2$HPO$_4$·3H$_2$O	[16788-57-1]	KH$_2$PO$_4$	[7778-77-0]
K$_2$HPO$_4$	[7758-11-4]	KH$_5$(PO$_4$)$_2$	[14887-42-4]

The conditions under which these hydrates exist and their transition points are discussed in the Critical Evaluation of the binary systems.

The incongruently soluble KH$_5$(PO$_4$)$_2$ has been observed in all studies which investigated strongly acid solutions (1,3-7,9) and its existence may be taken as proved. A study of the temperature dependence of the solubility of this compound (13) has proved that it is incongruently soluble up to 373 K.

K/P ratios between 1 and 2. Some compound may also exist in the region between KH$_2$PO$_4$ and K$_2$HPO$_4$ but the solubility results that have been reported (1,3-9) differ substantially. Parker (1) did not observe K$_2$HPO$_4$ and his solid phase formulations in this region are probably incorrect. Berg (3,4) reports the presence of K$_2$HPO$_4$·KH$_2$PO$_4$·3H$_2$O at 298 K in a very limited concentration interval: 35.53 to 34.48% K$_2$O and 29.09 to 29.39% P$_2$O$_5$, i.e., 53.40 to 52.85% K$_3$PO$_4$ and 15.54 to 15.69% H$_3$PO$_4$. Flatt, et al. (9) reported the presence of K$_5$H$_4$(PO$_4$)$_3$·H$_2$O or 2K$_2$HPO$_4$·KH$_2$PO$_4$·H$_2$O at 298 K in the concentration range 51.5 to 52.4% K$_3$PO$_4$ and 16.3 to 15.7% H$_3$PO$_4$. This compound was also reported by Staudenmayer (16).

A careful study of the transition between KH$_2$PO$_4$ and K$_2$HPO$_4$ at 298 K has been made (8). The existence of the above reported complexes was not observed, nor was the presence of K$_2$HPO$_4$·3H$_2$O. These authors suggest that the crystallization of K$_2$HPO$_4$ and KH$_2$PO$_4$ occurs somewhere between 20.5 to 31.3% K$_3$PO$_4$ and 13.7 to 14.8% H$_3$PO$_4$, i.e., at a significantly lower K$_3$PO$_4$ content.

Ravich (6,7) studied the system at 273 K and reports no solid phase with a K/P ratio between 1 and 2. Berg (5) reported the presence of K$_2$HPO$_4$·KH$_2$PO$_4$·2H$_2$O and 3K$_2$HPO$_4$·KH$_2$PO$_4$·2H$_2$O at 323 K. The latter compound was also observed by Staudenmayer (16). However, Berg (5) was not certain of the identity of these compounds.

In conclusion, the Evaluator believes that no decision about the existence of the solid phases in the K$_2$O-P$_2$O$_5$-H$_2$O system can be made until further studies are reported. In making such investigations the existence of very viscous solutions will be encountered. These solutions become supersaturated very readily. Meanwhile, the data of Berg (3-5) and Ravich (6,7) are to be considered as tentative values.

The solubility curves for KH$_2$PO$_4$ (1,3-14) agree satisfactorily. Punin's smoothing equation (15) may be useful for temperature extrapolation, but its use for the binary KH$_2$PO$_4$-H$_2$O system shows a systematic error of about +10% when compared with accepted experimental values.

(continued next page)

COMPONENTS:	EVALUATOR:
(1) Tripotassium phosphate; K$_3$PO$_4$; [7778-53-2]	J. Eysseltová Charles University Prague, Czechoslovakia
(2) Phosphoric acid; H$_3$PO$_4$; [7664-38-2]	
(3) Potassium hydroxide; KOH; [1310-58-3]	July, 1986
(4) Water; H$_2$O; [7732-18-5]	

CRITICAL EVALUATION:

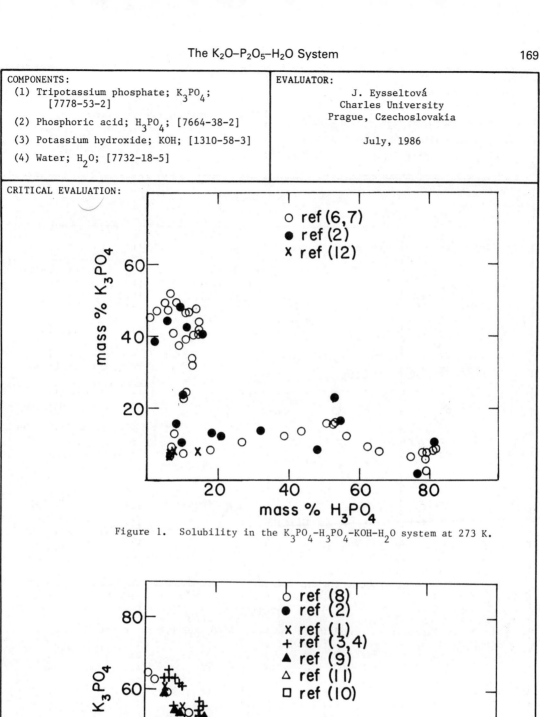

Figure 1.   Solubility in the K$_3$PO$_4$–H$_3$PO$_4$–KOH–H$_2$O system at 273 K.

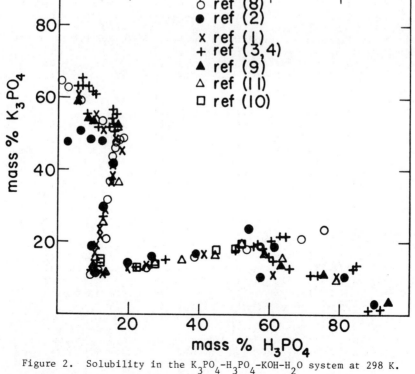

Figure 2.   Solubility in the K$_3$PO$_4$–H$_3$PO$_4$–KOH–H$_2$O system at 298 K.

COMPONENTS:	EVALUATOR:
(1) Tripotassium phosphate; K_3PO_4; [7778-53-2]	J. Eysseltová Charles University Prague, Czechoslovakia
(2) Phosphoric acid; H_3PO_4; [7664-38-2]	
(3) Potassium hydroxide; KOH; [1310-58-3]	July, 1986
(4) Water; H_2O; [7732-18-5]	

CRITICAL EVALUATION:

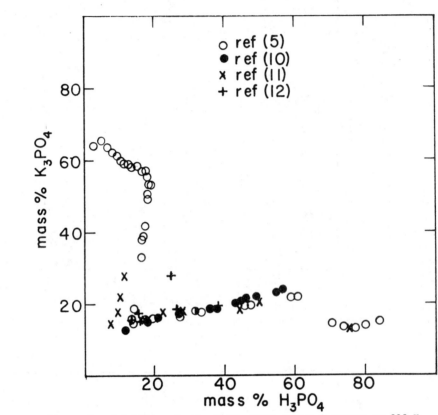

Figure 3. Solubility in the K_3PO_4-H_3PO_4-KOH-H_2O system at 323 K.

COMPONENTS:	EVALUATOR:
(1) Tripotassium phosphate; K$_3$PO$_4$; [7778-53-2] (2) Phosphoric acid; H$_3$PO$_4$; [7664-38-2] (3) Potassium hydroxide; KOH; [1310-58-3] (4) Water; H$_2$O; [7732-18-5]	J. Eysseltová Charles University Prague, Czechoslovakia July, 1986

CRITICAL EVALUATION: (cont'd)

Solubilities in some segments of this system have been determined at elevated temperatures (17,18). The system is similar to the Na$_2$O–P$_2$O$_5$–H$_2$O system (see chap. 2, p. 11). The presence of liquid-liquid immiscibility is characteristic of this system at high temperatures.

References

1. Parker, E.G. *J. Phys. Chem.* 1914, *18*, 653.
2. Jänecke, E. *Z. Phys. Chem.* 1927, *127*, 71.
3. Berg, A.G. *Izv. Akad. Nauk SSSR* 1933, 167.
4. Berg, A.G. *Izv. Akad. Nauk SSSR* 1938, 147.
5. Berg, A.G. *Izv. Akad. Nauk SSSR* 1938, 161.
6. Ravich, M.I. *Kaliy* 1936, *10*, 33.
7. Ravich, M.I. *Izv. Akad. Nauk SSSR* 1938, 167.
8. D'Ans, J.; Schreiner, O. *Z. Anorg. Chem.* 1911, *75*, 95.
9. Flatt, R.; Brunisholz, G.; Bourgeois, J. *Helv. Chim. Acta* 1956, *39*, 841.
10. Beremzhanov, B.A.; Voronina, L.V.; Savich, R.F. *Prikl. Teor. Khim.* 1978, 3.
11. Myl, J.; Solc, Z. *Coll. Czech. Chem. Commun.* 1960, *25*, 2414.
12. Krasil'shtschikov, A.I. *Izv. In-ta Fiz.-khim. An.* 1933, *6*, 159.
13. Paravano, N.; Mieli, A. *Gaz. Chim. Ital.* 1908, *38*, 535.
14. Orekhov, I.I.; Tereshchenko, L.Ya.; Balabanovich, Ya.K.; Vlasova, T.L. *Zh. Neorg. Khim.* 1969, *14*, 1637.
15. Punin, Yu.O.; Mirenkova, T.F.; Artamanova, O.I.; Ul'yanova, T.P. *Zh. Neorg. Khim.* 1975, *20*, 2813.
16. Staudenmayer, L. *Z. Anorg. Chem.* 1894, 5, 383.
17. Marshall, W.L.; Hall, C.E.; Mesmer, R.E. *J. Inorg. Nucl. Chem.* 1981, *43*, 449.
18. Marshall, W.L. *J. Chem. Eng. Data* 1982, *27*, 15.

COMPONENTS:	ORIGINAL MEASUREMENTS:
(1) Dipotassium hydrogenphosphate; K$_2$HPO$_4$; [7758-11-4] (2) Water; H$_2$O; [7732-18-5]	Marshall, W.L.; Hall, C.E.; Mesmer, R.E. J. Inorg. Nucl. Chem. <u>1981</u>, 43, 449-55.

VARIABLES:	PREPARED BY:
Temperature and composition.	J. Eysseltová

EXPERIMENTAL VALUES:

Part 1. Smoothed values for the solubility of K$_2$HPO$_4$ in H$_2$O.

t/°C	mass%	mol/kg^a
100	74.0	16.34
150	75.5	17.69
200	76.5	18.69
250	77.0	19.22
300	77.5$_b$	19.77
350	78.0$_b$	20.35
400	78.5^b	20.96

^{a}The mol/kg H$_2$O values were calculated by the compiler.

^{b}Based on experiments at temperature and extrapolation from lower temperatures. The accuracy is ± 1.5%.

(continued next page)

AUXILIARY INFORMATION

METHOD/APPARATUS/PROCEDURE:	SOURCE AND PURITY OF MATERIALS:
Gold-plated stainless steel high pressure vessels were used. The samples and a small amount of water (to counterbalance the vapor pressure) were rocked at constant high temperature between 100° and 400°C for period of time of 2 to 5 hr. The vanado-molybdate method for quantitative spectrophotometric determination was used. The reagent used was that of Bridger, et al. (1). The procedure was modified slightly.	No information is given.
	ESTIMATED ERROR: The accuracy of the temperature was ± 0.5 to 1 K. The accuracy of the smoothed values was ± 0.5%.
	REFERENCES: 1. Bridger, G.L.; Boylan, D.R.; Markey, J.W. Anal. Chem. <u>1953</u>, 25, 336.

COMPONENTS:	ORIGINAL MEASUREMENTS:
(1) Dipotassium hydrogenphosphate; K$_2$HPO$_4$; [7758-11-4] (2) Water; H$_2$O; [7732-18-5]	Marshall, W.L.; Hall, C.E.; Mesmer, R.E. J. Inorg. Nucl. Chem. 1981, 43, 449-55.

EXPERIMENTAL VALUES cont'd:

Part 2. Two liquid phase regions and critical phenomena in the K$_2$HPO$_4$-H$_2$O system at 360-400°C.

K$_2$HPO$_4$ mass%	mol/kg[a]	t^b/°C	the phase appearing[e]	t^c/°C	upper temperature reached/°C[d]
4.58	0.28	368.0	L$_2$	378.5[f]	390
10.0	0.64	360.0	L$_2$	379.0[f]	390
20.0	1.43	361.8	L$_2$	-----	---
30.0	2.46	360.0	L$_2$	377.4[f]	390
40.0	3.83	361.0	L$_1$	377.0[f]	400
50.0	5.74	366.0	L$_1$	g	400
60.0	8.61	>400	–	–	400

[a] The mol/kg H$_2$O values were calculated by the compiler.

[b] The temperature of two-liquid phase appearance with rising temperature.

[c] The liquid-vapor critical temperature of very dilute phase.

[d] The concentrated liquid phase was always clear at the highest temperature reached.

[e] L$_1$ is the more dilute phase; L$_2$ is the more concentrated phase.

[f] The actual composition of the dilute phase undergoing the critical phenomenon is estimated to be 3 mass%.

[g] Could not detect this.

COMPONENTS:	ORIGINAL MEASUREMENTS:
(1) Potassium dihydrogenphosphate; KH$_2$PO$_4$; [7778-77-0] (2) Phosphoric acid; H$_3$PO$_4$; [7664-38-2] (3) Water; H$_2$O; [7732-18-5]	Paravano, N.; Mieli, A. *Gaz. Chim. Ital.* <u>1908</u>, *11*, 535-44.

VARIABLES:	PREPARED BY:
Composition and temperature	J. Eysseltová

EXPERIMENTAL VALUES:

Saturation temperatures of the solutions of KH$_2$PO$_4$·H$_3$PO$_4$ in water.

KH$_2$PO$_4$·H$_3$PO$_4$		KH$_2$PO$_4$ [a]		H$_3$PO$_4$ [a]		t/°C.	solid phase
mass%	mol%	mass%	mol/kg	mass%	mol/kg		
0	0	0	0	0	0	0	ice
3.337	0.27	1.94	0.147	1.40	0.147	-0.6	"
8.824	0.69	4.82	0.385	3.47	0.385	-1.7	"
12.13	1.05	7.05	0.589	5.08	0.589	-2.5	"
20.50	1.94	11.92	1.101	8.58	1.101	-5.7	"
29.43	3.11	17.11	1.781	12.32	1.781	-6.7	"
36.98	4.32	21.50	2.506	15.48	2.506	-9.2	"
45.80	6.10	26.63	3.609	19.17	3.609	∼0	KH$_2$PO$_4$
50.33	7.21	29.26	4.328	21.07	4.328	10.9	"
68.44	14.30	39.79	9.263	28.65	9.263	65.2	"
72.43	16.81	42.11	11.222	30.32	11.222	78.0	"
77.60	21.05	45.11	14.798	32.48	14.798	87.5	"
85.88	31.86	49.93	25.981	35.95	25.981	105.5	"
92.18	47.57	53.59	50.353	38.59	50.353	120.0	"
95.73	63.31	55.65	95.767	40.07	95.676	134.5	"
96.10	65.47	55.87	100	40.22	100	135.0	"
98.85	88.88	57.47	100	41.39	100	137.5	"
100	100	58.14		41.86		139.0	"

[a] These values were calculated by the compiler.

AUXILIARY INFORMATION

METHOD/APPARATUS/PROCEDURE:	SOURCE AND PURITY OF MATERIALS:
There was a constant ratio of KH$_2$PO$_4$/H$_3$PO$_4$. Saturation temperatures were determined visually as the temperature of disappearance of the last crystal.	The KH$_2$PO$_4$·H$_3$PO$_4$ was prepared from an equimolar mixture of concentrated solutions of KH$_2$PO$_4$ and H$_3$PO$_4$ by slow crystallization. Analysis: observed calculated P$_2$O$_5$ 60.44% 60.64% K$_2$O 20.38% 20.13%

ESTIMATED ERROR:

No information is given.

REFERENCES:

COMPONENTS:	ORIGINAL MEASUREMENTS:
(1) Tripotassium phosphate; K$_3$PO$_4$; [7778-53-2] (2) Phosphoric acid; H$_3$PO$_4$; [7664-38-2] (3) Water; H$_2$O; [7732-18-5]	D'Ans, J.; Schreiner, O. Z. Anorg. Chem. <u>1911</u>, 75, 95-102.

VARIABLES:	PREPARED BY:
Composition at 25°C.	J. Eysseltová

EXPERIMENTAL VALUES: Solubility in the K$_3$PO$_4$–H$_3$PO$_4$–H$_2$O system at 25°C.

K$^+$ conc^a	PO$_4^{3-}$ concn^a	K$_3$PO$_4$^{b} mass%	mol/kg	H$_3$PO$_4$^{b} mass%	mol/kg	solid phase
9.14	3.13	64.71	8.84	0.84	0.25	K$_3$PO$_4$
8.84	3.22	62.58	8.49	2.70	0.79	"^{c}
8.42	3.44	59.61	8.23	6.23	1.68	"
7.52	3.78	53.24	7.32	12.50	3.72	"
6.90	4.15	48.85	6.97	18.15	5.61	"
6.88	4.12	48.70	6.87	17.92	5.48	"
6.80	4.08	48.14	6.60	17.79	5.32	K$_2$HPO$_4$^{c}
6.80	4.05	48.14	6.60	17.49	5.19	"
6.76	3.96	47.85	6.36	16.74	4.82	"^{c}
6.50	3.85	46.01	5.72	16.12	4.34	"^{c}
6.16	3.61	43.61	4.99	15.27	3.79	"
5.24	3.25	37.09	3.62	14.74	3.12	KH$_2$PO$_4$^{c}
4.42	2.94	31.29	2.71	14.38	2.70	"
2.90	2.36	20.53	1.46	13.66	2.11	"
1.70	1.71	12.03	0.73	11.21	1.49	"
1.60	1.67	11.32	0.68	11.14	1.46	"
1.48	1.46	10.46	0.61	9.51	1.21	"
1.78	3.15	12.60	0.95	25.06	4.10	"
2.18	4.65	15.43	1.57	38.45	8.51	"
2.54	6.32	17.98	2.98	53.64	19.29	"
2.66	6.76	18.83	3.75	57.56	24.89	"
2.98	8.03	21.09	10.00	68.96	70.85	"
3.32	8.80	23.50	101.6	75.40	706.4	"^{c}

^{a}The concentration unit is: mol/1000 g soln.
^{b}These values were calculated by the compiler.
^{c}These solid phases were analyzed.

AUXILIARY INFORMATION

METHOD/APPARATUS/PROCEDURE:	SOURCE AND PURITY OF MATERIALS:
The solid phases were separated by pressing them between two porous plates. H$_3$PO$_4$ was precipitated as NH$_4$MgPO$_4$·6H$_2$O and weighed as Mg$_2$P$_2$O$_7$. Potassium was determined gravimetrically as KClO$_4$.	Commercial materials were used and were recrystallized before use.
	ESTIMATED ERROR:
	The temperature was controlled to within ± 0.05 K.
	REFERENCES:

COMPONENTS:	ORIGINAL MEASUREMENTS:
(1) Tripotassium phosphate; K$_3$PO$_4$; [7778-53-2] (2) Phosphoric acid; H$_3$PO$_4$; [7664-38-2] (3) Water; H$_2$O; [7732-18-5]	Flatt, R.; Brunisholz, G.; Bourgeois, J. *Helv. Chim. Acta* <u>1956</u>, 39, 841-53.

VARIABLES:	PREPARED BY:
Composition at 25°C.	J. Eysseltová

EXPERIMENTAL VALUES:

Solubility in the K$_3$PO$_4$–H$_3$PO$_4$–H$_2$O system at 25°C.

K$^+$	H$^+$	K$_3$PO$_4$[a]		H$_3$PO$_4$[a]		H$_2$O		solid[c]
eq%	eq%	mass%	mol/kg	mass%	mol/kg	M[b]	mass%[a]	phase
0	100.0	0	0	95.6	220	8.4	4.4	A
1.6	98.4	3.3	5.5	93.9	337	5.4	2.8	"
6.0	94.0	11.1	5.8	80.0	91.6	19.0	8.9	B
6.4	93.6	11.2	3.9	75.3	56.8	30.5	13.5	"
8.0	92.0	12.3	2.6	65.5	30.2	56.5	22.2	"
9.1	90.9	13.7	2.8	63.3	28.1	60.0	23.0	"
11.7	88.3	16.9	3.3	58.7	24.6	66.6	24.4	"
13.0	87.0	19.0	4.0	58.7	26.9	60.0	22.3	"
13.4	86.6	19.9	4.6	59.5	29.5	54.4	20.6	B + C
13.3	86.7	18.3	3.2	55.1	21.2	75.8	26.6	C
19.8	80.2	13.0	1.0	24.4	3.97	374.2	62.6	"
22.5	77.5	11.9	0.8	19.0	2.81	511.4	69.1	"
27.2	72.8	11.2	0.7	13.8	1.89	715.0	74.9	"
31.4	68.6	10.4	0.6	10.5	1.35	942.0	79.2	"
33.2	66.7	10.4	0.6	9.7	1.24	1000.0	79.9	"
38.6	61.4	13.5	0.8	9.9	1.32	863.4	76.6	"
43.6	56.4	19.2	1.3	11.4	1.68	620.4	69.4	"
49.0	51.0	27.7	2.2	13.3	2.30	410.5	59.0	"
56.0	44.0	41.8	4.6	15.2	3.59	226.7	43.0	"
59.3	40.7	51.5	7.5	16.3	5.18	145.3	32.2	C + D
59.0	41.0	50.9	7.3	16.3	5.09	149.3	32.8	D
60.6	39.4	52.4	7.7	15.7	5.03	145.0	31.9	D + E
62.1	37.9	50.8	6.9	14.3	4.19	167.5	34.9	E

(continued next page)

AUXILIARY INFORMATION

METHOD/APPARATUS/PROCEDURE:	SOURCE AND PURITY OF MATERIALS:
No information is given.	No information is given.
	ESTIMATED ERROR: No information is given.
	REFERENCES:

COMPONENTS:	ORIGINAL MEASUREMENTS:
(1) Tripotassium phosphate; K$_3$PO$_4$; [7778-53-2]	Flatt, R.; Brunisholz, G.; Bourgeois, J.
(2) Phosphoric acid; H$_3$PO$_4$; [7664-38-2]	*Helv. Chim. Acta* 1956, 39, 841-53.
(3) Water; H$_2$O; [7732-18-5]	

EXPERIMENTAL VALUES cont'd:

Solubility in the K$_3$PO$_4$–H$_3$PO$_4$–H$_2$O system at 25°C.

K$^+$	H$^+$	K$_3$PO$_4$[a]		H$_3$PO$_4$[a]		H$_2$O		solid[c]
eq%	eq%	mass%	mol/kg	mass%	mol/kg	M[b]	mass%[a]	phase
66.1	33.9	50.9	6.5	12.1	3.33	188.8	37.0	E
66.8	33.2	51.1	6.5	11.7	3.22	191.1	37.2	"
71.5	28.5	52.8	6.6	9.7	2.6	199.8	37.5	"
74.4	25.6	53.7	6.7	8.5	2.3	205.4	37.7	"
82.3	17.7	58.7	7.8	5.8	1.7	195.2	35.4	E + F
100	0	48.1	4.4	0	0	424.4	51.9	F

[a] These values were calculated by the compiler.

[b] The concentration unit is: mol g H$_2$O/100 eq of solute.

[c] The solid phases are: A = H$_3$PO$_4$; B = KH$_5$(PO$_4$)$_2$; C = KH$_2$PO$_4$; D = K$_5$H$_4$(PO$_4$)$_3$·H$_2$O;

E = K$_2$HPO$_4$·3H$_2$O; F = K$_3$PO$_4$·7H$_2$O.

COMPONENTS:	ORIGINAL MEASUREMENTS:
(1) Potassium dihydrogenphosphate; KH_2PO_4; [7778-77-0] (2) Phosphoric acid; H_3PO_4; [7664-38-2] (3) Water; H_2O; [7732-18-5]	Orekhov, I.I.; Tereshchenko, L.Ya.; Balabanovich, Ya.K.; Vlasova, T.L. *Zh. Neorg. Khim.* <u>1969</u>, *14*, 1637-40.

VARIABLES:	PREPARED BY:
Temperature and composition.	J. Eysseltová

EXPERIMENTAL VALUES:

Part 1. Definition of the sections:

section	1	2	3	4	5	6
mass% H_3PO_4	4	13.5	20	30	40	50

Part 2. Eutectic points:

section	KH_2PO_4 mass%	H_3PO_4 mass%	H_2O mass%	t_s/°C.[a]	t_c/°C.[b]	solid phase
1	14.0	3.44	82.56	-2.5	-10.0	ice + KH_2PO_4
2	14.38	11.54	74.08	-5	-13	"
3	19.35	16.14	64.25	-7.5	-13	"
4	20.6	24.3	55.1	-13	-14.5	"
5	21.9	31.3	46.8	-21.5	-24	"
6	20.5	39.74	39.76	-35	-39	"

[a] Eutectic temperature measured by heating.

[b] Eutectic temperature measured by cooling.

(continued next page)

AUXILIARY INFORMATION

METHOD/APPARATUS/PROCEDURE:	SOURCE AND PURITY OF MATERIALS:
A visual polythermic method (1) was used as well as an isothermal method which involved conductivity measurements. Standard methods of analysis were used but no specific details are given.	Chemically pure H_3PO_4 and reagent grade KH_2PO_4 were used.

ESTIMATED ERROR:

No information is given.

REFERENCES:

1. Bergman, A.G.; Luzhnaya, N.P. *Fiziko-Khimicheskie Osnovy Izucheniya i Ispol'zovaniya solyanykh Mestorozhdeniy Khlorid-sul'fatnogo Tipa*, Moscow, IAN SSSR, <u>1951</u>.
2. Babayan, S.G.; Pokhomov, B.G.; Melichov, I.V.; Merkulova, M.S. *Radiokhimiya* <u>1961</u>, *3*, 391.

COMPONENTS:	ORIGINAL MEASUREMENTS:
(1) Potassium dihydrogenphosphate; KH_2PO_4; [7778-77-0]	Orekhov, I.I.; Tereshchenko, L. Ya.; Balabanovich, Ya.K.; Vlasova, T.L.
(2) Phosphoric acid; H_3PO_4 [7664-38-2]	Zh. Neorg. Khim. 1969, 14, 1637-40.
(3) Water; H_2O; [7732-18-5]	

EXPERIMENTAL VALUES cont'd:

Part 3.   Solubility of KH_2PO_4 in aqueous solutions of H_3PO_4.[a]

temperature in °C.

section	10	20	30	40	50	60	70
1	----	17.1	20.6	23.1	----	----	----
2	17.33	20.28	23.4	26.21	29.1	32.46	34.71
3	20.6	23.54	25.7	29.1	32.29	34.88	37.37
4	23.22	25.62	29.25	31.7	33.82	36.1	37.62
5	26.56	28.22	31.72	33.63	35.64	37.4	38.57
6	29.6	30.74	33.0	35.3	37.5	39.4	41.25

[a]The solubility values are given as mass%.

The authors emphasize that supersaturated solutions are formed very easily.

Part 4.   The relation of composition of saturated solutions to pH is given in graphical form only.

COMPONENTS:	ORIGINAL MEASUREMENTS:
(1) Potassium dihydrogenphosphate; KH_2PO_4; [7778-77-0]	Beremzhanov, B.A.; Voronina, L.V.; Savich, R.F.
(2) Phosphoric acid; H_3PO_4; [7664-38-2]	*Prikl. Teor. Khim.* 1978, 3-7.
(3) Water; H_2O; [7732-18-5]	

VARIABLES:	PREPARED BY:
Temperature and composition.	J. Eysseltová

EXPERIMENTAL VALUES:

Composition of saturated solutions in the KH_2PO_4-H_3PO_4-H_2O system.

KH_2PO_4		H_3PO_4		K_2O	P_2O_5		refr.	solid
mass%	mol/kg[a]	mass%	mol/kg[a]	mass%	mass%	pH	index	phase

temp = 25°C.

24.8	2.43	0.30	0.04	8.65	13.10	8.25	1.355	KH_2PO_4
25.8	2.69	3.75	0.54	8.92	22.02	2.38	1.361	"
25.0	2.86	10.9	1.73	8.67	20.78	1.97	1.381	"
25.2	2.90	11.1	1.77	8.72	21.37	1.86	1.387	"
27.2	3.42	14.4	2.51	9.4	22.41	1.61	1.397	"
30.8	4.59	23.7	4.40	10.92	26.70	1.12	1.417	"
30.4	4.86	23.7	5.26	10.67	31.57	0.56	1.417	"
34.7	6.89	28.3	7.80	12.0	33.65	0.21	1.419	"
36.7	8.74	32.8	10.97	12.7	35.58	0.20	1.422	"
37.9	9.83	33.8	12.18	13.3	44.32	0.16	1.423	"

temp = 35°C.

27.2	2.74	0.0	0.00	9.4	14.3	3.5	1.360	KH_2PO_4
28.0	3.02	4.0	0.60	9.5	17.3	2.76	1.373	"
30.1	3.85	12.5	2.22	10.4	24.7	1.82	1.389	"
31.9	4.47	15.7	3.05	11.1	28.0	1.20	1.400	"
33.9	5.71	22.5	5.26	11.81	34.4	0.88	1.414	"
38.0	6.10	27.3	7.01	13.3	39.8	0.30	1.417	"
39.2	9.86	31.6	11.04	13.6	43.0	0.15	1.423	"
41.5	12.44	34.0	14.16	14.3	46.6	0.02	1.424	"

(continued next page)

AUXILIARY INFORMATION

METHOD/APPARATUS/PROCEDURE:	SOURCE AND PURITY OF MATERIALS:
Crystalline KH_2PO_4 was dissolved in phosphoric acid solutions of different concentrations. Four days were allowed for equilibration. No further details are given.	No details are given.
	ESTIMATED ERROR:
	No details are given.
	REFERENCES:

COMPONENTS:	ORIGINAL MEASUREMENTS:
(1) Potassium dihydrogenphosphate; KH_2PO_4; [7778-77-0]	Beremzhanov, B.A.; Voronina, L.V.; Savich, R.F.
(2) Phosphoric acid; H_3PO_4; [7664-38-2]	*Prikl. Teor. Khim.* <u>1978</u>, 3-7.
(3) Water; H_2O; [7732-18-5]	

EXPERIMENTAL VALUES cont'd:

Composition of saturated solutions in the KH_2PO_4-H_3PO_4-H_2O system.

KH_2PO_4		H_3PO_4		K_2O	P_2O_5			
mass%	mol/kg[a]	mass%	mol/kg[a]	mass%	mass%	pH	refr. index	solid phase
				temp = 50°C.				
23.6	2.26	----	0.00	10.25	15.41	3.70	1.365	KH_2PO_4
29.9	3.28	3.3	0.50	10.35	17.99	3.10	1.370	"
30.0	3.48	6.8	1.09	10.96	20.58	2.19	1.372	"
31.5	4.09	12.0	2.16	10.90	25.11	1.78	1.373	"
33.6	4.81	15.1	3.00	11.52	28.40	1.56	1.384	"
35.2	5.59	18.6	4.10	12.19	31.81	1.12	1.398	"
36.3	6.25	20.0	4.72	12.55	33.42	0.89	1.394	"
38.7	7.56	23.7	6.43	13.39	37.2	0.69	1.401	"
39.5	8.29	25.5	7.43	13.65	39.10	0.44	1.409	"
40.9	9.10	26.1	8.07	13.75	40.25	0.34	1.410	"
41.9	10.72	29.4	10.45	14.50	43.12	0.25	1.413	"
42.5	11.69	30.8	11.77	14.70	44.50	0.23	1.415	"
44.4	14.30	32.8	14.67	15.35	46.89	-0.04	1.419	"
46.0	17.33	34.5	18.05	15.90	49.01	-0.08	1.423	"

[a] The mol/kg H_2O values were calculated by the compiler.

COMPONENTS:	ORIGINAL MEASUREMENTS:
(1) Potassium dihydrogenphosphate; KH_2PO_4; [7778-77-0] (2) Potassium hydroxide; KOH; [1310-58-3] (3) Water; H_2O; [7732-18-5]	Marshall, W.L. J. Chem. Eng. Data <u>1982</u>, 27, 175-80.
VARIABLES:	PREPARED BY:
Temperature and five K/P ratios.	J. Eysseltová

EXPERIMENTAL VALUES:

Immiscibility and liquid-vapor critical phenomena for
aqueous potassium phosphate solutions.

solute stoichiometry			immiscibility boundary		critical phenomenon		
K/PO_4 ratio	mass%	mol/kg^a	t^b/°C	phase^c	t/°C	mass%^{d}
1	5.05	0.39	386.8 ± 0.2	L_2	389.0 ± 0.5	2.2
1	10.09	0.82	385.6 ± 0.1	L_2	388.5 ± 0.5	2.2
1	14.8	1.28	385.7 ± 0.1	L_2	389.0 ± 0.5	2.2
1	20.0	1.84	386.1 ± 0.2	L_2^x	388.9 ± 0.2	2.2
1	25.0	2.45	386.9 ± 0.4	L_1	388.7 ± 0.5	2.2
1	29.5	3.07	387.8 ± 0.4	L_1	389.0 ± 0.5	2.2
1	33^d	3.62	389^d	L_1	389^d	2.2
1	41.5	5.21	e	e	e	e
1.2	5.07	0.37	376.0 ± 0.3	L_2	384.0 ± 0.5	1.8
1.2	10.3	0.79	371.8 ± 0.2	L_2	383.9 ± 0.3	1.8
1.2	15.1	1.23	371.0 ± 0.1	L_2	384.9 ± 0.3	1.8
1.2	20.2	1.75	370.9 ± 0.2	L_2	384.0 ± 0.5	1.8
1.2	25.1	2.32	372.0 ± 0.2	L_2^x	385.0 ± 0.2	1.8
1.2	30.2	2.99	373.0 ± 0.1	L_1	384.5 ± 0.5	1.8
1.2	35.1	3.74	374.9 ± 0.1	L_1	385.1 ± 0.1	1.8
1.2	39^d	4.42	384.5^d	L_1	384.5^d	1.8
1.2	40.2	4.64	e	e	e	e
1.2	50.1	6.94	e	e	e	e

(continued next page)

AUXILIARY INFORMATION

METHOD/APPARATUS/PROCEDURE:	SOURCE AND PURITY OF MATERIALS:
The synthetic method was used. Samples of known composition were sealed in fused silica capillary tubes and heated. A chromel-alumel thermocouple was used with a digital readout unit. The experimental details are described in ref. (1).	Analytical reagent grade K_2HPO_4 and K_3PO_4 and certified ACS grade KH_2PO_4 were used.
	ESTIMATED ERROR: The temperature at which immiscibility occurs had a precision of ± 0.1 K and an accuracy of 0.5 - 1.0 K. The critical temperature had a precision of ± 0.1-0.2 K and an accuracy of 1.0-1.5 K.
	REFERENCES: 1. Marshall, W.L.; Hall, C.E.; Mesmer, R.E. J. Inorg. Nucl. Chem. <u>1981</u>, 43, 449.

COMPONENTS:	ORIGINAL MEASUREMENTS:
(1) Potassium dihydrogenphosphate; KH_2PO_4; [7778-77-0] (2) Potassium hydroxide; KOH; [1310-58-3] (3) Water; H_2O; [7732-18-5]	Marshall, W.L. J. Chem. Eng. Data <u>1982</u>, 27, 175-80.

EXPERIMENTAL VALUES cont'd:

K/PO_4 ratio	mass%	mol/kg[a]	t[b]/°C	phase[c]	t/°C	mass%[d]
1.5	5.02	0.34	369.0 ± 2.0	L_2	380.1 ± 0.1	1.5
1.5	10.05	0.72	364.0 ± 1.0	L_2	380.0 ± 0.5	1.5
1.5	15.1	1.14	362.7 ± 0.1	L_2	380.0 ± 0.5	1.5
1.5	20.1	1.61	361.7 ± 0.1	L_2	380.0 ± 0.5	1.5
1.5	30.2	2.77	362.5 ± 0.1	L_x	380.0 ± 0.5	1.5
1.5	39.8	4.23	365.0 ± 0.5	L_1	380.5 ± 0.5	1.5
1.5	45.2	5.28	372.5 ± 0.5	L_1	380.0 ± 0.5	1.5
1.5	49[d]	6.15	380[d]	L_1	380[d]	1.5
1.5	50.3	6.48	e	e	e	e
2[f]	10.3	0.66	362.4 ± 0.4	L_2	375.9 ± 0.2	1
2[f]	20.0	1.44	360.0 ± 0.2	L_2	376.8 ± 0.5	1
2[f]	56[d]	7.31	377[d]	L_1	377[d]	1
2.12	4.73	0.28	369.4 ± 0.2	L_2	377.3 ± 0.2	2
2.12	9.88	0.61	361.4 ± 0.5	L_2	378.0 ± 0.2	2
2.12	16.7	1.12	360.9 ± 0.2	L_2	379.2 ± 0.4	2
2.12	30.3	2.42	362.1 ± 0.2	L_2	378.5 ± 0.4	2

[a] The mol/kg H_2O values were calculated by the compiler.

[b] Lower boundary of observation (appearance of second liquid phase with rising temperature).

[c] L_1 = dilute liquid phase; L_2 = concentrated liquid phase; L_x = liquid phase near the consolute solution composition (where composition L_1, equals composition L_2).

[d] The mass% of solute was estimated graphically; values at upper temperature limit of immiscibility.

[e] No second liquid or critical phenomenon is observed at temperatures up to 410°C.

[f] Additional values are given in ref. (1).

COMPONENTS:	ORIGINAL MEASUREMENTS:
(1) Tripotassium phosphate; K_3PO_4; [7778-53-2]	Parker, E.G.
(2) Phosphoric acid; H_3PO_4; [7664-38-2]	J. Phys. Chem. 1914, 18, 653-61.
(3) Potassium hydroxide; KOH; [1310-58-3]	
(4) Water; H_2O; [7732-18-5]	

VARIABLES:	PREPARED BY:
Composition at 25°C.	J. Eysseltová

EXPERIMENTAL VALUES:

Solubility in the KOH-H_3PO_4-H_2O system at 25°C.

K^+ conc[b]	PO_4^{3-} conc[b]	K_3PO_4[a] mass%	mol/kg	KOH[a] mass%	mol/kg	H_3PO_4[a] mass%	mol/kg	H_2O[a] mass%	solid phase[c]
1.40	8.56	9.90	4.32	----	----	79.31	75.08	18.79	A
1.47	6.74	10.40	1.72	----	----	61.25	22.04	28.35	"
2.31	5.00	16.34	1.82	----	----	41.45	10.02	42.21	B
1.89	3.20	13.37	1.02	----	----	25.18	4.18	61.45	"
1.78	2.60	12.59	0.87	----	----	19.66	2.96	67.75	"
1.51	1.81	10.68	0.65	----	----	12.80	1.70	76.52	"
1.46	1.46	10.33	0.60	----	----	9.53	1.21	80.14	"
2.31	1.84	16.34	1.05	----	----	10.48	1.46	73.18	"
2.61	1.99	18.46	1.23	----	----	10.97	1.58	70.57	"
3.06	2.25	21.65	1.53	----	----	12.05	1.85	66.30	"
3.20	2.28	22.64	1.62	----	----	11.89	1.85	65.47	"
3.98	2.67	28.16	2.26	----	----	13.16	2.28	58.68	"
5.22	3.24	36.93	3.59	----	----	14.70	3.10	48.37	"
5.33	3.33	37.71	3.77	----	----	15.22	3.30	47.07	"
5.67	3.41	40.11	4.20	----	----	14.89	3.37	45.00	"
6.38	3.69	45.14	5.37	----	----	15.32	3.95	39.54	"
6.80	3.92	48.11	6.35	----	----	16.20	4.63	35.69	"
7.23	3.73	51.15	6.71	----	----	12.93	3.67	35.92	C
7.79	3.66	55.11	7.53	----	----	10.42	3.08	34.47	"
8.56	3.42	60.56	8.42	----	----	5.55	1.67	33.89	"
8.81	2.92	61.98	7.73	0.28	0.13	----	----	37.74	D
7.14	2.07	43.94	4.07	5.21	1.82	----	----	50.85	"
7.18	2.09	44.36	4.11	4.88	1.71	----	----	50.83	"
9.19	0.48	10.18	1.03	43.48	16.72	----	----	46.34	"

(continued next page)

AUXILIARY INFORMATION

METHOD/APPARATUS/PROCEDURE:	SOURCE AND PURITY OF MATERIALS:
Bottles containing various amounts of phosphoric acid and potassium hydroxide in solution and in contact with a solid phase were placed in a thermostat and allowed to rotate until equilibrium was established. Phosphorus was analyzed according to ref. (1) and potassium was determined as $K_2[PtCl_6]$.	No information is given.
	ESTIMATED ERROR:
	No information is given.
	REFERENCES:
	1. Treadwell, F.P.; Hall, W.T. *Analytical Chemistry*, Vol. II, 1913, p. 434.

COMPONENTS:	ORIGINAL MEASUREMENTS:
(1) Tripotassium phosphate; K_3PO_4; [7778-53-2]	Parker, E.G.
(2) Phosphoric acid; H_3PO_4; [7664-38-2]	J. *Phys. Chem.* 1914, *18*, 653-61.
(3) Potassium hydroxide; KOH; [1310-58-3]	
(4) Water; H_2O; [7732-18-5]	

EXPERIMENTAL VALUES cont'd:

Solubility in the KOH–H_3PO_4–H_2O system at 25°C.

K^+ conc[b]	PO_4^{3-} conc[b]	K_3PO_4[a] mass%	mol/kg	KOH[a] mass%	mol/kg	H_3PO_4[a] mass%	mol/kg	H_2O[a] mass%	solid phase[c]
9.23	0.46	9.76	0.99	44.04	16.99	----	----	46.20	D
9.41	0.38	8.06	0.83	46.40	18.16	----	----	45.54	"
9.79	0.23	4.88	0.52	51.06	20.65	----	----	44.06	"
9.80	0.24	5.09	0.54	50.94	20.65	----	----	43.97	"
9.48	0.32	6.79	0.70	47.80	18.76	----	----	45.41	"
9.76	0.24	5.09	0.54	50.72	20.46	----	----	44.21	E
9.76	0.22	4.67	0.49	51.06	20.55	----	----	44.27	"
9.77	0.12	2.54	0.26	52.79	21.07	----	----	44.67	"

[a] All these values were calculated by the compiler.

[b] The concentration unit is: mol/1000 g of the solution.

[c] The solid phases are: A = $KH_2PO_4 \cdot H_3PO_4$; B = KH_2PO_4; C = K_3PO_4; D = $K_3PO_4 \cdot 3H_2O$;

E = $KOH \cdot 2H_2O$.

COMPONENTS:	ORIGINAL MEASUREMENTS:
(1) Tripotassium phosphate; K$_3$PO$_4$; [7778-53-2]	Jänecke, E.
(2) Phosphoric acid; H$_3$PO$_4$; [7664-38-2]	Z. Phys. Chem. <u>1927</u>, 127, 71-92.
(3) Potassium hydroxide; KOH; [1310-58-3]	
(4) Water; H$_2$O; [7732-18-5]	

VARIABLES:	PREPARED BY:
Composition and temperature.	J. Eysseltová

EXPERIMENTAL VALUES: Solubility in the K$_3$PO$_4$–H$_3$PO$_4$–KOH–H$_2$O system.

K$_2$O conc^a	H$_2$O conc^a	K$_3$PO$_4^b$ mass%	K$_3$PO$_4^b$ mol/kg	KOH^b mass%	KOH^b mol/kg	H$_3$PO$_4^b$ mass%	H$_3$PO$_4^b$ mol/kg	H$_2$O mass%
					temp. = 0°C.			
96.0	107.0	5.77	0.62	50.66	20.73	----	----	43.61
90.5	146.0	11.55	1.01	34.66	11.48	----	----	53.83
86.5	156.0	15.77	1.31	27.74	8.75	----	----	56.49
82.2	173.0	19.50	1.52	20.40	6.05	----	----	60.10
79.5	171.0	22.62	1.76	17.00	5.02	----	----	60.38
76.3	153.0	28.01	2.26	13.71	4.19	----	----	58.28
73.5	153.0	31.32	2.50	9.76	2.95	----	----	58.92
70.8	160.0	33.59	2.61	5.80	1.70	----	----	60.61
68.5	153.0	37.24	2.92	2.72	0.80	----	----	60.04
66.5	152.0	39.66	3.10	----	----	0.07	0.01	60.27
63.9	147.0	38.88	3.11	----	----	2.25	0.39	58.87
61.0	104.0	44.94	4.28	----	----	5.67	1.17	49.39
58.5	82.0	48.31	5.35	----	----	9.21	2.21	42.48
56.0	97.1	42.70	4.35	----	----	11.13	2.46	46.17
52.4	92.0	41.02	4.42	----	----	15.32	3.58	43.66
50.5	219.0	23.79	1.70	----	----	10.45	1.62	65.76
48.3	363.0	15.68	0.97	----	----	8.19	1.09	76.13
39.8	460.0	10.68	0.63	----	----	9.91	1.27	79.41
33.0	280.0	13.05	0.89	----	----	18.32	2.72	68.63
30.0	260.0	12.52	0.88	----	----	21.07	3.23	66.41
25.0	63.0	23.05	4.51	----	----	52.90	22.45	24.05
24.7	172.0	13.64	1.18	----	----	31.93	5.98	54.43
19.5	80.0	16.28	2.60	----	----	54.24	18.28	29.48

(continued next page)

AUXILIARY INFORMATION

METHOD/APPARATUS/PROCEDURE:	SOURCE AND PURITY OF MATERIALS:
No information is given.	No information is given.
	ESTIMATED ERROR:
	No information is given.
	REFERENCES:

COMPONENTS:	ORIGINAL MEASUREMENTS:
(1) Tripotassium phosphate; K$_3$PO$_4$; [7778-53-2]	Jänecke, E.
(2) Phosphoric acid; H$_3$PO$_4$; [7664-38-2]	Z. Phys. Chem. 1927, 127, 71-92.
(3) Potassium hydroxide; KOH; [1310-58-3]	
(4) Water; H$_2$O; [7732-18-5]	

EXPERIMENTAL VALUES cont'd:

Solubility in the K$_3$PO$_4$-H$_3$PO$_4$-KOH-H$_2$O system.

K$_2$O conc ^{a}	H$_2$O conc ^{a}	K$_3$PO$_4$ ^{b} mass%	mol/kg	KOH ^{b} mass%	mol/kg	H$_3$PO$_4$ ^{b} mass%	mol/kg	H$_2$O ^{b} mass%
				temp. = 0°C.				
13.1	131.0	8.52	0.92	----	----	48.01	11.27	43.47
10.0	44.0	10.43	6.08	----	----	81.48	102.97	8.09
2.4	73.5	2.07	0.46	----	----	76.71	36.91	30.22
				temp. = 25°C.				
96.0	97.0	6.07	0.70	53.23	23.31	----	----	40.70
90.5	98.0	14.35	1.58	43.07	18.02	----	----	42.58
86.5	102.0	19.98	2.09	35.16	13.97	----	----	44.86
82.2	130.0	23.14	2.07	24.22	8.20	----	----	52.64
79.5	121.0	26.54	2.33	19.95	6.64	----	----	53.51
76.3	119.0	32.36	2.94	15.83	5.45	----	----	51.81
73.5	127.0	34.91	3.03	10.88	3.57	----	----	54.21
70.8	118.0	40.06	3.55	6.92	2.32	----	----	53.02
68.5	119.0	43.02	3.76	3.14	1.04	----	----	53.84
66.5	117.5	45.95	4.01	-----	----	0.08	0.01	53.97
63.9	101.5	47.66	4.53	-----	----	2.76	0.56	49.58
61.0	80.2	50.88	5.61	-----	----	6.42	1.53	42.70
58.5	82.0	48.31	5.35	-----	----	9.21	2.21	42.48
56.0	76.0	47.82	5.67	-----	----	12.47	3.20	39.71
52.4	88.0	41.89	4.64	-----	----	15.64	3.76	22.47
50.5	159.0	29.30	2.38	-----	----	12.87	2.27	57.83
48.3	292.0	18.52	1.21	-----	----	9.67	1.37	71.81
39.8	435.0	11.18	0.67	-----	----	10.38	1.35	78.42
33.0	249.0	14.21	1.01	-----	----	19.95	3.09	65.84
30.0	185.0	15.82	1.29	-----	----	26.62	4.71	57.56
25.0	59.0	23.63	5.03	-----	----	54.23	25.00	22.14
24.7	121.0	16.79	1.80	-----	----	39.30	9.13	43.91
19.5	58.0	18.55	4.44	-----	----	61.79	32.09	19.66
13.1	92.0	10.25	1.51	-----	----	57.76	18.43	31.99
10.0	44.0	10.43	6.08	-----	----	81.48	102.87	12.44
2.4	47.8	2.44	1.53	-----	----	90.05	122.44	7.51

^{a}The concentration unit is: g/100 g of K$_2$O + P$_2$O$_5$.

^{b}These concentrations were calculated by the compiler.

COMPONENTS:	ORIGINAL MEASUREMENTS:
(1) Potassium dihydrogenphosphate; KH_2PO_4; [7778-77-0]	Krasil'shtschikov, A.I.
(2) Potassium hydroxide; KOH; [1310-58-3]	*Izv. In-ta Fiz.-khim. An.* <u>1933</u>, *6*, 159-68.
(3) Phosphoric acid; H_3PO_4; [7664-38-2]	
(4) Water; H_2O; [7732-18-5]	

VARIABLES:	PREPARED BY:
Temperature and composition.	J. Eysseltová

EXPERIMENTAL VALUES:

Part 1. The following data are given in the paper.

soln. no.	t/°C	density g/cm^3	K_2O concn.^{a}	concn.^{b}	concn.^{c}	P_2O_5 concn.^{a}	concn.^{b}	concn.^{c}	H_2O concn.^{c}
1	0	1.131	14.5	6.47	44.2	12.1	8.16	55.8	583.6
2	0	1.101	10.9	5.01	41.2	10.3	7.16	55.8	721.6
3	0	1.094	9.44	4.39	39.85	9.44	6.62	60.15	808.2
4	0	1.096	9.6	4.44	38.9	10.0	6.97	61.1	776.4
5	0	1.102	9.9	4.57	37.5	11.0	7.63	62.5	719.6
6	0	1.157	12.4	5.30	28.8	20.4	13.10	71.2	443.5
7	25	1.247	32.9	12.5	46.0	25.5	14.65	54.0	268.3
8	25	1.193	23.5	9.63	43.8	20.1	12.35	56.2	355.0
9	25	1.176	20.3	8.45	42.6	18.0	11.41	57.4	403.6
10	25	1.157	17.9	7.65	41.2	17.1	10.90	58.8	439.1
11	25	1.151	16.6	7.17	40.5	16.2	10.55	59.5	464.3
12	25	1.148	16.0	6.95	40.2	15.9	10.34	59.8	478.4
13	25	1.147	15.9	6.89	39.85	15.9	10.40	60.15	478.5
14	25	1.148	15.9	6.88	39.5	16.1	10.56	60.5	473.4
15	25	1.152	16.5	7.03	39.2	16.9	10.92	60.8	467.1
16	25	1.154	16.8	7.13	38.8	17.4	11.26	61.2	443.7
17	25	1.160	16.8	7.18	38.2	18.2	11.63	61.8	431.6
18	25	1.162	16.9	7.18	37.5	18.7	11.95	62.5	422.8
19	25	1.168	17.3	7.21	36.3	20.0	12.65	63.7	403.6
20	25	1.206	18.8	7.56	32.5	26.1	15.70	67.5	330.0
21	25	1.242	21.9	8.30	30.1	33.9	19.30	69.9	262.3
22	50		45.7	12.68	27.2	80.8	33.96	72.8	114.4
23	50		36.0	11.85	32.1	50.6	25.10	67.9	170.7

(continued next page)

AUXILIARY INFORMATION

METHOD/APPARATUS/PROCEDURE:	SOURCE AND PURITY OF MATERIALS:
The mixtures were allowed to equilibrate for 12-15 hours in a water thermostat. The phosphorus content was determined gravimetrically as $Mg_2P_2O_7$, the hydrogen or hydroxide ion content was determined by titration using methylorange as indicator.	Kahlbaum KH_2PO_4 was used. No other information is given.
	ESTIMATED ERROR:
	The temperature was controlled to within ± 0.1 K.
	REFERENCES:

COMPONENTS:	ORIGINAL MEASUREMENTS:
(1) Potassium dihydrogenphosphate; KH$_2$PO$_4$; [7778-77-0] (2) Potassium hydroxide; KOH; [1310-58-3] (3) Phosphoric acid; H$_3$PO$_4$; [7664-38-2] (4) Water; H$_2$O; [7732-18-5]	Krasil'shtschikov, A.I. *Izv. In-ta Fiz.-khim. An.* <u>1933</u>, *6*, 159-68

EXPERIMENTAL VALUES cont'd:

Part 1. The following data are given in the paper.

soln. no.	t/°C	density g/cm^3	K$_2$O concn.^{a}	concn.^{b}	concn.^{c}	P$_2$O$_5$ concn.^{a}	concn.^{b}	concn.^{c}	H$_2$O concn.^{c}
24	50		30.5	11.0	35.9	36.2	19.70	64.1	225.8
25	50		28.6	10.78	39.0	29.6	16.84	61.0	262.1
26	50		27.7	10.50	38.9	28.6	16.50	61.1	270.3
27	50		25.9	10.12	39.5	26.4	15.53	60.5	289.9
28	50		25.8	10.09	39.85	25.8	15.22	60.15	295.1
29	50		28.6	10.91	40.6	27.6	15.95	59.4	272.4
30	50		32.0	11.99	42.2	29.0	16.41	57.8	252.1
31	50		125.0	28.69	51.2	79.1	27.37	48.8	78.4

^{a}The concentration unit is: mol/1000 mol H$_2$O.

^{b}The concentration unit is: g/100 g of solution.

^{c}The concentration unit is: g/100 g of oxides.

Part 2. The compiler has recalculated the data in Part 1 to give the following values.

soln. no.	t/°C	KH$_2$PO$_4$ mass%	mol/kg	H$_3$PO$_4$ mass%	mol/kg	KOH mass%	mol/kg
1	0	15.65	1.363	----	----	0.013	0.002
2	0	13.73	1.169	----	----	0.003	0.000
3	0	12.68	1.067	0.006	0.000	----	----
4	0	12.83	1.086	0.385	0.045	----	----
5	0	13.21	1.131	1.03	0.122	----	----
6	0	15.31	1.449	7.06	0.928	----	----
7	25	28.09	2.871	----	----	0.035	0.008
8	25	23.68	2.280	----	----	0.018	0.004
9	25	21.88	2.056	----	----	0.011	0.002
10	25	20.90	1.941	----	----	0.005	0.001
11	25	20.23	1.863	----	----	0.002	0.000
12	25	19.83	1.817	----	----	0.001	0.000
13	25	19.91	1.827	0.024	0.003	----	----
14	25	19.88	1.829	0.265	0.033	----	----
15	25	20.31	1.883	0.45	0.058	----	----
16	25	20.60	1.923	0.712	0.092	----	----
17	25	20.75	1.962	1.12	0.146	----	----
18	25	20.75	1.962	1.56	0.205	----	----
19	25	20.83	1.995	2.46	0.327	----	----
20	25	21.85	2.222	5.95	0.840	----	----
21	25	23.98	2.644	9.38	1.436	----	----
22	50	36.64	6.28	20.51	4.88	----	----
23	50	34.24	4.51	10.00	1.83	----	----
24	50	31.79	3.65	4.31	0.689	----	----
25	50	31.15	3.36	0.82	0.123	----	----
26	50	30.34	3.24	0.94	0.139	----	----
27	50	29.24	3.05	0.39	0.056	----	----
28	50	29.15	3.04	0.44	0.063	----	----
29	50	30.59	3.24	----	----	0.004	0.001
30	50	31.45	3.37	----	----	0.013	0.003
31	50	52.49	8.14	----	----	0.13	0.050

COMPONENTS:	ORIGINAL MEASUREMENTS:
(1) Tripotassium phosphate; K_3PO_4; [7778-53-2]	1. Berg, A.G. *Izv. Akad. Nauk SSSR* <u>1933</u>, 167-82.
(2) Phosphoric acid; H_3PO_4; [7664-38-2]	2. Berg, A.G. *Izv. Akad. Nauk SSSR* <u>1938</u>, 147-60.
(3) Potassium hydroxide; KOH; [1310-58-3]	
(4) Water; H_2O; [7732-18-5]	

VARIABLES:	PREPARED BY:
Composition at 25°C.	J. Eysseltová

EXPERIMENTAL VALUES:

Part 1. Solubility isotherm in the K_3PO_4-H_3PO_4-KOH-H_2O system at 25°C.

soln. no.	K_2O mass%	mol%	P_2O_5 mass%	mol%	H_2O mass%	mol%	solid phase[a]
1	47.25	14.62	----	----	52.72	83.38	A1, A2(1)
2	45.51	13.78	----	----	54.49	86.22	A2
3	47.49	17.92	1.37	0.74	41.14	81.34	A1 + B3
4	46.18	14.44	1.73	0.36	52.09	85.20	B3, A1(1)
5	45.92	14.32	1.78	0.37	52.30	85.31	" "
6	45.90	14.28	1.86	0.38	52.24	85.34	A2 + B3
7	45.75	14.25	1.85	0.38	52.40	85.37	B3(1) A2 + B3
8	43.20	13.37	3.75	0.77	53.05	85.86	B3
9	41.69	13.65	9.04	1.97	49.27	84.38	B3
10	41.53	13.67	9.52	2.08	48.95	84.25	B3 + B7
11	41.67	13.67	9.61	2.11	48.72	84.13	B3[b]
12	41.63	14.20	11.78	2.67	46.59	83.13	"
13	41.55	14.55	13.51	3.14	44.94	82.31	"
14	41.50	14.78	14.56	3.36	43.94	81.86	"
15	41.70	15.03	15.19	3.67	43.11	81.30	"
16	41.58	14.98	15.23	3.67	43.19	81.35	"
17	41.56	15.03	15.45	3.70	42.99	81.27	"
18	41.87	15.49	16.54	4.06	41.59	80.45	"
19	42.01	15.81	17.44	4.39	40.55	79.80	"
20	42.12	15.95	17.69	4.45	40.19	79.60	"
21	42.44	17.08	20.78	5.54	36.78	77.38	"

(continued next page)

AUXILIARY INFORMATION

METHOD/APPARATUS/PROCEDURE:	SOURCE AND PURITY OF MATERIALS:
The isothermal method was used. The mixtures were equilibrated in a water thermostat by agitation (900 rpm) for at least 20 hours. The solid and liquid phases were separated from each other by centrifuging at 1500-2000 rpm. The potassium content was determined as $KClO_4$, the phosphorus content as $Mg_2P_2O_7$, and the water content by difference. The nature of the solid phases was determined microscopically and by the use of Schreinemakers' method.	Kahlbaum reagent grade KOH was used. The KH_2PO_4 was recrystallized two or three times. The H_3PO_4 was imported and had a content of 90%.

	ESTIMATED ERROR:
	The temperature was kept constant to within ± 0.05 K. For further information see the critical evaluation.

	REFERENCES:

COMPONENTS:	ORIGINAL MEASUREMENTS:
(1) Tripotassium phosphate; K$_3$PO$_4$; [7778-53-2]	1. Berg. A.G. *Izv. Akad. Nauk SSSR* <u>1933</u>, 167-82.
(2) Phosphoric acid; H$_3$PO$_4$; [7664-38-2]	2. Berg. A.G. *Izv. Akad. Nauk SSSR* <u>1938</u>, 147-60.
(3) Potassium hydroxide; KOH; (1310-58-3]	
(4) Water; H$_2$O; [7732-18-5]	

EXPERIMENTAL VALUES cont'd:

Part 1. Solubility isotherm in the K$_3$PO$_4$–H$_3$PO$_4$–KOH–H$_2$O system at 25°C.

soln. no.	K$_2$O mass%	mol%	P$_2$O$_5$ mass%	mol%	H$_2$O mass%	mol%	solid phase[a]
22	43.34	19.93	26.75	8.16	29.91	71.91	B3 + C
23	41.53	13.67	9.52	2.08	48.95	84.25	B3 + B7
24	41.50	13.66	9.57	2.10	48.93	84.24	B3 + B7
25	37.18	11.32	8.15	1.65	54.67	87.03	B7
26	34.39	10.51	10.99	2.23	54.62	87.26	"
27	34.06	11.25	16.70	3.66	49.21	85.09	"
28	39.42	16.03	24.17	6.52	36.41	77.45	B7 + C3
29	41.48	17.99	25.62	7.37	32.90	74.64	B7[b]
30	41.64	18.71	27.21	8.10	31.15	73.19	C[c]
31	41.07	18.35	27.46	8.14	31.47	73.51	"[b]
32	40.40	----	28.12	----	31.48	----	C[b]
33	40.22	17.80	27.79	8.16	31.99	74.04	C[c]
34	37.59	16.57	30.03	8.78	32.39	74.65	"
35	37.20	16.51	30.72	9.04	32.08	74.45	"
36	36.96	16.31	30.65	8.96	32.39	74.43	"
37	41.90	18.06	24.89	7.11	32.21	74.83	C3[c]
38	40.33	16.81	24.62	6.80	35.05	76.39	"
39	36.69	14.22	24.37	6.28	38.94	79.45	C3[d]
40	34.15	13.03	25.48	6.45	40.37	80.52	"
41	34.05	12.98	25.52	6.45	40.43	80.57	"
42	34.22	13.19	26.02	6.69	39.76	80.12	C3
43	34.13	13.22	26.39	6.82	39.48	79.96	"
44	34.22	13.36	26.90	7.26	38.88	79.38	"
45	34.55	13.76	27.57	7.32	37.88	78.92	"
46	34.92	14.16	28.90	7.77	36.82	78.07	"
47	34.97	14.42	29.04	7.95	35.99	77.63	C3 + D[b,e]
48	37.46	16.29	29.47	8.50	33.07	75.21	D[b]
49	35.84	14.98	28.95	8.03	35.21	76.99	"
50	35.53	14.81	29.09	8.08	35.38	77.11	C3 + D[b]
51	35.51	14.76	28.96	7.99	35.53	77.25	C3(1)[b]
52	35.16	14.54	29.01	7.96	35.83	77.50	C3 + D
53	35.01	14.49	29.19	8.02	35.80	77.49	D
54	34.48	14.19	29.39	8.02	36.13	77.78	"
55	33.32	13.18	28.30	7.42	38.38	79.40	E
56	32.65	12.65	27.75	7.13	39.60	80.22	"
57	31.30	11.98	27.12	7.72	41.58	81.30	"
58	26.33	8.75	24.23	5.34	49.44	85.91	"
59	20.12	5.78	19.67	3.75	60.21	90.47	"
60	17.68	4.84	18.13	3.29	64.19	91.87	"
61	14.20	3.64	15.89	2.70	69.91	93.66	"
62	11.66	2.77	13.37	2.10	74.97	95.13	"
63	7.77	1.76	10.73	1.61	81.50	96.63	"
64	7.14	1.60	10.45	1.56	82.41	96.84	"
65	6.87	1.53	10.17	1.51	82.96	96.96	"
66	6.90	1.55	11.00	1.65	82.10	96.80	"
67	7.06	1.60	11.44	1.72	81.50	96.68	"
68	7.08	1.62	12.12	1.84	80.80	96.54	"
69	8.49	2.15	20.35	3.43	71.16	94.42	"
70	9.52	2.65	26.89	4.95	63.59	92.40	"
71	10.19	3.10	32.90	6.48	56.91	90.42	"
72	11.69	4.18	42.42	10.05	45.89	85.77	E[d]
73	12.50	4.84	46.38	11.91	41.12	83.25	"
74	12.68	5.01	47.40	12.43	39.92	82.56	E[b]
75	12.91	5.38	49.90	13.49	37.19	81.13	"
76	13.67	5.86	50.76	14.43	35.57	79.71	"
77	14.27	6.44	52.78	15.80	32.95	77.76	"
78	14.29	6.49	53.94	16.26	31.77	77.25	"
79	12.77	5.04	47.16	12.33	40.07	82.63	E + F

(continued next page)

COMPONENTS:	ORIGINAL MEASUREMENTS:
(1) Tripotassium phosphate; K_3PO_4; [7778-53-2]	1. Berg, A.G. *Izv. Akad. Nauk SSSR* <u>1933</u>, 167-82.
(2) Phosphoric acid; H_3PO_4; [7664-38-2]	
(3) Potassium hydroxide; KOH; [1310-58-3]	2. Berg, A.G. *Izv. Akad. Nauk SSSR* <u>1938</u>, 147-60.
(4) Water; H_2O; [7732-18-5]	

EXPERIMENTAL VALUES cont'd:

Part 1. Solubility isotherm in the K_3PO_4-H_3PO_4-KOH-H_2O system at 25°C.

soln. no.	K_2O mass%	mol%	P_2O_5 mass%	mol%	H_2O mass%	mol%	solid phase[a]
80	11.83	4.64	47.73	12.42	41.07	82.94	F
81	10.27	3.99	48.66	12.54	41.07	83.47	"
82	9.69	3.78	49.40	12.78	40.91	83.44	"
83	8.36	3.32	51.58	13.57	40.06	83.11	"
84	8.10	3.22	51.91	13.68	39.99	83.10	"
85	6.99	2.90	55.23	15.19	37.78	81.91	"
86	6.85	2.93	56.80	15.75	36.35	81.32	"
87	8.36	4.39	65.08	22.67	26.56	72.94	"
88	8.76	4.74	65.94	23.67	25.30	71.59	"
89	----	----	64.41	18.67	35.59	81.33	G
90	0.46	0.20	64.40	18.82	35.14	80.98	"
91	1.13	0.51	64.83	19.37	34.01	80.12	"
92	1.27	0.57	64.86	19.47	33.87	80.00	"
93	1.93	0.87	65.03	19.80	33.04	79.43	"
94	2.35	1.09	65.16	20.06	32.49	78.85	"
95	0.89	0.42	67.55	21.26	31.56	78.32	"

[a] The solid phases are: A1 = $KOH \cdot H_2O$; A2 = $KOH \cdot 2H_2O$; B3 = $K_3PO_4 \cdot 3H_2O$; B7 = $K_3PO_4 \cdot 7H_2O$;

C = K_2HPO_4; C3 = $K_2HPO_4 \cdot 3H_2O$; D = $K_2HPO_4 \cdot KH_2PO_4 \cdot 3H_2O$; E = KH_2PO_4; F = $KH_5(PO_4)_2$;

G = $2H_3PO_4 \cdot H_2O$. All the data concerning these phases are from source paper (2).

Those from source paper (1) are indicated by A(1), for example.

[b] This is a metastable equilibrium.

[c] This is given as stable in source paper (1), but as metastable in source paper (2).

[d] This is given as metastable in source paper (1), but as stable in source paper (2).

[e] These data appear in source paper (1) only.

Part 2. The compiler has calculated the following results from the data given in Part 1 above.

soln. no.	K_3PO_4 mass%	mol/kg	KOH mass%	mol/kg	H_3PO_4 mass%	mol/kg
1	----	----	56.28	22.95	----	----
2	----	----	54.21	21.10	----	----
3	4.09	0.45	53.32	22.32	----	----
4	5.17	0.55	50.91	20.66	----	----
5	5.32	0.56	50.48	20.35	----	----
6	5.38	0.58	51.60	21.38	----	----
7	5.53	0.58	50.11	20.13	----	----
8	11.21	1.14	42.57	16.41	----	----
9	27.03	2.84	28.22	11.24	----	----
10	28.47	3.00	26.89	10.74	----	----
11	28.74	3.04	26.84	10.77	----	----
12	35.23	3.84	21.65	8.95	----	----
13	40.40	4.51	17.45	7.38	----	----
14	43.54	4.93	14.90	6.39	----	----
15	45.43	5.23	13.65	5.94	----	----
16	45.55	5.22	13.41	5.82	----	----
17	46.21	5.31	12.86	5.60	----	----
18	49.47	5.84	10.65	4.76	----	----
19	52.16	6.27	8.68	3.95	----	----
20	52.91	6.41	8.22	3.77	----	----

(continued next page)

COMPONENTS:	ORIGINAL MEASUREMENTS:
(1) Tripotassium phosphate; K₃PO₄; [7778-53-2]	1. Berg, A.G. *Izv. Akad. Nauk SSSR* <u>1933</u>, 167-82.
(2) Phosphoric acid; H₃PO₄; [7664-38-2]	2. Berg, A.G. *Izv. Akad. Nauk SSSR* <u>1938</u>, 147-60.
(3) Potassium hydroxide; KOH; [1310-58-3]	
(4) Water; H₂O; [7732-18-5]	

EXPERIMENTAL VALUES cont'd:

Part 2. The compiler has calculated the following results from the data given in Part 1 above.

soln. no.	K₃PO₄ mass%	K₃PO₄ mol/kg	KOH mass%	KOH mol/kg	H₃PO₄ mass%	H₃PO₄ mol/kg
21	62.15	8.00	1.27	0.62	----	----
22	65.14	10.98	----	----	6.90	2.52
23	28.47	3.00	26.89	10.74	----	----
24	28.62	3.02	26.74	10.67	----	----
25	24.37	2.26	24.96	8.78	----	----
26	32.87	2.96	14.90	5.08	----	----
27	49.95	4.79	0.96	0.35	----	----
28	59.25	8.04	----	----	6.06	1.78
29	62.35	9.47	----	----	6.63	2.18
30	62.59	10.27	----	----	8.71	3.10
31	61.73	10.09	----	----	9.45	3.35
32	60.72	10.05	----	----	10.83	3.88
33	60.45	9.80	----	----	10.50	3.69
34	56.50	9.47	----	----	15.41	5.60
35	55.91	9.59	----	----	16.64	6.18
36	55.55	9.43	----	----	16.71	6.14
37	62.98	9.36	----	----	5.33	1.71
38	60.62	8.56	----	----	6.05	1.85
39	55.15	7.09	----	----	8.22	2.29
40	51.33	6.50	----	----	11.52	3.16
41	51.18	6.48	----	----	11.64	3.19
42	51.43	6.50	----	----	12.21	3.43
43	51.30	6.73	----	----	12.79	3.63
44	51.43	6.89	----	----	13.43	3.90
45	51.93	7.20	----	----	14.12	4.24
46	52.49	7.77	----	----	15.70	5.04
47	52.56	7.84	----	----	15.86	5.12
48	56.30	9.16	----	----	14.73	5.19
49	53.87	8.18	----	----	15.14	4.98
50	53.40	8.10	----	----	15.54	5.11
51	53.37	8.04	----	----	15.38	5.02
52	52.85	7.91	----	----	15.69	5.09
53	52.62	7.91	----	----	16.04	5.22
54	51.82	7.75	----	----	16.69	5.41
55	50.08	6.95	----	----	15.98	4.80
56	49.07	6.56	----	----	15.69	4.54
57	47.04	5.96	----	----	15.75	4.32
58	39.57	4.12	----	----	15.21	3.43
59	30.24	2.51	----	----	13.21	2.38
60	26.57	2.06	----	----	12.78	2.15
61	21.34	1.51	----	----	12.10	1.85
62	17.52	1.14	----	----	10.38	1.46
63	11.67	0.69	----	----	9.43	1.22
64	10.73	0.63	----	----	9.48	1.21
65	10.32	0.60	----	----	9.28	1.17
66	10.37	0.61	----	----	10.40	1.34
67	10.61	0.63	----	----	10.90	1.41
68	10.64	0.64	----	----	11.82	1.55
69	12.76	0.92	----	----	22.21	3.48
70	14.31	1.22	----	----	30.53	5.64
71	15.31	1.55	----	----	38.36	8.45
72	17.57	2.59	----	----	50.47	16.11
73	18.78	3.42	----	----	55.37	21.87
74	19.06	3.69	----	----	56.66	23.81
75	19.40	4.42	----	----	59.95	29.64
76	20.54	5.13	----	----	60.61	32.83
77	21.45	6.49	----	----	62.98	41.30

(continued next page)

COMPONENTS:	ORIGINAL MEASUREMENTS:
(1) Tripotassium phosphate; K_3PO_4; [7778-53-2]	1. Berg, A.G. *Izv. Akad. Nauk SSSR* <u>1933</u>, 167-82.
(2) Phosphoric acid; H_3PO_4; [7664-38-2]	2. Berg, A.G. *Izv. Akad. Nauk SSSR* <u>1938</u>, 147-60.
(3) Potassium hydroxide; KOH; [1310-58-3]	
(4) Water; H_2O; [7732-18-5]	

EXPERIMENTAL VALUES cont'd:

Part 2. The compiler has calculated the following results from the data given in Part 1 above.

soln. no.	K_3PO_4 mass%	K_3PO_4 mol/kg	KOH mass%	KOH mol/kg	H_3PO_4 mass%	H_3PO_4 mol/kg
78	21.48	7.25	----	----	64.57	47.26
79	19.19	3.68	----	----	56.26	23.40
80	17.78	3.41	----	----	57.70	24.02
81	15.43	2.96	----	----	60.07	25.02
82	14.56	2.86	----	----	61.49	26.21
83	12.56	2.68	----	----	65.42	30.33
84	12.17	2.63	----	----	66.06	30.97
85	10.50	2.73	----	----	71.41	40.31
86	10.29	3.02	----	----	73.68	46.92
87	12.56	17.58	----	----	84.06	254.88
88	13.16	33.45	----	----	84.97	467.70
89	----	----	----	----	88.93	82.01
90	0.69	0.30	----	----	88.60	84.45
91	1.69	0.83	----	----	88.73	94.61
92	1.90	0.95	----	----	88.67	96.11
93	2.90	1.58	----	----	88.45	104.40
94	3.53	2.04	----	----	88.34	110.94
95	1.33	1.04	----	----	92.65	157.36

COMPONENTS:	ORIGINAL MEASUREMENTS:
(1) Tripotassium phosphate; K$_3$PO$_4$; [7778-53-2] (2) Phosphoric acid; H$_3$PO$_4$; [7664-38-2] (3) Potassium hydroxide; KOH; [1310-58-3] (4) Water; H$_2$O; [7732-18-5]	1. Ravich, M.I. *Kaliy* <u>1936</u>, *10*, 33-7. 2. Ravich, M.I. *Izv. Akad. Nauk SSSR* <u>1938</u>, 167-76.

VARIABLES:	PREPARED BY:
Composition at 0°C.	J. Eysseltová

EXPERIMENTAL VALUES:

Part 1. Solubility isotherm for the K$_3$PO$_4$–H$_3$PO$_4$–KOH–H$_2$O system at 0°C.

soln. no.	K$_2$O		P$_2$O$_5$		H$_2$O		solid phase[a]
	mass%	mol%	mass%	mol%	mass%	mol%	
1	----	----	57.17	14.48	42.83	85.52	A
2	1.64	0.65	58.01	15.31	40.35	84.04	"
3	3.78	1.58	58.81	16.37	37.78	82.05	"
4	4.89	2.12	59.57	17.15	35.54	80.73	"
5	5.06	2.20	59.50	17.16	35.44	80.64	A +[b] B
6	5.75	2.67	62.01	19.09	32.24	78.24	B
7	5.51	2.52	61.60	18.70	32.89	78.78	"
8	5.27	2.37	61.11	18.23	33.62	79.40	"
9	4.93	2.12	58.92	16.76	36.15	81.12	B
10	4.38	1.76	56.12	14.99	39.50	83.25	"
11	4.33	1.73	55.87	14.85	39.80	83.42	"
12	5.30	1.95	50.19	12.26	44.51	85.79	"
13	6.05	2.17	47.93	11.41	46.02	86.42	"
14	8.15	2.88	45.02	10.55	46.83	86.57	"
15	10.37	3.70	43.6	10.33	46.01	85.97	B + C
16	10.54	3.77	43.69	10.31	45.97	85.92	"
17	10.92	4.02	45.04	11.01	44.04	84.97	C[b]
18	10.22	3.53	41.76	9.58	48.02	86.69	C
19	8.93	2.78	35.92	7.42	55.15	89.80	"
20	8.17	2.39	32.30	6.28	59.53	91.33	"
21	6.85	1.75	22.48	3.81	70.67	99.44	"
22	5.75	1.35	15.75	2.42	78.50	96.23	"

(continued next page)

AUXILIARY INFORMATION

METHOD/APPARATUS/PROCEDURE:	SOURCE AND PURITY OF MATERIALS:
The isothermal method was used. At least 24 hours with constant agitation was allowed for equilibration. For the more viscous solutions the time of equilibration was 3 days. The solid and liquid phases were separated from each other by centrifuging. Schreinemakers' method was used to identify the solid phases. K$_2$O was determined by the chloride method (probably weighed as KCl--compiler), P$_2$O$_5$ was determined by normal gravimetry but no details are given. The compiler supposes that water was determined by difference.	Kahlbaum reagent grade KOH was used. The H$_3$PO$_4$ was imported. The KH$_2$PO$_4$ was recrystallized twice. The K$_3$PO$_4$·7H$_2$O and K$_2$HPO$_4$·3H$_2$O were prepared by a method to be published by L.G. Berg.
	ESTIMATED ERROR: No information is given.
	REFERENCES:

COMPONENTS:	ORIGINAL MEASUREMENTS:
(1) Tripotassium phosphate; K$_3$PO$_4$; [7778-53-2]	1. Ravich, M.I. *Kaliy* <u>1936</u>, *10*, 33-7.
(2) Phosphoric acid; H$_3$PO$_4$; [7664-38-2]	2. Ravich, M.I. *Izv. Akad. Nauk SSSR* <u>1938</u>, 167-76.
(3) Potassium hydroxide; KOH [1310-58-3]	
(4) Water; H$_2$O; [7732-18-5]	

EXPERIMENTAL VALUES cont'd:

Part 1. Solubility isotherm for the K$_3$PO$_4$–H$_3$PO$_4$–KOH–H$_2$O system at 0°C.

soln. no.	K$_2$O mass%	K$_2$O mol%	P$_2$O$_5$ mass%	P$_2$O$_5$ mol%	H$_2$O mass%	H$_2$O mol%	solid phase[a]
23	4.89	1.07	9.75	1.41	85.36	97.52	C
24	4.41	0.93	6.66	0.93	88.93	98.14	"
25	6.35	1.38	8.05	1.16	85.60	97.46	"
26	8.50	1.92	9.80	1.49	81.70	96.59	"
27	15.18	3.89	15.10	2.57	69.72	93.54	"
28	16.20	4.24	15.98	2.78	67.82	92.98	"
29	21.18	6.20	19.86	3.84	58.96	89.96	"
30	22.49	6.72	20.45	4.05	57.06	89.23	"
31	27.10	9.06	24.00	5.33	48.90	85.61	"
32	27.63	9.31	23.97	5.36	48.40	85.33	C + F[c]
33	29.24	10.29	25.23	5.89	45.53	83.82	?
34	30.54	11.04	25.69	6.16	43.77	82.80	?
35	31.50	11.54	25.58	6.14	42.92	82.32	D
36	31.03	11.00	24.05	5.66	44.92	83.84	"
37	30.86	10.78	23.26	5.39	45.88	83.83	"
38	32.70	11.54	22.23	5.20	45.07	83.26	"
39	34.38	12.38	21.90	5.23	43.72	87.39	D + E
40	34.49	12.47	22.06	5.29	43.45	82.24	"
41	27.05	8.91	23.01	5.03	49.94	86.06	F
42	25.94	8.15	20.80	4.33	53.26	87.52	"
43	24.91	7.48	18.82	3.76	56.27	88.76	"
44	26.92	8.30	18.60	3.80	54.48	87.90	"
45	29.35	9.36	18.72	3.96	51.93	86.68	"
46	31.29	10.41	19.75	4.36	48.96	85.23	"
47	32.62	11.07	19.80	4.36	47.58	84.48	E + F[c]
48	32.83	11.17	19.80	4.47	47.37	84.36	E
49	32.28	10.85	19.53	4.35	48.19	84.80	"
50	31.26	10.07	17.55	3.6	51.19	86.27	"
51	30.20	9.34	15.81	3.24	53.99	87.42	"
52	29.46	8.89	14.80	2.96	55.74	88.15	"
53	29.60	8.81	13.59	2.68	56.81	88.51	"
54	29.28	8.66	13.41	2.63	57.31	88.71	E[c]
55	29.18	8.46	11.98	2.30	58.84	89.24	"
56	29.20	7.97	7.25	1.32	63.55	90.71	E
57	29.16	7.90	7.76	1.22	64.08	90.88	"
58	33.19	8.89	2.12	0.38	64.69	90.73	"
59	38.16	10.78	1.74	0.33	60.10	88.89	"
60	40.16	11.68	2.11	0.41	57.73	87.91	"
61	41.82	12.52	2.70	0.54	55.48	86.94	E + G[c]
62	36.23	13.23	21.04	5.10	42.73	81.67	H[b,c]
63	33.65	11.40	18.74	4.21	47.61	84.39	"
64	32.65	10.69	17.45	3.79	49.90	85.52	"
65	31.70	10.02	15.92	3.34	52.38	86.64	"
66	31.18	9.58	14.40	2.93	54.42	87.49	"
67	31.40	9.39	12.26	2.43	56.34	88.18	"
68	31.52	9.41	12.00	2.38	56.48	88.21	"
69	32.24	9.60	11.16	2.20	56.60	88.20	"
70	36.82	11.61	11.00	2.30	52.18	86.09	"
71	41.75	15.44	16.67	4.09	41.58	80.47	I[b,c]
72	40.93	13.98	12.52	2.84	46.55	83.18	"
73	41.14	12.77	5.92	1.22	52.94	86.01	"
74	41.86	12.71	3.66	0.74	54.48	86.55	"
75	42.30	12.79	2.99	0.60	54.71	86.61	G
76	41.95	12.57	2.62	0.52	55.43	86.91	"
77	41.80	12.42	2.17	0.43	56.03	87.15	"
78	41.68	12.26	1.50	0.29	56.82	87.45	"
79	41.01	11.73	-----	----	58.99	88.27	"

[a] The solid phases are: A = 2H$_3$PO$_4$·H$_2$O; B = KH$_5$(PO$_4$)$_2$; C = KH$_2$PO$_4$; D = K$_2$HPO$_4$·3H$_2$O; E = K$_3$PO$_4$·7H$_2$O; F = K$_2$HPO$_4$·6H$_2$O; G = KOH·2H$_2$O; H = K$_3$PO$_4$·9H$_2$O; I = K$_3$PO$_4$·3H$_2$O.
[b] Metastable equilibrium.
[c] These data appear in source paper (2) only.

(continued next page)

COMPONENTS:	ORIGINAL MEASUREMENTS:
(1) Tripotassium phosphate; K$_3$PO$_4$; [7778-53-2]	1. Ravich, M.I. *Kaliy* <u>1936</u>, *10*, 33-7.
(2) Phosphoric acid; H$_3$PO$_4$; [7664-38-2]	2. Ravich, M.I. *Izv. Akad. Nauk SSSR* <u>1938</u>, 167-76.
(3) Potassium hydroxide; KOH; [1310-58-3]	
(4) Water; H$_2$O; [7732-18-5]	

EXPERIMENTAL VALUES cont'd:

Part 2. The compiler has calculated the following values from the data given above in Part 1.

soln. no.	K$_3$PO$_4$ mass%	mol/kg	KOH mass%	mol/kg	H$_3$PO$_4$ mass%	mol/kg
1	----	----	----	----	78.93	38.24
2	2.46	0.62	----	----	78.96	43.38
3	5.68	1.70	----	----	78.62	51.12
4	7.35	2.51	----	----	78.86	58.37
5	7.60	2.60	----	----	78.64	58.39
6	8.64	4.18	----	----	81.63	85.70
7	8.28	3.72	----	----	81.23	79.09
8	7.92	3.28	----	----	80.72	72.57
9	7.41	2.38	----	----	77.93	54.28
10	6.58	1.63	----	----	74.45	40.06
11	6.50	1.58	----	----	74.14	39.10
12	7.96	1.42	----	----	65.62	25.36
13	9.09	1.48	----	----	61.98	21.87
14	12.25	1.84	----	----	56.51	18.46
15	15.58	2.33	----	----	53.01	17.23
16	15.84	2.37	----	----	52.74	17.13
17	16.41	2.66	----	----	54.62	19.24
18	15.36	2.12	----	----	50.57	15.15
19	13.42	1.46	----	----	43.40	10.26
20	12.28	1.18	----	----	38.80	8.09
21	10.29	0.76	----	----	26.69	4.23
22	8.64	0.55	----	----	17.76	2.46
23	7.35	0.41	----	----	10.07	1.24
24	6.62	0.35	----	----	6.14	0.71
25	9.54	0.53	----	----	6.71	0.81
26	12.77	0.75	----	----	7.64	0.97
27	22.81	1.60	----	----	10.33	1.57
28	24.35	1.76	----	----	10.84	1.70
29	31.83	2.70	----	----	12.74	2.34
30	33.80	2.97	----	----	12.65	2.41
31	40.73	4.27	----	----	14.36	3.26
32	41.53	4.39	----	----	13.95	3.19
33	43.95	4.99	----	----	14.57	3.58
34	45.90	5.43	----	----	14.31	3.67
35	47.34	5.69	----	----	13.49	3.51
36	46.64	5.27	----	----	11.70	2.86
37	46.38	5.09	----	----	10.73	2.55
38	49.15	5.40	----	----	8.03	1.91
39	51.67	5.80	----	----	6.41	1.56
40	51.84	5.87	----	----	6.56	1.61
41	40.66	4.13	----	----	13.02	2.87
42	38.99	3.65	----	----	10.74	2.18
43	37.44	3.27	----	----	8.72	1.65
44	40.46	3.63	----	----	7.03	1.36
45	44.11	4.12	----	----	5.51	1.11
46	47.03	4.67	----	----	5.59	1.20
47	49.03	4.73	----	----	4.73	1.04
48	49.34	5.04	----	----	4.59	1.01
49	48.52	4.87	----	----	4.60	1.00
50	46.98	4.38	----	----	2.57	0.52
51	45.39	3.98	----	----	0.90	0.17
52	44.28	3.74	----	----	0.02	0.00
53	40.64	3.39	3.03	0.95	----	----
54	40.11	3.32	3.07	0.96	----	----
55	35.83	2.91	6.34	1.95	----	----
56	21.68	1.68	17.59	5.16	----	----
57	23.21	1.80	16.33	4.81	----	----

(continued next page)

COMPONENTS:	ORIGINAL MEASUREMENTS:
(1) Tripotassium phosphate; K_3PO_4; [7778-53-2] (2) Phosphoric acid; H_3PO_4; [7664-38-2] (3) Potassium hydroxide; KOH; [1310-58-3] (4) Water; H_2O; [7732-18-5]	1. Ravich, M.I. *Kaliy* <u>1936</u>, *10*, 33-7. 2. Ravich, M.I. *Izv. Akad. Nauk SSSR* <u>1938</u>, 167-76.

EXPERIMENTAL VALUES cont'd:

Part 2. The compiler has calculated the following values from the data given above in Part 1.

soln. no.	K_3PO_4		KOH		H_3PO_4	
	mass%	mol/kg	mass%	mol/kg	mass%	mol/kg
58	6.34	0.50	34.51	10.39	----	----
59	5.20	0.45	41.33	13.77	----	----
60	6.31	0.58	42.83	15.01	----	----
61	8.07	0.78	43.41	15.95	----	----
62	54.45	6.16	----	-----	3.94	0.96
63	50.58	5.08	----	-----	2.56	0.55
64	49.07	4.67	----	-----	1.47	0.30
65	47.61	4.28	----	-----	----	----
66	43.07	3.76	2.99	0.98	----	----
67	36.67	3.14	8.32	2.69	----	----
68	35.89	3.07	9.08	2.94	----	----
69	33.38	2.87	11.93	3.89	----	----
70	32.90	3.14	17.77	6.42	----	----
71	49.86	5.88	10.20	4.55	----	----
72	37.44	4.05	19.06	7.81	----	----
73	17.70	1.76	34.96	13.16	----	----
74	10.94	1.07	41.18	15.33	----	----
75	8.94	0.88	43.30	16.15	----	----
76	7.83	0.76	43.76	16.11	----	----
77	6.49	0.62	44.64	16.28	----	----
78	4.48	0.42	46.09	16.62	----	----
79	0.00	0.00	48.85	17.02	----	----

COMPONENTS:	ORIGINAL MEASUREMENTS:
(1) Tripotassium phosphate; K_3PO_4; [7778-53-2] (2) Phosphoric acid; H_3PO_4; [7664-38-2] (3) Potassium hydroxide; KOH; [1310-58-3] (4) Water; H_2O; [7732-18-5]	Berg. A.G. *Izv. Akad. Nauk SSSR* <u>1938</u>, 161-6.

VARIABLES:	PREPARED BY:
Composition at 50°C.	J. Eysseltová

EXPERIMENTAL VALUES:

Part 1. The solutility isotherm in the K_3PO_4-H_3PO_4-KOH-H_2O system at 50°C.

soln. no.	K_2O mass%	mol%	P_2O_5 mass%	mol%	H_2O mass%	mol%	solid phase[a]
1	48.43	15.42	----	----	51.17	84.57	A
2	49.44	16.60	2.13	0.47	48.43	82.93	A +[b] B
3	50.19	16.80	2.60	0.58	47.21	82.62	B
4	46.60	14.85	2.64	0.56	50.76	84.59	B
5	46.50	14.87	2.95	0.63	50.55	84.51	"
6	44.88	14.40	4.70	1.00	50.42	84.60	"
7	43.12	14.14	7.80	1.70	49.08	84.16	"
8	42.35	14.74	12.36	2.85	45.29	82.41	"
9	42.30	15.52	15.64	3.80	42.06	80.68	"
10	42.25	17.03	21.07	5.63	36.68	77.34	"
11	42.89	18.33	23.55	6.67	33.56	75.00	"
12	43.80	19.85	25.64	7.71	30.56	72.44	B + C
13	42.39	18.90	26.14	7.73	31.47	73.37	C
14	41.72	18.66	26.92	7.99	31.36	73.35	"
15	41.39	18.38	26.87	7.92	31.74	73.70	"
16	40.98	18.29	27.49	8.14	31.53	73.57	"
17	40.84	18.26	27.59	8.18	31.47	73.56	"
18	40.01	17.68	27.90	8.18	32.09	74.14	"
19	39.68	17.63	28.48	8.39	31.84	73.98	"
20	39.73	17.71	28.59	8.45	31.68	73.84	"
21	39.52	17.55	28.59	8.42	31.89	74.03	"
22	39.18	17.38	28.86	8.49	31.96	74.13	"
23	38.98	17.38	29.29	8.86	31.73	73.96	C
24	38.72	17.30	29.64	8.78	31.64	73.92	"

(continued next page)

AUXILIARY INFORMATION

METHOD/APPARATUS/PROCEDURE:	SOURCE AND PURITY OF MATERIALS:
The isothermal method was used. At least 20 hours with agitation at 900 rpm was allowed for equilibration. The identity of the solid phase was determined microscopically and by the use of the Schreinemakers' method. The solid and liquid phases were separated from each other by centrifuging at 1500-2000 rpm. Potassium was determined as $KClO_4$, phosphorus as $Mg_2P_2O_7$, and water by difference.	Kahlbaum reagent grade KOH was used. The KH_2PO_4 was recrystallized two or three times. The phosphoric acid was imported and had a content of 90%.

ESTIMATED ERROR:
The temperature was kept constant within 0.05 K. For additional information see the Critical Evaluation.

REFERENCES:

COMPONENTS:	ORIGINAL MEASUREMENTS:
(1) Tripotassium phosphate; K_3PO_4; [7778-53-2]	Berg, A.G.
(2) Phosphoric acid; H_3PO_4; [7664-38-2]	*Izv. Akad. Nauk SSSR* <u>1938</u>, 161-6.
(3) Potassium hydroxide; KOH; [1310-58-3]	
(4) Water; H_2O; [7732-18-5]	

EXPERIMENTAL VALUES cont'd:

Part 1. The solubility isotherm in the K_3PO_4-H_3PO_4-KOH-H_2O system at 50°C.

soln. no.	K_2O mass%	K_2O mol%	P_2O_5 mass%	P_2O_5 mol%	H_2O mass%	H_2O mol%	solid phase[a]
25	38.70	17.41	30.00	8.95	31.30	73.64	metastable
26	38.12	17.36	31.13	9.41	30.75	73.23	C + D(?)
27	37.96	17.27	31.24	9.43	30.80	73.30	C + D(?)[b]
28	37.97	17.49	31.79	9.71	30.24	72.80	C[b]
29	38.50	17.12	29.60	8.73	31.90	74.15	E(?)
30	36.71	16.38	31.42	9.29	31.87	74.33	E(?)[b]
31	38.93	17.80	30.57	9.27	30.62	72.93	D(?)[b]
32	38.83	17.78	30.74	9.34	30.43	72.88	D(?)[b]
33	38.06	17.19	30.78	9.22	31.16	73.59	C + D(?)
34	36.65	16.23	31.02	9.12	32.23	74.65	D
35	36.61	16.28	31.38	9.26	32.01	74.46	"
36	35.69	15.77	31.91	9.36	32.40	74.87	"
37	35.77	15.94	31.10	9.19	32.13	74.87	F
38	35.58	15.45	31.16	8.98	33.28	75.57	"
39	35.54	15.47	31.30	9.04	33.16	75.49	"
40	35.35	15.38	31.44	9.07	33.21	75.55	"
41	34.00	14.12	30.30	8.35	35.70	77.53	"
42	33.79	13.89	29.93	8.16	36.28	77.95	"
43	33.70	13.91	30.25	8.28	36.05	77.81	"
44	32.80	13.19	29.69	7.92	37.51	78.89	"
45	31.80	12.44	29.08	7.54	39.12	80.02	"
46	27.91	9.88	26.78	6.28	45.31	83.84	F
47	25.60	8.54	25.20	5.58	49.20	85.88	"
48	25.25	8.34	24.80	5.43	49.95	86.23	"
49	21.78	6.66	22.71	4.61	55.51	88.73	"
50	12.56	3.20	17.01	2.88	70.43	93.92	"
51	10.63	2.60	15.31	2.49	74.06	94.91	"
52	10.06	2.45	15.15	2.45	74.79	95.11	"
53	10.15	2.47	15.30	2.45	74.55	95.08	"
54	10.32	2.53	15.56	2.53	74.12	94.94	"
55	10.48	2.64	17.87	2.99	71.65	94.37	"
56	10.71	2.75	19.42	3.31	69.87	93.34	"
57	11.11	3.09	25.51	4.71	63.38	92.20	"
58	12.03	3.61	30.40	6.05	57.57	90.34	"
59	12.25	3.77	31.94	6.51	55.82	89.72	"
60	12.96	4.52	39.74	9.20	47.30	86.28	"
61	13.22	4.71	40.85	9.66	45.93	85.63	"
62	14.21	6.02	49.71	13.98	36.08	80.00	"
63	14.30	6.09	49.87	14.09	35.83	79.82	"
64	14.34	6.27	51.15	14.83	34.51	78.90	G
65	9.65	4.25	55.85	16.30	34.51	79.45	"
66	9.53	4.18	55.72	16.19	34.75	79.63	"
67	8.84	3.59	57.92	17.45	33.24	78.96	"
68	8.71	3.95	58.17	17.50	33.12	78.55	"
69	8.68	3.50	60.24	16.13	38.08	80.37	"
70	9.10	4.56	62.48	20.75	28.52	74.69	"
71	10.05	5.60	65.93	24.38	24.02	70.02	"

[a] The solid phases are: A = $KOH \cdot H_2O$; B = $K_3PO_4 \cdot 3H_2O$; C = K_2HPO_4; D = $K_2HPO_4 \cdot KH_2PO_4 \cdot 2H_2O$; E = $3K_2HPO_4 \cdot KH_2PO_4 \cdot 2H_2O$; F = KH_2PO_4; G = $KH_5(PO_4)_2$.

[b] A metastable equilibrium.

COMPONENTS:	ORIGINAL MEASUREMENTS:
(1) Tripotassium phosphate; K$_3$PO$_4$; [7778-53-2]	Berg, A.G.
(2) Phosphoric acid; H$_3$PO$_4$; [7664-38-2]	*Izv. Akad. Nauk SSSR* 1938, 161-6.
(3) Potassium hydroxide; KOH; [1310-58-3]	
(4) Water; H$_2$O; [7732-18-5]	

EXPERIMENTAL VALUES cont'd:

Part 2. The compiler has calculated the following values from the data given in
        Part 1 above.

soln. no.	K$_3$PO$_4$ mass%	K$_3$PO$_4$ mol/kg	KOH mass%	KOH mol/kg	H$_3$PO$_4$ mass%	H$_3$PO$_4$ mol/kg
1	----	----	58.17	24.78	----	----
2	6.37	0.75	53.48	24.12	----	----
3	7.77	0.94	53.62	20.76	----	----
4	7.89	0.86	49.25	20.48	----	----
5	8.82	0.97	48.39	20.16	----	----
6	14.05	1.51	42.31	17.28	----	----
7	23.33	2.50	32.86	13.37	----	----
8	36.96	4.15	21.13	8.99	----	----
9	46.78	5.51	13.29	5.93	----	----
10	63.02	8.10	0.36	0.17	----	----
11	64.47	9.27	----	----	2.80	0.87
12	65.83	10.65	----	----	5.05	1.77
13	63.71	10.15	----	----	6.72	2.32
14	62.71	10.17	----	----	8.26	2.90
15	62.21	9.98	----	----	8.42	2.92
16	61.59	10.06	----	----	9.56	3.38
17	61.38	10.03	----	----	9.79	3.47
18	60.14	9.74	----	----	10.80	3.79
19	59.64	9.85	----	----	11.83	4.23
20	59.72	9.92	----	----	11.94	4.30
21	59.40	9.81	----	----	12.09	4.32
22	58.89	9.76	----	----	12.70	4.56
23	58.59	9.86	----	----	13.43	4.90
24	58.20	9.89	----	----	14.09	5.19
25	58.18	10.07	----	----	14.60	5.47
26	57.30	10.33	----	----	16.57	6.47
27	57.05	10.29	----	----	16.83	6.57
28	57.07	10.60	----	----	17.57	7.07
29	57.87	9.75	----	----	14.19	5.18
30	55.18	9.67	----	----	17.94	6.81
31	58.51	10.50	----	----	15.23	5.92
32	58.36	10.53	----	----	15.54	6.07
33	57.21	10.10	----	----	16.12	6.17
34	55.09	9.44	----	----	17.43	6.47
35	55.03	9.59	----	----	17.96	6.78
36	53.64	9.35	----	----	19.33	7.30
37	53.76	9.02	----	----	18.15	6.59
38	53.48	8.95	----	----	18.37	6.66
39	53.42	8.99	----	----	18.59	6.77
40	53.13	8.95	----	----	18.91	6.90
41	51.10	7.86	----	----	18.27	6.09
42	50.79	7.64	----	----	17.91	5.84
43	50.65	7.71	----	----	18.41	6.07
44	49.30	7.16	----	----	18.26	5.74
45	47.80	6.60	----	----	18.11	5.42
46	41.95	4.89	----	----	17.63	4.45
47	38.48	4.07	----	----	17.05	3.91
48	37.95	3.94	----	----	16.74	3.77
49	32.73	3.02	----	----	16.26	3.25
50	18.87	1.34	----	----	14.78	2.27
51	15.97	1.07	----	----	13.77	2.00
52	15.12	1.00	----	----	13.94	2.00
53	15.25	1.01	----	----	14.09	2.03
54	15.51	1.04	----	----	14.33	2.08
55	15.75	1.11	----	----	17.41	2.65
56	16.09	1.17	----	----	19.39	3.06
57	16.70	1.41	----	----	27.52	5.03
58	18.08	1.76	----	----	33.64	7.11

(continued next page)

COMPONENTS:	ORIGINAL MEASUREMENTS:
(1) Tripotassium phosphate; K_3PO_4; [7778-53-2]	Berg, A.G.
(2) Phosphoric acid; H_3PO_4; [7664-38-2]	*Izv. Akad. Nauk SSSR* <u>1938</u>, 161-6.
(3) Potassium hydroxide; KOH; [1310-58-3]	
(4) Water; H_2O; [7732-18-5]	

EXPERIMENTAL VALUES cont'd:

Part 2. The compiler has calculated the following values from the data given in Part 1 above.

soln. no.	K_3PO_4 mass%	K_3PO_4 mol/kg	KOH mass%	KOH mol/kg	H_3PO_4 mass%	H_3PO_4 mol/kg
59	18.39	1.88	----	----	35.62	7.90
60	19.48	2.65	----	----	45.89	13.52
61	19.87	2.84	----	----	47.24	14.66
62	21.35	5.06	----	----	58.79	30.22
63	21.49	5.17	----	----	58.95	30.76
64	21.55	5.71	----	----	60.69	34.88
65	14.50	4.53	----	----	70.41	47.65
66	14.32	4.39	----	----	70.33	46.78
67	13.28	4.86	----	----	73.84	58.58
68	13.09	4.88	----	----	74.28	60.04
69	13.04	6.27	----	----	77.16	80.43
70	13.67	10.13	----	----	79.96	128.37
71	15.10	86.51	----	----	84.07	1042.9

COMPONENTS:	ORIGINAL MEASUREMENTS:
(1) Potassium dihydrogenphosphate; KH$_2$PO$_4$; [7778-77-0] (2) Potassium hydroxide; KOH; [1310-58-3] (3) Phosphoric acid; H$_3$PO$_4$; [7664-38-2] (4) Water; H$_2$O; [7732-18-5]	Mýl. J.; Šolc. Z. *Collection Czechoslov. Chem. Comm.* 1960, 25, 2414-8.

VARIABLES:	PREPARED BY:
Temperature and composition	J. Eysseltová

EXPERIMENTAL VALUES:

Composition of saturated solutions in the KH$_2$PO$_4$-H$_3$PO$_4$-KOH-H$_2$O system.

K$_2$O	P$_2$O$_5$	KH$_2$PO$_4$[a]		KOH[a]		H$_3$PO$_4$[a]		H$_2$O[a]
mass%	mass%	mass%	mol/kg	mass%	mol/kg	mass%	mol/kg	mass%

temp = 25°C.

K$_2$O	P$_2$O$_5$	KH$_2$PO$_4$ mass%	KH$_2$PO$_4$ mol/kg	KOH mass%	KOH mol/kg	H$_3$PO$_4$ mass%	H$_3$PO$_4$ mol/kg	H$_2$O mass%
10.0	12.4	23.78	1.18	2.11	0.25	----	----	74.11
12.19	13.65	26.17	1.37	3.73	0.47	----	----	70.09
16.7	16.65	31.93	1.91	6.73	0.98	----	----	61.34
24.25	21.7	41.61	3.28	11.73	2.24	----	----	46.65
6.85	10.4	19.79	0.91	----	----	0.11	0.01	80.10
7.79	13.85	22.51	1.11	----	----	2.92	0.20	74.58
8.78	18.75	25.37	1.39	----	----	7.62	0.58	67.01
8.97	23.18	25.92	1.57	----	----	13.34	1.12	60.74
9.84	30.6	28.43	2.10	----	----	21.78	2.23	49.79
10.86	37.6	31.38	2.93	----	----	29.32	3.81	39.30
11.55	41.85	33.37	3.73	----	----	33.75	5.24	32.87
9.9	51.3	28.61	4.97	----	----	50.23	12.11	21.16
6.3	59.8	18.20	5.42	----	----	69.46	28.73	12.33

(continued next page)

AUXILIARY INFORMATION

METHOD/APPARATUS/PROCEDURE:	SOURCE AND PURITY OF MATERIALS:
A modification of Toepler's method (1) was used.	Reagent grade materials were used.

ESTIMATED ERROR:

No information is given.

REFERENCES:

1. Mýl, J.; Kvapil, J. *Colln. Czechoslov. Chem. Commun.* 1960, 25, 194.

COMPONENTS:	ORIGINAL MEASUREMENTS:
(1) Potassium dihydrogenphosphate; KH$_2$PO$_4$; [7778-77-0] (2) Potassium hydroxide; KOH; [1310-58-3] (3) Phosphoric acid; H$_3$PO$_4$; [7664-38-2] (4) Water; H$_2$O; [7732-18-5]	Mýl. J.; Šolc. Z. *Collection Czechoslov. Chem. Comm.* 1960, 25, 2414-8.

EXPERIMENTAL VALUES cont'd:

Composition of saturated solutions in the KH$_2$PO$_4$-H$_3$PO$_4$-KOH-H$_2$O system.

K$_2$O mass%	P$_2$O$_5$ mass%	KH$_2$PO$_4$[a] mass%	KH$_2$PO$_4$[a] mol/kg	KOH[a] mass%	KOH[a] mol/kg	H$_3$PO$_4$[a] mass%	H$_3$PO$_4$[a] mol/kg	H$_2$O[a] mass%
				temp = 30°C.				
10.2	12.9	24.74	1.24	1.95	0.24	----	----	73.31
12.5	14.21	27.25	1.45	3.66	0.47	----	----	69.09
17.5	17.22	33.02	2.03	7.23	1.08	----	----	59.75
24.4	22.0	42.19	3.34	11.44	2.20	----	----	46.37
7.15	10.8	20.66	0.96	----	----	0.04	0.00	79.30
8.4	14.6	24.27	1.22	----	----	2.68	0.19	73.04
9.35	19.5	27.02	1.52	----	----	7.47	0.58	65.51
9.5	23.8	27.45	1.70	----	----	13.10	1.12	59.45
10.37	30.8	29.96	2.24	----	----	20.95	2.18	49.08
11.3	37.8	32.65	3.10	----	----	28.68	3.78	38.67
11.9	42.0	34.39	3.90	----	----	33.23	5.24	32.38
10.3	51.0	29.76	5.14	----	----	49.00	11.76	21.25
6.76	59.9	19.53	6.07	----	----	68.64	29.62	11.82
				temp = 40°C.				
10.8	14.2	27.23	1.41	1.64	0.21	----	----	71.13
13.3	15.6	29.91	1.65	3.51	0.47	----	----	66.57
18.1	18.5	35.48	2.26	6.94	1.07	----	----	57.59
24.5	22.6	43.34	3.51	11.32	2.22	----	----	45.34
8.43	12.7	24.35	1.18	0.00	0.00	----	----	75.65
9.4	16.15	27.16	1.42	----	----	2.74	0.20	70.10
10.45	20.95	30.20	1.77	----	----	7.18	0.59	62.62
10.5	25.0	30.34	1.96	----	----	12.67	1.13	56.99
11.4	31.8	32.94	2.58	----	----	20.19	2.20	46.87
12.1	38.3	34.96	3.44	----	----	27.71	3.79	37.33
12.6	42.4	36.41	4.28	----	----	32.33	5.28	31.26
8.68	59.3	22.16	7.00	----	----	66.20	29.02	11.64
				temp = 50°C.				
11.55	15.8	30.30	1.63	1.27	0.16	----	----	68.43
14.3	17.25	33.08	1.91	3.40	0.48	----	----	63.52
18.7	19.7	37.78	2.50	6.70	1.08	----	----	55.52
9.77	14.8	28.23	1.45	----	----	0.11	0.01	71.66
10.5	17.75	30.34	1.66	----	----	2.66	0.20	67.00
11.56	22.4	33.40	2.05	----	----	6.88	0.59	59.72
11.69	26.25	33.78	2.28	----	----	11.92	1.12	54.30
12.4	32.7	35.83	2.94	----	----	19.35	2.20	44.82
13.1	38.9	37.85	3.90	----	----	26.45	3.78	35.69
13.42	42.6	38.78	4.70	----	----	30.90	5.20	30.23
8.68	59.3	25.08	8.30	----	----	63.82	29.34	11.10

[a]These values were calculated by the compiler.

COMPONENTS:	ORIGINAL MEASUREMENTS:
(1) Potassium dihydrogenphosphate; KH$_2$PO$_4$; [7778-77-0] (2) Phosphoric acid; H$_3$PO$_4$; [7664-38-2] (3) Potassium hydroxide; KOH; [1310-58-3] (4) Water; H$_2$O; [7732-18-5]	Punin, Yu.O.; Mirenkova, T.F.; Artamanova, O.I.; Ul'yanova, T.P. *Zh. Neorg. Khim.* 1975, *20*, 2813-5.

VARIABLES:	PREPARED BY:
Temperature and composition.	J. Eysseltová

EXPERIMENTAL VALUES:

Parameters a_i of the equation: $c_t = a_o + a_1 t + a_2 t^2 + \ldots + a_5 t^5$ were calculated and are given in the Table below. c_t is the concentration of the saturated solution (as g/100 g of solvent) at the temperature $t°$C.

solvent	concn[b]	a_o	$a_1 \times 10$	$a_2 \times 10^3$	$a_3 \times 10^5$	$a_4 \times 10^7$	$a_5 \times 10^9$	σ[a] (g/100 g)
H$_2$O		14.958	2.881	4.914	1.826	2.311	0	0.03
H$_3$PO$_4$	1.72	16.478	2.229	5.354	0	0	0	0.16
"	5.00	17.509	3.008	6.239	6.051	4.889	0	0.06
"	9.94	20.184	4.177	1.906	1.784	0	0	0.13
"	14.90	32.498	8.209	64.574	135.678	137.658	51.195	0.16
KOH	2.00	14.682	12.533	52.068	147.490	178.247	80.992	0.14
"	4.92	30.370	2.826	6.213	6.207	4.963	0	0.03
"	10.04	48.563	0.694	15.135	21.332	13.776	0	0.08
"	15.68	68.154	3.231	38.031	80.746	84.653	31.888	0.08

[a]This is the mean quadratic error.

[b]The concentration unit is: g/100 g.

AUXILIARY INFORMATION

METHOD/APPARATUS/PROCEDURE:	SOURCE AND PURITY OF MATERIALS:
Nine solubility polytherms were studied. They differed from each other in the composition of the solvent. The pH of each solvent was measured. For each polytherm, 10-14 samples were prepared by precise weighing. The saturation temperature of each sample was determined by using an apparatus constructed for measuring crystal growth rate (1).	All the components were of a "special purity" grade.

ESTIMATED ERROR:

The accuracy of the saturation temperature was ± 0.1 K.

REFERENCES:

1. Petrov, T.G.; Trejbus, E.B.; Kosatkin, A.P. *"Vyrashchivanie Kristallov iz Rastvorov"*, Nedra, Leningrad, 1967.

COMPONENTS:	EVALUATOR:
(1) Potassium dihydrogenphosphate; KH_2PO_4; [7778-77-0] (2) Water; H_2O; [7732-18-5]	J. Eysseltová Charles University Prague, Czechoslovakia May, 1985

CRITICAL EVALUATION:

THE BINARY SYSTEM

Several papers report solubility data for this system (1-6). Empirical smoothing equations describing the temperature dependence of this solubility have also been reported (6,7). Solubility data for this system have also been reported as part of a study of multicomponent systems having KH_2PO_4 as one component (8-24). Some of these values are clearly incorrect and were rejected immediately (8-10). The same is true for some of the solubility values reported for 323 K (11,12). In another report (13) the solubility of KH_2PO_4 at 298 K is given as 20.3% and also as 21.6%. The latter value is an obvious error. All the other data from the studies of multicomponent systems were evaluated together with the data in refs. (1-6).

The evaluation procedure was the same as that described in chapter 3. It was possible to use only one equation because no hydrate formation has been observed. The assumptions concerning the precision of the data were the same as those used in chapter 3. The general solubility equation [1] was used. Equation [2] was used for the

$$\ln x/x_o = A \cdot (1/T - 1/T_o) + B \cdot \ln (T/T_o) + C \cdot (T - T_o) \qquad [1]$$

selection of experimental points during the iterative procedure.

$$\left| \frac{xj - x(Tj)}{x(Tj)} \right| \leq 0.015 \qquad [2]$$

Table I is a summary of the solubility data used in the evaluation procedure. All the data from only two reports (22,24) were eliminated during the iteration procedure.

The results of this evaluation procedure are summarized in Table II, which gives the values obtained for the parameters in equation [1].

Table III is a list of the recommended solubility values calculated from equation [1]. The other smoothing equations that have been suggested (6,7) give values that are 8-10% larger than the recommended values (Table III) in the 273-288 K temperature interval. Above 288 K the one equation (6) gives values that agree with those in Table III while the other equation (7) gives values that have a constant error of about +10%.

(continued next page)

COMPONENTS:	EVALUATOR:
(1) Potassium dihydrogenphosphate; KH_2PO_4; [7778-77-0] (2) Water; H_2O; [7732-18-5]	J. Eysseltová Charles University Prague, Czechoslovakia May 1985

CRITICAL EVALUATION: (cont'd)

Table I. Solubility of KH_2PO_4 in water.

T/K	mass%	ref.	weight init/final	T/K	mass%	ref.	weight init/final
273.2	10.48	1	1/0	298.2	20.30	13	1/0
273.2	12.48	2	1/0	298.2	19.92	4	1/1
273.2	12.68	4	1/0	298.2	19.93	19	1/1
273.2	12.88	6	1/0	298.2	20.3	21	1/0
273.2	12.15	15	1/0	298.2	20.21	23	1/1
273.2	11.8	5	1/1	298.2	20.42	12,13	2/0
273.2	12.2	3	1/0	299.2	20.00	3	1/0
273.2	12.7	4	1/0	300.2	20.00	5	1/0
273.2	12.41	19	1/0	303.2	21.90	6	1/1
273.2	12.30	24	1/0	303.2	21.0	3	1/0
274.0	12.00	5	1/1	305.0	22.00	5	1/0
278.2	14.00	6	1/0	305.4	22.00	3	1/0
283.2	15.50	6	1/0	308.2	23.65	6	1/1
283.2	14.95	15	1/1	308.2	22.90	15	1/0
283.2	15	5	1/1	313.2	25.10	6	1/1
283.2	14.9	3	1/1	313.2	25.00	20	1/1
283.2	15.20	24	1/0	313.2	27.15	22	1/0
286.8	16.00	5	1/1	313.2	27.12	22	1/0
287.3	16.00	3	1/1	318.2	26.90	6	1/1
288.2	16.78	2	1/0	323.2	29.26	1	1/1
293.0	18.00	5	1/1	323.2	29.15	4,16	2/2
293.2	18.45	6	1/1	323.2	29.00	6,21	2/2
293.2	18.20	15	1/1	323.2	29.42	14	1/0
293.2	18.2	5	1/1	323.2	29.1	4	1/1
293.2	18.50	18	1/0	323.2	28.70	19	1/1
293.2	17.8	3	1/0	333.2	33.40	6	1/0
293.2	17.73	2	1/0	333.2	32.37	20	1/0
293.2	18.04	20	1/0	343.2	36.68	1	1/1
293.2	18.00	24	1/0	343.2	36.65	1	1/1
293.8	18.00	3	1/0	343.2	37.05	6	1/1
298.2	19.87	1	1/1	348.2	39.20	21	1/1
298.2	20.15	1	1/1	348.2	38.61	19	1/1
298.2	19.90	4	1/1	353.2	41.30	6	1/1
298.2	20.04	6	1/1	353.2	40.75	20	1/1
298.2	19.80	15	1/1	356.2	41.38	1	1/0
298.2	20.07	2	1/1	356.2	41.92	1	1/1
298.2	20.49	16	1/0	363.2	45.5	6	1/1
				378.3	52.7	17	1/1

Table II. Values for the parameters of equation [1].

A		B		C		x_o	T_o
value	σ^a	value	σ^a	value	σ^a	value	value
-2.121×10^4	10	-1.224×10^2	0.5	0.1929	0.0005	0.031948	298.2

a Standard deviation for the respective parameter.

(continued next page)

COMPONENTS:	EVALUATOR:
(1) Potassium dihydrogenphosphate; KH_2PO_4; [7778-77-0] (2) Water; H_2O; [7732-18-5]	J. Eysseltová Charles University Prague, Czechoslovakia May 1985

CRITICAL EVALUATION: (cont'd)

Table III. Recommended values calculated from equation [1] for the solubility of KH_2PO_4 in water.

T/K	mole fraction	mol/kg	mass%
273.2	0.017287	0.98	11.74
278.2	0.019886	1.13	13.30
283.2	0.022653	1.29	14.91
288.2	0.025586	1.46	16.56
293.2	0.028684	1.64	18.25
298.2	0.031948	1.83	19.97
303.2	0.035385	2.04	21.71
308.2	0.039006	2.25	23.48
313.2	0.042825	2.48	25.28
318.2	0.046861	2.73	27.10
323.2	0.051139	2.99	28.95
328.2	0.055690	3.28	30.84
333.2	0.060548	3.58	32.76
338.2	0.065757	3.91	34.73
343.2	0.071367	4.27	36.75
348.2	0.077434	4.66	38.82
353.2	0.084028	5.10	40.96
358.2	0.091224	5.58	43.15
363.2	0.099113	6.11	45.41
368.2	0.10780	6.71	47.74
373.2	0.11740	7.39	50.14
378.2	0.12806	8.16	52.62

MULTICOMPONENT SYSTEMS

Solubility measurements have been made for a variety of multicomponent systems that include potassium dihydrogenphosphate as a component. However, an evaluation cannot be made of most of these studies because of a lack of corroborating results. The systems may be differentiated according to the type of solid phases that are in equilibrium with the saturated solutions.

The formation of compounds has been reported for two of these systems: the compounds $KH_2PO_4 \cdot KHSeO_4$ and $3KH_2PO_4 \cdot KHSeO_4$ were identified in the KH_2PO_4-$KHSeO_4$-H_2O system at 298 K (18); the compound $Ca_9K_4H_{32}(PO_4)_{18} \cdot 10H_2O$ was found to be present in the KH_2PO_4-$Ca(H_2PO_4)_2$-H_3PO_4-H_2O system at 298 K (25).

Solid solutions were found to be present in several systems. A continuous series of solid solutions is present in the KH_2PO_4-$NH_4H_2PO_4$-H_2O system (10-13, 23, 26, 27). Some of the solubility data at 298 K (11-13, 23) are compared on Figure 1. All the values agree fairly well with each other and these are accepted as tentative values. However, the discontinuity around 2.9-3 molal $NH_4H_2PO_4$ and 1 molal KH_2PO_4 cannot be explained. Further work is needed in this part of the system. The solid solutions, which have been designated β-solid solutions (26), are also found in the K^+, $NH_4^+||Cl^-$, $H_2PO_4^-$-H_2O system at 273 K (10, 29). Two forms of solid solutions are said to exist in the KH_2PO_4-$NH_4H_2PO_4$-$CO(NH_2)_2$-H_2O system (28). In addition, there are also the α-, β- and γ-modifications of urea. This is true also of the KH_2PO_4-$CO(NH_2)_2$-H_2O system (15).

A series of solid solutions was also observed in the KH_2PO_4-KH_2AsO_4-H_2O system at 280 K (9). These results cannot be evaluated, but it should be noted that the values reported for the solubility of KH_2PO_4 in water are in error by about 50%.

In all the other systems having two or more saturating components, no solid solutions or compounds were found to be present. These systems are referred to individually.

1. The KH_2PO_4-$[(C_2H_5)_3N]H_3PO_4$-H_2O system. Solubility measurements were made at 293, 313, 333 and 353 K (20).

2. The KH_2PO_4-KBO_2-H_2O system. Solubility values have been reported for 298 and 323 K (8). These values cannot be evaluated but mention should be made of the fact that the value given for the solubility of KH_2PO_4 in water is incorrect.

3. The KH_2PO_4-KNO_3-H_2O system. There are four reports of solubility values for this system (1, 5, 16, 17). The results are compared on Figure 2.

(continued next page)

COMPONENTS:	EVALUATOR:
(1) Potassium dihydrogenphosphate; KH_2PO_4; [7778-77-0] (2) Water; H_2O; [7732-18-5]	J. Eysseltová Charles University Prague, Czechoslovakia May 1985

CRITICAL EVALUATION:

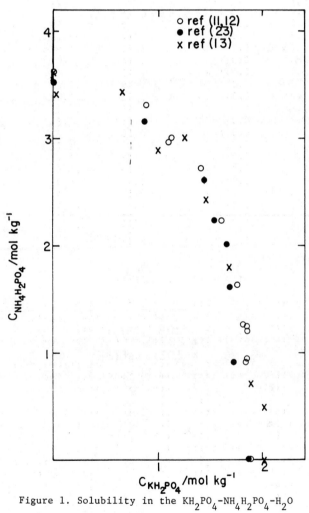

Figure 1. Solubility in the $KH_2PO_4-NH_4H_2PO_4-H_2O$
system at 298 K.

COMPONENTS:	EVALUATOR:
(1) Potassium dihydrogenphosphate; KH_2PO_4; [7778-77-0] (2) Water; H_2O; [7732-18-5]	J. Eysseltová Charles University Prague, Czechoslovakia May 1985

CRITICAL EVALUATION:

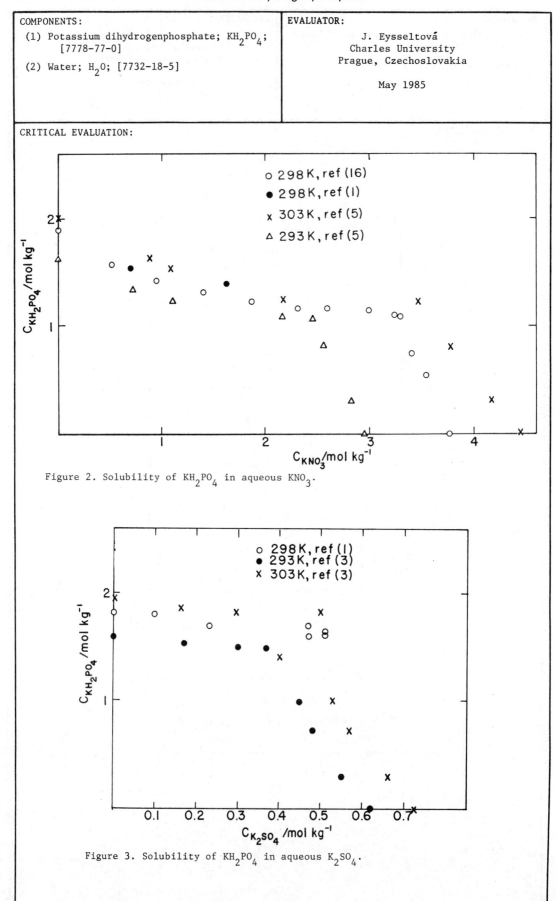

Figure 2. Solubility of KH_2PO_4 in aqueous KNO_3.

Figure 3. Solubility of KH_2PO_4 in aqueous K_2SO_4.

COMPONENTS:

(1) Potassium dihydrogenphosphate; KH_2PO_4;
      [7778-77-0]

(2) Water; H_2O; [7732-18-5]

EVALUATOR:

J. Eysseltová
Charles University
Prague, Czechoslovakia

May 1985

CRITICAL EVALUATION:

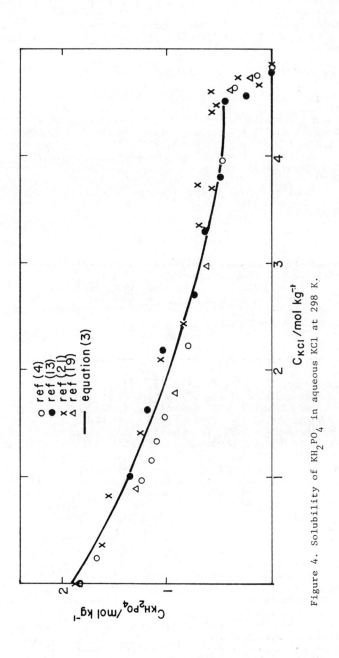

Figure 4. Solubility of KH_2PO_4 in aqueous KCl at 298 K.

COMPONENTS:	EVALUATOR:
(1) Potassium dihydrogenphosphate; KH_2PO_4; [7778-77-0] (2) Water; H_2O; [7732-18-5]	J. Eysseltová Charles University Prague, Czechoslovakia May 1985

CRITICAL EVALUATION: (cont'd)

4. The KH_2PO_4-K_2SO_4-H_2O system. The solubility values reported for this system (1, 3, 30) are compared on Figure 3.

5. The KH_2PO_4-KCl-H_2O system. Several investigators have published solubility values for this system (4, 10, 13, 19-22). The solubility of KH_2PO_4 at 298 K may be expressed by the smoothing equation of Kirgintsev (31) where m_1 is the molality of KH_2PO_4 and y_1 is its solute mole

$$\log m_1 = \log 1.87 + (0.59 \pm 0.07) \log y_1 \qquad [3]$$

fraction. The results calculated from equation [3] are compared with the experimental values on Figure 4.

In addition to the solubility values reported above, there are values given for the KH_2PO_4-H_2O_2-H_2O system at 273 K (2); some values for the KH_2PO_4-$HCONH_2$-H_2O system at 298 K (32), at 323 K (14) and a polytherm in ref. (33); two values for the solubility of KH_2PO_4 in aqueous potassium acetate and in aqueous potassium carbonate at 298 K (1). The latter system cannot be treated as a ternary system because of the observed decomposition of the K_2CO_3.

No solid solutions or compounds were observed in a study of the $2K^+$, $Ca^{2+}||2Cl^-$, $2H_2PO_4^-$-H_2O system at 298 K (34).

References

1. Apfel, O. Dissertation, Technical University, Darmstadt 1911.
2. Menzel, H.; Gabler, C. Z. Anorg. Chem. 1929, 177, 187.
3. Bel'tschev, F.V. Trudy Beloruss. S.-Kh. Akad. 1953, 19, 145.
4. Krasil'shtschikov, A.I. Izv. In-ta Fiz. Khim. Anal. 1933, 6, 159.
5. Bergman, A.G.; Bochkarev, N.F. Izv. Akad. Nauk SSSR 1938, 237.
6. Kazantsev, A.A. Zh Obshch Khim. 1938, 13, 1230.
7. Punin, Yu. O.; Mirenkova, T.F.; Artamanova, O.I.; Ul'yanova, T.P. Zh. Neorg. Khim. 1975, 20, 2813.
8. Beremzhanov, B.A.; Voronina, L.V.; Savich, R.F. Khim. Khim. Tekhnol. (Alma Ata) 1978, 173.
9. Muthmann, F.; Kuntze, O. Z. Kryst. 1894, 23, 368.
10. Askenasy, P.; Nessler, F. Z. Anorg. Chem. 1930, 189, 305.
11. Dombrovskaya, N.S.; Zvorykin, A.Y. Kaliy 1937, 2, 24.
12. Zvorykin, A.Y.; Kuznetsov, V.G. Izv. AN SSSR, Ser. Khim. 1938, 195.
13. Solov'ev, A.P.; Balashova, E.F.; Verendyakina, N.A.; Zyuzina, L.F. Vzaimodeistvie Khloridov Kaliya, Magniya, Ammoniya s ich Nitratami i Fosfatami, 1977, 3.
14. Yugai, M.R.; Tukhtaev, S.; Beglov, B.M. Uzb. Khim. Zh. 1981, 6, 15.
15. Polosin, V.A.; Shakhparonov, M.I. Zh. Obshch. Khim. 1947, 17, 397.
16. Girich, T.E.; Gulyamov, Yu. M.; Ganz, S.N. Zh. Neorg. Khim. 1979, 24, 1084.
17. Shenkin, Ya. S.; Gorozhankin, E.V. Zh. Neorg. Khim. 1976, 21, 2293.
18. Zbořilová, L.; Krejčí, J. Scripta Fac. Sci. Nat. UJEP Brunensis, Chemia 1 1972, 77.
19. Brunisholz, G.; Bodmer, M. Helv. Chim. Acta 1963, 46, 288, 2566.
20. Filipescu, L. Rev. Chim. (Bucharest) 1971, 22, 533.
21. Mráz, R.; Srb, V.; Tichý, S.; Vosolsobě, J. Chem. Prům. 1976, 26, 511.
22. Khallieva, Sh. D.; Izv. AN Turkm.-SSR, Ser. Khim. 1978, 3, 125.
23. Kuznetsov, D.I.; Kozhukhovskij, A.A.; Borovaya, F.E. Zh. Prikl. Khim. 1948, 21, 1278.
24. Babenko, A.M.; Vorob'eva, T.A. Zh. Prikl. Khim. (Leningrad) 1975, 48, 2437.
25. Flatt, R.; Brunisholz, G.; Bourgeois, J. Helv. Chim. Acta 1956, 39, 841.
26. Bergman, A.G.; Gladkovskaya, A.A.; Galushkina, R.A. Zh. Neorg. Khim. 1972, 17, 2055.
27. Shenkin, Ya. S.; Ruchnova, S.A.; Rodionova, N.A.; Zh. Neorg. Khim. 1972, 17, 3368.
28. Bergmanm, A.G.; Gladkovskaya, A.A.; Galushkina, N.A. Zh. Neorg. Khim. 1973, 18, 1978.
29. Iovi, A.; Haiduc, C. Rev. Roum. Chim. 1971, 16, 1743.
30. Gladkovskaya, A.A.; Bergman, A.G. Tr. Kuban. S.-Kh. In-ta 1975, 102, 31, 130.
31. Kirgintsev. A.N. Izv. Akad. Nauk SSSR, Ser. Khim. 1965, 8, 1591.
32. Becker, B. J. Chem. Eng. Data 1969, 14, 431.
33. Beglov, B.M.; Tukhtaev, S.; Yugai, M.R. Zh. Neorg. Khim. 1980, 25, 2283.
34. Timoshenko, Yu. M.; Gilyazova, G.N. Zh. Neorg. Khim. 1981, 26, 1104.

COMPONENTS:	ORIGINAL MEASUREMENTS:
(1) Potassium dihydrogenphosphate; KH_2PO_4; [7778-77-0] (2) Water; H_2O; [7732-18-5]	Apfel, O. Dissertation, Technical University, Darmstadt, <u>1911</u>.
VARIABLES: Temperature and composition.	**PREPARED BY:** J. Eysseltová

EXPERIMENTAL VALUES:

Composition of the saturated liquid phase.

$t/°C$.	PO_4^{3-} concn[b]	K^+ concn[b]	KH_2PO_4[a] mass%	mol/kg
0	0.77	0.77	10.48	0.86
25	1.47	1.48	19.87	1.82
			20.15[c]	1.85[c]
50	2.15	2.15	29.26	3.04
70	2.695	2.693	36.68	4.26
			36.65[c]	4.25[c]
83	3.04	3.08	41.38	5.19
			41.92[c]	5.30[c]

[a] These values were calculated by the compiler.

[b] The concentration unit is: mol/1000 g of solution.

[c] In these calculations the potassium content was taken as the starting point of the calculation. In the other calculations it was the PO_4^{3-} content.

AUXILIARY INFORMATION

METHOD/APPARATUS/PROCEDURE:	SOURCE AND PURITY OF MATERIALS:
All the experiments were carried out in a water thermostat. Equilibrium was checked by repeated analysis. The solid and liquid phases were separated from each other by filtration through a platinum wire mat. The phosphate ion content was determined gravimetrically as $Mg_2P_2O_7$, and potassium was determined as $KClO_4$.	No information is given.
	ESTIMATED ERROR: No information is given.
	REFERENCES:

COMPONENTS:	ORIGINAL MEASUREMENTS:
(1) Potassium dihydrogenphosphate; KH_2PO_4; [7778-77-0] (2) Water; H_2O; [7732-18-5]	Menzel, H.; Gabler, C. Z. Anorg. Chem. 1929, 177, 187-214.

VARIABLES:	PREPARED BY:
Temperature and composition.	J. Eysseltová

EXPERIMENTAL VALUES:

Composition of saturated solutions of KH_2PO_4 in water.

$t/°C.$	in 1000 cm^3 of solution		in 1000 g of solution		in 1000 g of water	
	mols	grams	mols	grams	mols	grams
0	1.001	136.3	0.917	124.8	1.047	142.6
15	1.359	185.0	1.233	167.8	1.481	201.7
18	1.433	195.2	1.302	177.3	1.583	215.5
25	1.699	231.3	1.474	200.7	1.845	251.2
-2.75^a					1.08	147

[a] This is the cryohydric point of the system under consideration. However, the authors are in doubt about the accuracy of their analytical results.

AUXILIARY INFORMATION

METHOD/APPARATUS/PROCEDURE:	SOURCE AND PURITY OF MATERIALS:
The apparatus was that described previously (1). The equilibrium was checked by repeated analysis. The $H_2PO_4^-$ ion content was determined gravimetrically as ammonium phosphomolybdate.	Kahlbaum KH_2PO_4 was used. This material had been prepared for enzyme investigations.

ESTIMATED ERROR:
The temperature was controlled to ± 0.1 K. The accuracy of the cryohydric temperature was ± 0.01 K. No other information is given.

REFERENCES:
1. Menzel, H. Z. Anorg. Allg. Chem. 1927, 164, 6.

COMPONENTS:	ORIGINAL MEASUREMENTS:
(1) Potassium dihydrogenphosphate; KH_2PO_4; [7778-77-0] (2) Water; H_2O; [7732-18-5]	Krasil'shtschikov, A.I. *Izv. In-ta Fiz-khim. An.* 1933, *6*, 159-68.

VARIABLES:	PREPARED BY:
Composition at 0, 25 and 50°C.	J. Eysseltová

EXPERIMENTAL VALUES:

Composition of saturated solutions of KH_2PO_4 in water.

$t/°C$	d g cm^{-3}	K$_2$O conc^a	conc^b	conc^c	conc^a	P$_2$O$_5$ conc^b	conc^c	H$_2$O conc^c	KH$_2$PO$_4$ mass%	^{d} mol/kg
0	1.094	9.44	4.39	39.85	9.44	6.62	60.15	808.2	12.68	1.06
25	1.147	15.9	6.89	39.85	15.9	10.40	60.15	478.5	19.90	1.82
50		25.8	10.09	39.85	25.8	15.22	60.15	295.1	29.15	3.02

^{a}The concentration unit is: mol/1000 mol H_2O.

^{b}The concentration unit is: g/100 g of solution.

^{c}The concentration unit is: g/100 g of oxides.

^{d}These values were calculated by the compiler.

AUXILIARY INFORMATION

METHOD/APPARATUS/PROCEDURE:	SOURCE AND PURITY OF MATERIALS:
The mixtures were allowed to equilibrate for 12-15 hours in a water thermostat. Phosphorus was determined gravimetrically as $Mg_2P_2O_7$.	Kahlbaum KH_2PO_4 was used.
	ESTIMATED ERROR: The temperature was controlled to within ± 0.1 K.
	REFERENCES:

COMPONENTS:	ORIGINAL MEASUREMENTS:
(1) Potassium dihydrogenphosphate; KH_2PO_4; [7778-77-0] (2) Water; H_2O; [7732-18-5]	Bergman, A.G.; Bochkarev, N.F. *Izv. Akad. Nauk SSSR* <u>1938</u>, 237-65.

VARIABLES:	PREPARED BY:
Temperature and composition.	J. Eysseltová

EXPERIMENTAL VALUES:

Composition and crystallization temperatures in the aqueous KH_2PO_4 system.

KH_2PO_4			solid
mass%	mol/kg[a]	$t/°C.$	phase
4	0.30	-0.7	ice
8	0.63	-1.5	"
10	0.81	-2.1	"
12	1.00	0.8	KH_2PO_4
16	1.39	13.6	"
18	1.61	19.8	"
20	1.83	26.0	"
22	2.07	31.8	"

[a]The mol/kg H_2O values were calculated by the compiler.

AUXILIARY INFORMATION

METHOD/APPARATUS/PROCEDURE:	SOURCE AND PURITY OF MATERIALS:
No details are given except that a visual polythermic method was used.	Chemically pure KH_2PO_4 was recrystallized twice before being used.
	ESTIMATED ERROR: No information is given.
	REFERENCES:

COMPONENTS:	ORIGINAL MEASUREMENTS:
(1) Potassium dihydrogenphosphate; KH_2PO_4; [7778-77-0] (2) Water; H_2O; [7732-18-5]	Kazantsev, A.A. *Zh. Obshch. Khim.* 1938, *13*, 1230-1.

VARIABLES:	PREPARED BY:
Temperature and composition.	J. Eysseltová

EXPERIMENTAL VALUES:

Composition of saturated solutions of KH_2PO_4 in water.

$t/°C$.	mass% KH_2PO_4 calcd.[a]	exptl	mol/kg KH_2PO_4[b]
0	12.79	12.88	1.086
5	14.05	14.00	1.196
10	15.46	15.50	1.347
15	16.93	16.87	1.491
20	18.46	18.45	1.662
25	20.09	20.04	1.841
30	21.77	21.90	2.060
35	23.51	23.65	2.275
40	25.31	25.10	2.462
45	27.17	26.90	2.703
50	29.07	29.00	3.001
60	33.01	33.40	3.684
70	37.10	37.05	4.324
80	41.29	41.30	5.169
90	45.53	45.5	6.134

[a]These values were calculated from the empirical formula:

$$a = 12.79 + 0.250 \, t + 0.00182 \, t^2 - 0.00000616 \, t^3$$

where a is given as g KH_2PO_4/100 g soln, and t = temperature.

[b]These values were calculated by the compiler from the experimental results reported by the author.

AUXILIARY INFORMATION

METHOD/APPARATUS/PROCEDURE:	SOURCE AND PURITY OF MATERIALS:
A small amount of solid salt was added to 10-15 ml of a solution saturated at a higher temperature. The mixtures were equilibrated in a thermostat. Equilibration times were 3 hours for temperatures above 30°C, and 6 hours for temperatures lower than 30°C. After equilibration, samples of the solution were analyzed for KH_2PO_4 by drying at 110°C and weighing.	KH_2PO_4 was synthesized from K_2CO_3 and H_3PO_4 and recrystallized. Its purity was said to be equivalent to "chemically pure".
	ESTIMATED ERROR:
	The temperature had a precision of ± 0.05°C. No other details are given.
	REFERENCES:

COMPONENTS:	ORIGINAL MEASUREMENTS:
(1) Potassium dihydrogenphosphate; KH_2PO_4; [7778-77-0] (2) Water; H_2O; [7732-18-5]	Bel'tschev. F.V. *Trudy Beloruss. S.-Kh. Akad.* <u>1953</u>, *19*, 145-9.

VARIABLES:	PREPARED BY:
Temperature and composition.	J. Eysseltová

EXPERIMENTAL VALUES:

Composition and mean crystallization temperature
in the aqueous KH_2PO_4 system.

KH_2PO_4			solid
mass%	mol/kg^a	$t/°C.$	phase
4	0.30	-0.8	ice
8	0.63	-1.6	"
10	0.81	-2.0	"
12	1.00	-0.2	"
14	1.19	-7.8	"
16	1.39	+14.1	KH_2PO_4
18	1.61	+20.6	"
20	1.83	+26.7	"
22	2.07	+32.2	"

^{a}The mol/kg H_2O values were calculated by the compiler.

AUXILIARY INFORMATION

METHOD/APPARATUS/PROCEDURE:	SOURCE AND PURITY OF MATERIALS:
A visual polythermic method (1) was used. No other details are given.	Chemically pure KH_2PO_4 was recrystallized twice before use.
	ESTIMATED ERROR: No information is given.
	REFERENCES: 1. Bel'tschev, F.V.; Bergman, A.G. *Zh. Prikl. Khim.* <u>1944</u>, *17*, 9.

COMPONENTS:	ORIGINAL MEASUREMENTS:
(1) Potassium dihydrogenphosphate; KH_2PO_4; [7778-77-0] (2) Water; H_2O; [7732-18-5]	Punin, Yu.O.; Mirenkova, T.F.; Artamanova, O.I.; Ul'yanova, T.P. *Zh. Neorg. Khim.* <u>1975</u>, *20*, 2813-5.
VARIABLES:	PREPARED BY:
Temperature and composition.	J. Eysseltová

EXPERIMENTAL VALUES:

Parameters a_i of the equation

$$c_t = a_o + a_1 t + a_2 t^2 + \text{---} + a_5 t^5$$

$a_o = 14.958$

$a_1 = 2.881 \times 10^{-1}$

$a_2 = 4.914 \times 10^{-3}$

$a_3 = 1.826 \times 10^{-5}$

$a_4 = 2.311 \times 10^{-7}$

$a_5 = 0$

c_t = concentration of the saturated solution as g/100 g H_2O at the temperature, t°C.

AUXILIARY INFORMATION

METHOD/APPARATUS/PROCEDURE:	SOURCE AND PURITY OF MATERIALS:
Mixtures of KH_2PO_4 and water were prepared by precise weighing. The saturation temperatures were measured by an apparatus constructed for the purpose of measuring the rate of crystal growth (1).	A special purity grade of KH_2PO_4 was used.
	ESTIMATED ERROR:
	No information is given.
	REFERENCES:
	1. Petrov, T.G.; Trejbus, E.B.; Kosatkin, A.P. *"Vyrashtchivanie kristallov iz rastvorov"*, Nedra, Leningrad, <u>1967</u>.

COMPONENTS:	ORIGINAL MEASUREMENTS:
(1) Potassium dihydrogenphosphate; KH_2PO_4; [7778-77-0]	Muthmann, W.; Kuntze, O.
(2) Potassium dihydrogenarsenate; KH_2AsO_4; [7784-41-0]	Z. *Kryst.* <u>1894</u>, *23*, 368-76.
(3) Water; H_2O; [7732-18-5]	

VARIABLES:	PREPARED BY:
Composition at 7°C.	J. Eysseltová

EXPERIMENTAL VALUES:

Composition of saturated solutions in the KH_2PO_4-KH_2AsO_4-H_2O system at 7°C.

d g cm^{-3}	KH_2PO_4 concn.^{a}	concn.^{b}	mass%^{c}	mol/kg^c	KH_2AsO_4 concn.^{a}	concn.^{b}	mass%^{c}	mol/kg^c	H_2O mass%^{c}
1.1634	249.86	1834.9	21.48	2.00	----	----	0	0	78.52
1.1720	220.02	1615.4	18.77	1.76	37.60	208.7	3.20	0.22	78.01
1.1773	204.83	1504.3	17.39	1.64	59.84	332.1	5.08	0.36	77.51
1.1848	181.08	1329.8	15.28	1.45	92.10	511.3	7.77	0.56	76.94
1.1903	160.24	1176.8	13.46	1.29	120.80	670.6	10.14	0.73	76.38
1.1971	137.61	1010.6	11.49	1.11	151.39	840.4	13.64	0.92	75.85
1.2004	111.36	815.6	9.27	0.89	179.74	997.8	14.97	1.09	75.74
1.1999	80.89	594.7	6.74	0.65	205.69	1141.6	17.13	1.25	76.11
1.2000	51.09	375.2	4.25	0.41	234.05	1299.3	19.50	1.42	76.23
1.2009	29.17	214.2	2.42	0.23	256.20	1425.5	21.33	1.55	76.25
1.1955	----	----	0	0	282.37	1567.5	23.61	1.71	76.39

^{a}The concentration unit is: g/1000 ml.

^{b}The concentration unit is: mg mol/1000 ml.

^{c}These values were calculated by the compiler.

COMMENT: The solid phases were solid solutions.

AUXILIARY INFORMATION

METHOD/APPARATUS/PROCEDURE:	SOURCE AND PURITY OF MATERIALS:
Cool saturated solutions were mixed in a volume ratio of 1:9, 2:8,..., 9:1. Solid components were added to the mixtures. The mixtures were then heated to dissolve the solid phase and placed in a cellar. Super-saturated solutions were formed and seeded with a residue obtained by evaporation of a drop of the respective solution. The contents of the pycnometers used for density measurements were evaporated and the residue was dried at 100°C in a dry box. The arsenic content of the residue was then determined gravimetrically as As_2S_5.	No information is given.
	ESTIMATED ERROR:
	No details are given except that the temperature interval was 6.8 to 7.2°C.
	REFERENCES:

COMPONENTS:	ORIGINAL MEASUREMENTS:
(1) Potassium dihydrogenphosphate; KH_2PO_4; [7778-77-0] (2) Potassium acetate; CH_3CO_2K; [127-08-2] (3) Water; H_2O; [7732-18-5]	Apfel, O. Dissertation, Technical University, Darmstadt <u>1911</u>.
VARIABLES: Composition at 25°C.	PREPARED BY: J. Eysseltová

EXPERIMENTAL VALUES:

Composition of saturated solutions in the KH_2PO_4-CH_3CO_2K-H_2O system at 25°C.

PO_4^{3-} [b] conc	$CH_3CO_2^-$ [b] conc	KH_2PO_4 [a] mass%	mol/kg	CH_3CO_2K [a] mass%	mol/kg
1.28	0.50	17.42	1.65	4.91	0.64
1.06	0.98	14.43	1.40	9.62	1.29

[a] These values were calculated by the compiler.

[b] The concentration unit is: mol/1000 g of solution.

AUXILIARY INFORMATION

METHOD/APPARATUS/PROCEDURE:	SOURCE AND PURITY OF MATERIALS:
The isothermal method was used. Equilibrium was checked by repeated analysis. The solid and liquid phases were separated from each other by filtration through a platinum mat. Phosphate content was determined gravimetrically as $Mg_2P_2O_7$, potassium was determined gravimetrically as $KClO_4$.	No information is given.
	ESTIMATED ERROR: No information is given.
	REFERENCES:

COMPONENTS:	ORIGINAL MEASUREMENTS:
(1) Potassium dihydrogenphosphate; KH_2PO_4; [7778-77-0] (2) Potassium carbonate; K_2CO_3; [584-08-7] (3) Water; H_2O; [7732-18-5]	Apfel, O. Dissertation, Technical University, Darmstadt <u>1911</u>.

VARIABLES:	PREPARED BY:
Composition at 25°C.	J. Eysseltová

EXPERIMENTAL VALUES:

Composition of saturated solutions in the KH_2PO_4-K_2CO_3-H_2O system at 25°C.

PO_4^{3-}	CO_3^{2-}	KH_2PO_4[a]		K_2CO_3[a]	
conc[b]	conc[b]	mass%	mol/kg	mass%	mol/kg
1.69	0.12	23.00	2.24	1.66	0.16
2.34	0.415	31.86	3.75	5.74	0.66

[a] These values were calculated by the compiler.

[b] The concentration unit is: mol/1000 g of solution.

COMMENT: The author observed a vigorous evolution of CO_2, and, therefore, expresses doubt about the establishment of equilibrium in the system.

AUXILIARY INFORMATION

METHOD/APPARATUS/PROCEDURE:	SOURCE AND PURITY OF MATERIALS:
The isothermal method was used. The solid and liquid phases were separated from each other by filtration through a platinum wire mat. Analyses were done gravimetrically: phosphorus as $Mg_2P_2O_7$, and potassium as $KClO_4$.	No information is given.
	ESTIMATED ERROR: No information is given.
	REFERENCES:

COMPONENTS:	ORIGINAL MEASUREMENTS:
(1) Potassium dihydrogenphosphate; KH_2PO_4; [7778-77-0] (2) Dipotassium sulfate; K_2SO_4; [7778-80-5] (3) Water; H_2O; [7732-18-5]	Apfel, O. Dissertation, Technical University, Darmstadt, 1911.
VARIABLES: Composition at 25°C.	PREPARED BY: J. Eysseltová

EXPERIMENTAL VALUES:

Composition of saturated solutions in the KH_2PO_4-K_2SO_4-H_2O system at 25°C.

PO_4^{3-} [b] conc.	SO_4^{2-} [b] concn.	KH_2PO_4 [a] mass%	mol/kg	K_2SO_4 [a] mass%	mol/kg
1.47	----	19.87	1.82	----	----
1.43	0.08	19.46	1.81	1.39	0.10
1.34	0.18	18.24	1.70	3.14	0.23
1.30	0.36	17.69	1.71	6.27	0.47
1.24	0.39	16.88	1.62	6.80	0.51
1.25	0.39	17.01	1.64	6.80	0.51
1.23	0.36	16.74	1.60	6.27	0.47

[a] These values were calculated by the compiler.

[b] The concentration unit is: mol/1000 g solution.

AUXILIARY INFORMATION

METHOD/APPARATUS/PROCEDURE:	SOURCE AND PURITY OF MATERIALS:
The isothermal method was used. Equilibrium was checked by repeated analysis. The liquid and solid phases were separated from each other by filtration through a platinum wire mat. Analysis was done gravimetrically. Phosphate was determined as $Mg_2P_2O_7$, potassium was determined as $KClO_4$, and sulfate was determined as $BaSO_4$.	No details are given.
	ESTIMATED ERROR: No information is given.
	REFERENCES:

COMPONENTS:	ORIGINAL MEASUREMENTS:
(1) Potassium dihydrogenphosphate; KH_2PO_4; [7778-77-0] (2) Potassium nitrate; KNO_3; [7757-79-1] (3) Water; H_2O; [7732-18-5]	Apfel, O. Dissertation, Technical University, Darmstadt, 1911.

VARIABLES:	PREPARED BY:
Composition at 25°C.	J. Eysseltová

EXPERIMENTAL VALUES:

Composition of saturated solutions in the KH_2PO_4-KNO_3-H_2O system at 25°C.

PO_4^{3-} concn.[b]	NO_3^{-} concn.[b]	KH_2PO_4[a] mass%	mol/kg	KNO_3[a] mass%	mol/kg
1.20	0.55	16.33	1.54	5.56	0.70
1.03	1.19	14.01	1.39	12.03	1.61

[a] These values were calculated by the compiler.

[b] The concentration unit is: mol/1000 g of solution.

AUXILIARY INFORMATION

METHOD/APPARATUS/PROCEDURE:	SOURCE AND PURITY OF MATERIALS:
The isothermal method was used. Equilibrium was checked by repeated analysis of the liquid phase. The liquid and solid phases were separated from each other by filtration through a platinum wire mat. Analyses were done gravimetrically: phosphate as $Mg_2P_2O_7$, and potassium as $KClO_4$.	No information is given.
	ESTIMATED ERROR: Nothing is stated.
	REFERENCES:

COMPONENTS:	ORIGINAL MEASUREMENTS:
(1) Potassium dihydrogenphosphate; KH_2PO_4; [7778-77-0] (2) Hydrogen peroxide; H_2O_2; [7722-84-1] (3) Water; H_2O; [7732-18-5]	Menzel, H.; Gabler, C. Z. Anorg. Chem. 1929, 177, 187-214.

VARIABLES:	PREPARED BY:
Composition at 0°C.	J. Eysseltová

EXPERIMENTAL VALUES:

Solubility in the KH_2PO_4-H_2O_2-H_2O system at 0°C.

mol P : mol O_2^{2-}	H_2O_2		KH_2PO_4	
	g/1000 g soln	mol/kg	g/1000 g soln	mol/kg
1 : 0	-----	----	124.8	1.047
1 : 1.69	65.90	2.506	160.7	1.527
1 : 1.78	82.42	3.313	186.7	1.876

AUXILIARY INFORMATION

METHOD/APPARATUS/PROCEDURE:	SOURCE AND PURITY OF MATERIALS:
Equilibrium was reached isothermally in an ice-water bath. Repeated analyses were made to check the equilibrium. The dihydrogenphosphate ion content was determined gravimetrically as ammonium phosphomolybdate. The hydrogen peroxide content was determined by titration with potassium permanganate.	Kahlbaum KH_2PO_4 intended for use in enzyme investigation according to Soerensen was used. The H_2O_2 was the purest Merck reagent grade.

	ESTIMATED ERROR: The temperature was controlled to within ± 0.1 K. No other details are given.
	REFERENCES:

COMPONENTS:	ORIGINAL MEASUREMENTS:
(1) Potassium dihydrogenphosphate; KH_2PO_4; [7778-77-0] (2) Potassium chloride; KCl; [7447-40-7] (3) Water; H_2O; [7732-18-5]	Askenasy, P.; Nessler, F. Z. Anorg. Chem. <u>1930</u>, *189*, 305-28.

VARIABLES:	PREPARED BY:
Composition at 0°C.	J. Eysseltová

EXPERIMENTAL VALUES:

The 0°C isotherm for the KH_2PO_4-KCl-H_2O system.

d g/cm^3	KH_2PO_4			KCl			H_2O		solid$_c$ phase
	conc^b	mass%^{a}	mol/kg^a	conc^b	mass%^{a}	mol/kg^a	conc^b	mass%^{a}	
1.1151	100	15.5	1.35	0	0	0	4125	84.5	A
1.1158	91.9	14.6	1.28	18.1	1.77	0.28	3540	83.6	"
1.1179	68.3	13.0	1.04	31.7	3.31	0.53	3320	83.7	"
1.1183	55.1	12.0	1.07	44.9	5.37	0.87	2860	82.6	"
1.1199	37.3	10.2	0.93	62.7	9.44	1.58	2210	80.3	"
1.1427	27.0	8.70	0.82	73.0	12.9	2.21	1840	78.4	"
1.1750	7.9	3.21	0.31	92.1	20.6	3.63	1415	76.2	A + B
1.1690	4.2	1.72	0.16	95.8	21.5	3.76	1420	76.8	B
1.1420	0	0	0	100	22.3	3.85	1440	77.7	"

^{a}These values were calculated by the compiler.

^{b}The concentration unit is: mol/100 mol of solute.

^{c}The solid phases are: A = KH_2PO_4; B = KCl.

AUXILIARY INFORMATION

METHOD/APPARATUS/PROCEDURE:	SOURCE AND PURITY OF MATERIALS:
The isothermal method was used. The mixtures were agitated in a thermostat for 2 to 4 days. No information is given about the analytical procedures. The solid phase was separated from the liquid by centrifuging and analyzed.	No information is given.
	ESTIMATED ERROR: The only information given is that the temperature was controlled to within ± 0.1 K.
	REFERENCES:

COMPONENTS:	ORIGINAL MEASUREMENTS:
(1) Potassium dihydrogenphosphate; KH_2PO_4; [7778-77-0]	Askenasy, P.; Nessler, F.
(2) Ammonium dihydrogenphosphate; $NH_4H_2PO_4$; [7722-76-1]	Z. Anorg. Chem. <u>1930</u>, 189, 305-28.
(3) Water; H_2O; [7732-18-5]	

VARIABLES:	PREPARED BY:
Composition at 0°C.	J. Eysseltová

EXPERIMENTAL VALUES:

Composition of saturated solutions in the KH_2PO_4-$NH_4H_2PO_4$-H_2O system at 0°C.

d g cm^{-3}	KH_2PO_4 conc.^{a}	mass%^{b}	mol/kg^b	$NH_4H_2PO_4$ conc.^{a}	mass%^{b}	mol/kg^b	H_2O conc.^{a}	mass%^{b}
1.1151	100	15.5	1.35	0	0	0	4125	85.5
1.1169	80.6	14.7	1.32	19.4	3.6	0.32	3400	81.7
1.1393	67.4	13.8	1.27	32.6	7.0	0.62	2950	79.2
1.1472	54.7	12.8	1.20	45.3	10.2	0.89	2540	77.0
1.1577	49.3	12.4	1.17	50.7	12.2	1.21	2325	75.4
1.1571	44.2	11.4	1.10	55.8	13.7	1.38	2240	74.9
1.1603	38.1	10.2	0.89	61.9	15.7	1.61	2130	74.1
1.1604	36.4	9.8	0.86	63.6	16.1	1.67	2120	74.1
1.1611	35.2	9.6	0.83	64.8	16.4	1.71	2105	74.0
1.1574	31.2	8.4	0.82	68.8	17.1	1.73	2130	74.5
1.1568	26.4	7.0	0.67	73.6	17.6	1.86	2200	76.4
1.1447	18.2	4.6	0.44	81.8	18.7	1.99	2280	76.7
1.1350	15.3	2.8	0.36	84.7	19.7	2.00	2350	77.5
1.1312	12.4	3.0	0.28	87.6	18.6	1.97	2450	78.4
1.1043	0	0	0	100	18.6	1.97	2815	81.4

^{a}The concentration unit is: mol/100 mol of solute.

^{b}These values were calculated by the compiler.

AUXILIARY INFORMATION

METHOD/APPARATUS/PROCEDURE:	SOURCE AND PURITY OF MATERIALS:
Mixtures of salts and water were shaken in a thermostat for 2-4 days. The solid phase was isolated by centrifuging and analyzed. No details about the analytical procedures are given.	No information is given.
	ESTIMATED ERROR:
	The temperature was controlled to within ± 0.1 K. No other information is given.
	REFERENCES:

COMPONENTS:	ORIGINAL MEASUREMENTS:
(1) Potassium dihydrogenphosphate; KH_2PO_4; [7778-77-0]	Krasil'shtschikov, A.I.
(2) Potassium chloride; KCl; [7747-40-7]	*Izv. In-ta Fiz.-khim. Anal.* 1933, 6, 159-68.
(3) Water; H_2O; [7732-18-5]	

VARIABLES:	PREPARED BY:
Temperature and composition.	J. Eysseltová

EXPERIMENTAL VALUES:

Solubility isotherms in the KH_2PO_4-KCl-H_2O system.

No	t/°C.	d g cm^{-3}	KH_2PO_4 conc[a]	conc[b]	mass%	KCl conc[a]	conc[b]	mass%	H_2O mass%	conc[b]	solid phase[c]
1	0	1.094	19.3	100	12.7	----	----	----	87.30	687.4	A
2	0	1.095	16.0	81.4	10.5	6.6	18.6	2.4	87.10	675.2	"
3	0	1.097	11.7	54.7	7.6	17.7	45.3	1.3	91.10	619.4	"
4	0	1.165	4.3	10.7	2.5	65.8	89.3	20.9	76.60	327.3	A + B
5	0	1.156	----	----	----	69.0	100	22.2	77.80	350.4	B
6	25	1.147	32.9	100	19.92	----	----	----	80.08	402.0	A
7	25	1.144	30.1	92.6	18.28	4.3	7.4	1.45	80.27	406.8	"
8	25	1.137	22.4	70.1	13.65	17.4	29.9	5.83	80.52	413.3	"
9	25	1.138	21.0	65.3	12.75	20.6	35.0	6.87	80.38	409.6	"
10	25	1.139	19.8	60.2	11.98	23.9	39.8	7.92	80.10	402.5	"
11	25	1.140	18.3	54.6	11.05	27.8	45.4	9.18	79.77	394.3	"
12	25	1.148	14.5	39.6	8.54	40.3	60.4	13.05	78.41	363.2	"
13	25	1.179	8.8	18.2	4.83	71.3	81.8	21.67	73.50	277.4	"
14	25	1.196	6.5	12.6	3.57	83.9	87.4	24.82	71.61	252.2	A + B
15	25	1.184	2.7	5.3	1.44	85.8	94.7	25.81	72.75	267.0	B
16	25	1.179	----	----	----	86.8	100	26.40	73.60	278.0	"
17	50	-----	54.3	100	29.10	----	----	----	60.90	243.7	A
18	50	-----	38.1	75.5	20.8	22.6	24.7	6.8	72.40	262.3	"
19	50	-----	23.2	45.7	12.7	50.4	54.3	15.1	72.20	260.0	"
20	50	-----	10.2	15.96	5.23	99.0	84.04	27.5	67.23	205.2	A + B
21	50	-----	----	----	----	104.0	100	30.1	69.90	232.2	B

[a] The concentration unit is: mol/1000 mol H_2O.
[b] The concentration unit is: g/100 g salts.
[c] The solid phases are: A = KH_2PO_4; B = KCl.

AUXILIARY INFORMATION

METHOD/APPARATUS/PROCEDURE:	SOURCE AND PURITY OF MATERIALS:
The isothermal method was used, with 12-15 hours being allowed for equilibration. Chloride ion content was determined argentimetrically, the amount of total salts was determined by evaporating and drying a sample of the saturated solution.	Kahlbaum KH_2PO_4 was used. The source of the KCl is not specified.
	ESTIMATED ERROR:
	The temperature was controlled to within ± 0.1 K.
	REFERENCES:

COMPONENTS:	ORIGINAL MEASUREMENTS:
(1) Potassium dihydrogenphosphate; KH_2PO_4; [7778-77-0]	Krasil'shtschikov, A.I.
(2) Potassium chloride; KCl; [7747-40-7]	*Izv. In-ta Fiz.-khim. Anal.* 1933, 6, 159-68.
(3) Water; H_2O; [7732-18-5]	

EXPERIMENTAL VALUES cont'd:

The compiler has recalculated the above results to give the following:

No	KH_2PO_4 mol/kg	KCl mol/kg	No	KH_2PO_4 mol/kg	KCl mol/kg	No	KH_2PO_4 mol/kg	KCl mol/kg
1	1.068	-----	6	1.827	-----	17	3.015	-----
2	0.885	0.369	7	1.673	0.242	18	2.110	1.259
3	0.612	0.191	8	1.245	0.971	19	1.292	2.805
4	0.239	3.659	9	1.166	1.146	20	0.571	5.494
5	-----	3.827	10	1.098	1.326	21	-----	5.775
			11	1.017	1.543			
			12	0.800	2.232			
			13	0.482	3.954			
			14	0.366	4.648			
			15	0.145	4.758			
			16	-----	4.810			

COMPONENTS:	ORIGINAL MEASUREMENTS:

(1) Potassium dihydrogenphosphate; KH_2PO_4; [7778-77-0]
 	
 (2) Ammonium dihydrogenphosphate; $NH_4H_2PO_4$; [7722-76-1]
 	
 (3) Water; H_2O; [7732-18-5]
	1. Dombrovskaya, N.S.; Zvorykin, A.J.
	     Kaliy <u>1937</u>, *2*, 24-8.
	
	 2. Zvorykin, A.J.; Kuznetsov, V.G.
	     Izv. AN SSSR. ser. khim. <u>1938</u>, 195-201.

VARIABLES:	PREPARED BY:
Temperature and composition.	J. Eysseltová

EXPERIMENTAL VALUES:

The solubility in the KH_2PO_4-$NH_4H_2PO_4$-H_2O system has been reported by Zvorykin and co-workers in 2 publications. Source paper (1) reports the solubility isotherms at 25 and 50°C. Source paper (2) repeats only the data at 50°C. The solubility data are:

$t/°C$	KH_2PO_4		$NH_4H_2PO_4$		$t/°C$	KH_2PO_4		$NH_4H_2PO_4$	
	mass%	mol/kg[a]	mass%	mol/kg[a]		mass%	mol/kg[a]	mass%	mol/kg[a]
25	20.42	1.88	----	----	25	7.98	0.88	25.42	3.31
25	19.12	1.92	7.87	0.93	25	----	----	29.45	3.62
25	18.52	1.85	8.04	0.95	50	28.09	2.87	----	----
25	18.14	1.85	9.98	1.20	50	25.23	2.83	9.49	1.26
25	17.66	1.80	10.44	1.26	50	18.97	2.52	25.91	4.08
25	17.94	1.84	10.52	1.27	50	15.92	2.12	29.12	4.60
25	16.78	1.75	13.09	1.62	50	14.38	1.93	30.96	4.92
25	14.80	1.60	17.47	2.24	50	11.79	1.57	33.22	5.25
25	12.74	1.40	20.82	2.72	50	7.69	0.98	35.19	5.35
25	10.17	1.12	23.16	3.01	50	3.45	0.43	38.42	5.74
25	10.02	1.09	22.93	2.97	50	----	----	39.88	5.76

[a] The mol/kg H_2O values were calculated by the compiler.

(continued next page)

AUXILIARY INFORMATION

METHOD/APPARATUS/PROCEDURE:	SOURCE AND PURITY OF MATERIALS:
The isothermal method was used. The mixtures were agitated continuously in a thermostat for 2-5 days. Equilibrium was checked by repeated analysis. Potassium was determined as $KClO_4$, nitrogen was determined by the Kjeldahl method, and phosphorus was determined as $Mg_2P_2O_7$.	No information is given.
	ESTIMATED ERROR: The temperature was controlled to within ± 0.1 K.
	REFERENCES:

COMPONENTS:	ORIGINAL MEASUREMENTS:
(1) Potassium dihydrogenphosphate; KH_2PO_4; [7778-77-0]	1. Dombrovskaya, N.S.; Zvorykin, A.J. *Kaliy* 1937, 2, 24-8.
(2) Ammonium dihydrogenphosphate; $NH_4H_2PO_4$; [7722-76-1]	2. Zvorykin, A.J.; Kuznetsov, V.G. *Izv. AN SSSR. ser. khim.* 1938, 195-201.
(3) Water; H_2O; [7732-18-5]	

EXPERIMENTAL VALUES cont'd:

COMMENTS: The authors also express the composition of the saturated solutions in units other than mass% and mol/kg. These are given below.

$t/°C$	mol%	KH_2PO_4 conc.[a]	conc.[b]	mol%	$NH_4H_2PO_4$ conc.[a]	conc.[b]	mass%	H_2O mol%	concn.[b]
25	3.28	33.91	100	----	----	----	79.58	96.72	2949
25	3.22	34.64	67.2	1.61	16.87	32.8	73.01	95.17	1941
25	3.17	33.37	66.07	1.63	17.14	33.93	73.44	95.20	1980
25	3.16	33.34	60.54	2.06	21.74	39.46	71.88	94.78	1816
25	3.08	32.50	58.84	2.15	22.73	41.16	71.90	94.77	1810
25	3.14	33.19	59.04	2.18	23.03	40.96	71.54	94.68	1779
25	2.98	31.66	52.02	2.75	29.22	47.98	70.13	94.27	1643
25	2.70	28.92	41.72	3.78	40.39	58.28	67.73	93.52	1443
25	2.36	25.37	34.09	4.56	49.05	65.91	66.44	93.08	1344
25	1.88	20.18	27.07	5.06	54.38	72.93	66.67	93.06	1344
25	1.83	19.63	26.88	4.98	53.45	73.12	67.05	93.19	1368
25	1.47	15.85	20.99	5.55	59.74	79.01	66.60	92.98	1325
25	----	-----	-----	6.13	65.36	100	70.55	93.87	1530
50	4.91	51.69	100	----	-----	-----	71.91	95.09	1935
50	4.76	51.15	69.23	2.12	22.76	30.77	65.28	93.12	1354
50	4.07	45.56	38.22	6.58	73.69	67.77	55.12	89.35	839
50	3.41	38.33	31.62	7.40	82.97	68.38	54.96	89.19	825
50	3.10	34.81	27.71	7.89	95.01	72.29	54.66	89.01	810
50	2.52	28.35	23.18	8.42	94.52	77.26	54.99	89.10	817.4
50	1.60	17.82	15.58	8.66	96.45	84.37	57.12	89.74	874.7
50	0.71	7.86	7.05	9.31	103.5	92.95	58.13	89.98	898
50	----	-----	-----	9.41	102.7	100	60.12	91.59	974

[a] The concentration unit is: mol/1000 mol H_2O.

[b] The concentration unit is: mol/100 mol solute.

The authors state that the composition of the solution which is in equilibrium with a solid phase of the same composition is:
30.41 mol KH_2PO_4/100 mol solute and 69.59 mol $MH_4H_2PO_4$/100 mol solute at 25°C.
and
26.19 mol KH_2PO_4/100 mol solute and 73.81 mol $NH_4H_2PO_4$/100 mol solute at 50°C.

COMPONENTS:	ORIGINAL MEASUREMENTS:
(1) Potassium dihydrogenphosphate; KH_2PO_4; [7778-77-0] (2) Potassium nitrate; KNO_3; [7757-79-1] (3) Water; H_2O; [7732-18-5]	Bergman, A.G.; Bochkarev, N.F. *Izv. Akad. Nauk SSSR* <u>1938</u>, 237-65.

VARIABLES:	PREPARED BY:
Temperature and composition.	J. Eysseltová

EXPERIMENTAL VALUES:

Solubility isotherms in the KH_2PO_4-KNO_3-H_2O system.

t/°C.	KH_2PO_4			KNO_3			H_2O		solid[c] phase
	mass%	conc.[a]	mol/kg[b]	mass%	conc.[a]	mol/kg[b]	mass%	conc.[a]	
0	0	0	0	11	100	1.22	89	4532	A
0	3.6	21.0	0.30	9.9	79.0	1.13	86.5	3872	"
0	6.0	32.4	0.52	9.3	67.6	1.08	84.7	3457	"
0	9.0	44.0	0.80	8.5	56.0	1.01	82.5	3053	"
0	9.8	46.5	0.88	8.4	53.5	1.01	81.8	2929	A + B
0	10.5	65.3	0.90	4.2	34.7	0.48	85.3	4013	B
0	11.8	100	0.98	0	0	0	88.2	5628	"
10	0	0	0	16.6	100	1.98	83.4	2823	A
10	3.4	14.2	0.30	15.3	85.8	1.86	81.3	2564	"
10	8.6	32.5	0.80	13.2	67.5	1.66	78.2	2238	"
10	10.2	37.1	0.97	12.8	62.9	1.64	77	2116	A + B
10	11.2	48.0	1.02	8.9	52.0	1.10	79.9	2594	B
10	12.6	65.5	1.12	5.0	34.5	0.60	82.4	3221	"
10	15	100	1.29	0	0	0	85	4290	"
20	0	0	0	23.0	100	2.95	77.0	1875	A
20	3.0	9.3	0.29	21.6	90.7	2.83	75.4	1773	"
20	8.0	23.9	0.80	19.0	76.1	2.57	73.0	1640	"
20	10.4	30.0	1.06	17.9	70.0	2.46	71.7	1573	A + B
20	10.8	33.2	1.08	16.1	66.8	2.17	73.1	1705	B
20	13.2	53.0	1.24	8.7	77.0	1.10	78.1	2369	"
20	14.6	65.2	1.34	5.8	34.8	0.72	79.6	2694	"
20	18.2	100	1.63	0	0	0	81.8	3388	"

(continued next page)

AUXILIARY INFORMATION

METHOD/APPARATUS/PROCEDURE:	SOURCE AND PURITY OF MATERIALS:
A visual polythermic method was used. The isotherms were obtained by interpolation. No details are given.	Chemically pure KH_2PO_4 and KNO_3 were recrystallized twice before being used.
	ESTIMATED ERROR: No details are given.
	REFERENCES:

COMPONENTS:	ORIGINAL MEASUREMENTS:
(1) Potassium dihydrogenphosphate; KH_2PO_4; [7778-77-0] (2) Potassium nitrate; KNO_3; [7757-79-1] (3) Water; H_2O; [7732-18-5]	Bergman, A.G.; Bochkarev, N.F. *Izv. Akad. Nauk SSSR* <u>1938</u>, 237-65.

EXPERIMENTAL VALUES cont'd:

Solubility isotherms in the KH_2PO_4-KNO_3-H_2O system.

$t/°C.$	KH_2PO_4			KNO_3			H_2O		solid phase[c]
	mass%	conc.[a]	mol/kg[b]	mass%	conc.[a]	mol/kg[b]	mass%	conc.[a]	
30	0	0	0	31.0	100	4.44	69.0	1248	A
30	2.8	6.8	0.30	28.8	93.1	4.16	68.4	1241	"
30	7.4	17.6	0.81	25.6	82.4	3.77	67.0	1211	"
30	10.9	26.1	1.21	23.2	73.9	3.48	65.9	1180	A + B
30	12.2	36.6	1.24	15.8	63.4	2.17	72	1624	B
30	15.8	58.3	1.53	8.4	41.7	1.09	75.8	2114	"
30	17	65.1	1.63	6.8	34.9	0.88	76.2	2203	"
30	21.4	100	2.00	0	0	0	78.6	2779	"

[a]The concentration unit is: mol/100 mol of solute.

[b]The mol/kg H_2O values were calculated by the compiler.

[c]The solid phases are: A = KNO_3; B = KH_2PO_4.

Ternary eutectic point: temperature is -3.8°C.

composition is 7 mass% KNO_3 (0.84 mol/kg-compiler)

9.3 mass% KH_2PO_4 (0.82 mol/kg-compiler)

COMPONENTS:	ORIGINAL MEASUREMENTS:
(1) Potassium dihydrogenphosphate; KH_2PO_4; [7778-77-0] (2) Urea; CH_4N_2O; [57-13-6] (3) Water; H_2O; [7732-18-5]	Polosin, V.A.; Shakhparonov, M.I. *Zh. Obshch. Khim.* <u>1947</u>, *17*, 397-401.
VARIABLES: Temperature and composition.	PREPARED BY: J. Eysseltová

EXPERIMENTAL VALUES:

Part 1. Solubility isotherms in the KH_2PO_4-urea-H_2O system.

t/°C.	KH_2PO mass%	conc^b	mol/kg^a	$CO(NH_2)_2$ mass%	conc^b	mol/kg^a	H_2O mass%	conc^b	solid phase
-10	0	0	0.00	30.70	100	7.37	69.30	752	ice
-10	2.00	1.50	0.21	29.70	98.50	7.24	69.30	765	"
-10	2.92	4.60	0.30	27.10	95.40	6.44	69.98	820	"
-10	6.86	11.20	0.72	23.80	88.80	5.71	69.34	859	"
-10	7.60	10.70	0.86	27.72	89.30	7.13	64.68	694	KH_2PO_4
-10	6.16	7.80	0.72	31.60	92.20	8.45	62.24	602	urea
-10	2.69	3.50	0.30	32.60	96.50	8.38	64.71	638	"
-10	0	0	0.00	33.50	100	8.38	66.50	660	"
0	12.15	100	1.01	0	0	0.00	87.85	5460	KH_2PO_4
0	10.65	34.50	0.97	8.93	65.50	1.84	80.42	1967	"
0	9.85	19.60	1.01	18.10	80.40	4.18	71.95	1065	"
0	8.80	12.50	1.01	27.36	87.50	7.13	63.84	680	"
0	5.60	6.10	0.72	37.70	93.90	11.07	56.70	470	urea
0	2.45	2.70	0.30	38.80	97.30	10.99	58.75	491	"
0	0	0	0.00	39.80	100	11.00	60.20	520	"
+10	14.95	100	1.29	0	0	0.00	85.06	4300	KH_2PO_4
+10	13.25	40.30	1.24	8.67	59.70	1.84	78.08	1793	"
+10	11.60	22.50	1.20	17.68	77.50	4.16	70.72	1033	"
+10	10.00	14.10	1.16	27.00	85.90	7.13	63.00	669	"
+10	8.20	8.90	1.09	36.70	91.10	11.08	55.10	455	"
+10	7.40	7.20	1.06	41.65	92.80	13.62	50.95	377	"

(continued next page)

AUXILIARY INFORMATION

METHOD/APPARATUS/PROCEDURE:	SOURCE AND PURITY OF MATERIALS:
A polythermic method was used in the temperature range -12.9 to 35°C.	No information is given.
	ESTIMATED ERROR: No information is given.
	REFERENCES:

COMPONENTS:	ORIGINAL MEASUREMENTS:
(1) Potassium dihydrogenphosphate; KH_2PO_4; [7778-77-0]	Polosin, V.A.; Shakhparonov, M.I.
(2) Urea; CH_4N_2O; [57-13-6]	*Zh. Obshch. Khim.* <u>1947</u>, *17*, 397-401.
(3) Water; H_2O; [7732-18-5]	

EXPERIMENTAL VALUES cont'd:

Part 1. Solubility isotherms in the KH_2PO_4-urea-H_2O system.

	KH_2PO_4			$CO(NH_2)_2$			H_2O		solid
t/°C.	mass%	concn[b]	mol/kg[a]	mass%	concn[b]	mol/kg[a]	mass%	concn[b]	phase
+10	5.06	4.80	0.72	43.70	95.20	14.19	51.24	371	urea
+10	2.20	2.10	0.30	45.10	97.90	14.24	52.70	380	"
+10	0	0	0.00	46.00	100	14.18	54.00	391	"
+20	18.20	100	1.63	0	0	0.00	81.80	3398	KH_2PO_4
+20	15.85	45.40	1.53	8.41	54.60	1.84	75.74	1639	"
+20	10.40	25.50	1.05	17.32	74.50	3.98	69.28	993	"
+20	11.40	15.80	1.35	26.58	84.20	7.13	62.02	654	"
+20	9.20	10.00	1.24	36.30	90.00	11.08	54.50	449	"
+20	8.00	7.90	1.16	41.40	92.10	13.62	50.60	375	"
+20	7.60	6.80	1.20	46.20	93.20	16.65	46.20	311	"
+20	4.57	3.80	0.72	49.20	96.20	17.71	46.23	300	urea
+20	1.97	1.70	0.30	50.60	98.30	17.76	47.43	306	"
+20	0	0	0.00	51.80	100	17.89	48.20	310	"
+25	19.80	100	1.81	0	0	0.00	80.20	3056	KH_2PO_4
+25	17.10	45.40	1.68	8.29	54.60	1.85	74.61	1639	"
+25	14.40	27.20	1.54	17.12	72.80	4.16	68.48	972	"
+25	12.20	17.00	1.45	26.34	83.00	7.18	61.46	645	"
+25	9.70	10.60	1.31	36.10	89.40	11.08	54.20	446	"
+25	8.50	8.40	1.24	41.15	91.60	13.60	50.45	374	"
+25	8.00	7.10	1.27	46.00	93.20	16.65	46.00	309	"
+25	4.32	3.40	0.72	52.00	96.60	19.82	43.69	269	urea
+25	1.86	1.50	0.30	53.40	98.50	19.87	44.74	275	"
+25	0	0	0.00	54.50	100	19.94	45.50	278	"
+35	22.90	100	2.18	0	0	0.00	77.10	2545	KH_2PO_4
+35	19.70	52.00	2.00	8.03	48.00	1.85	72.27	1440	"
+35	16.70	30.80	1.84	16.66	69.20	4.16	66.64	924	"
+35	14.40	19.39	1.76	25.80	80.70	7.13	60.20	627	"
+35	11.20	12.20	1.54	35.50	87.80	11.09	53.30	435	"
+35	9.80	9.60	1.45	40.55	90.40	13.61	49.65	368	"
+35	8.70	7.80	1.40	40.65	92.20	16.65	45.65	306	"
+35	8.00	6.30	1.41	50.60	93.70	20.35	41.40	256	"
+35	3.82	2.80	0.72	57.50	97.20	24.75	39.68	226	urea
+35	1.64	1.20	0.30	59.00	98.80	24.95	39.36	218	"
+35	0	0	0.00	59.50	100	24.46	40.50	227	"

Part 2. Monovariant and invariant points.

	KH_2PO_4			$CO(NH_2)_2$			H_2O		
t/°C	mass%	concn[b]	mol/kg[a]	mass%	concn[b]	mol/kg[a]	mass%	concn[b]	solid phases
-5.1	9.40	31.40	0.84	9.06	68.60	1.85	81.50	2057	ice
-8.5	8.30	16.70	0.83	18.34	83.30	4.16	73.36	1111	"
-12.0	7.30	10.40	0.82	27.81	89.60	7.13	64.89	696	"
+0.6	7.20	7.80	0.94	37.10	92.20	11.09	55.70	460	urea + KH_2PO_4
+8.6	7.31	7.10	1.05	41.70	92.90	13.61	51.00	377	"
+18.2	7.60	6.80	1.20	46.20	93.20	16.65	46.20	311	"
+27.5	7.80	6.10	1.38	50.70	93.90	20.34	41.50	255	"
-11.7	2.74	3.70	0.30	31.15	96.30	7.84	65.76	670	urea + ice
+10.0	2.20	2.10	0.30	45.10	97.90	14.24	52.70	380	α-urea + β-urea
-12.8	6.31	8.50	0.72	29.90	91.50	7.80	63.79	649	urea + ice
+2.0	5.48	5.80	0.72	39.20	54.20	11.79	55.32	442	α-urea + β-urea
-10.8	0	0	0.00	33.30	100	8.20	67.00	676	urea + ice
+22.5	0	0	0.00	53.30	100	19.00	46.70	292	α-urea + β-urea
-2.4	11.70	100	0.97	0	0	0.00	88.30	5700	ice + KH_2PO_4
-12.9	7.00	90.60	0.81	29.70	9.40	7.81	63.30	643	ice+urea+KH_2PO_4
+1.0	7.20	7.90	0.85	37.50	92.10	11.29	55.90	482	α-urea + β-urea + KH_2PO_4

[a] The mol/kg H_2O values were calculated by the compiler.

[b] The concentration unit is: mol/100 mol of salts.

COMPONENTS:	ORIGINAL MEASUREMENTS:
(1) Potassium dihydrogenphosphate; KH_2PO_4; [7778-77-0]	Kuznetsov, D.I.; Kozhukhovskij, A.A.; Borovaya, F.E.
(2) Ammonium dihydrogenphosphate; $NH_4H_2PO_4$; [7722-76-1]	Zh. Prikl. Khim. 1948, 21, 1278-81
(3) Water; H_2O; [7732-18-5]	

VARIABLES:	PREPARED BY:
Composition at 25°C.	J. Eysseltová

EXPERIMENTAL VALUES:

Solubility in the KH_2PO_4-$NH_4H_2PO_4$-H_2O system at 25°C.

KH_2PO_4		$NH_4H_2PO_4$		H_2O	vapor press
mass%	mol/kg^a	mass%	mol/kg^a	mass%	mm Hg
20.21	1.86	----	0	79.79	22.66
17.53	1.72	7.85	0.91	74.62	22.08
16.19	1.68	13.05	1.60	70.76	21.90
15.48	1.65	15.84	2.00	69.68	21.84
14.27	1.53	17.53	2.23	67.20	----
13.08	1.43	20.12	2.61	66.80	21.61
7.97	0.86	24.48	3.15	67.55	21.95
----	0.00	28.85	3.52	71.15	22.00

^{a}The mol/kg H_2O values were calculated by the compiler.

AUXILIARY INFORMATION

METHOD/APPARATUS/PROCEDURE:	SOURCE AND PURITY OF MATERIALS:
The method has been described elsewhere (1). The nitrogen content was determined by using the Kjeldahl method, potassium was determined as $KClO_4$ after removal of the NH_3, and phosphorus was weighed as $NH_4MgPO_4 \cdot 6H_2O$. The vapor pressure was measured using the apparatus described by Vrevskiy (2).	The KH_2PO_4 and $NH_4H_2PO_4$ were recrystallized three times before use and dried at 100°C.

ESTIMATED ERROR:

No information is given.

REFERENCES:

1. Kuznetsov, D.I.; Kozhukhovaskij, A.A. Zh. Prikl. Khim. 1936, 9, 185.

2. Vrevskiy, M.S.; Zavaritskiy, N.N.; Sharlova, L.E. Zh. Russ. Fiz. Khim. Obshch. 1923, 54, 360.

COMPONENTS:	ORIGINAL MEASUREMENTS:
(1) Potassium dihydrogenphosphate; KH_2PO_4; [7778-77-0] (2) Dipotassium sulfate; K_2SO_4; [7778-80-5] (3) Water; H_2O; [7732-18-5]	Bel'teschev, F.V. *Trudy Beloruss. S.-Kh. Akad.* <u>1953</u>, *19*, 145-9.

VARIABLES:	PREPARED BY:
Temperature and composition.	J. Eysseltová

EXPERIMENTAL VALUES:

The following sections were studied in the binary salt-water systems:

No 1   (4% KH_2PO_4 + 96% H_2O) $-K_2SO_4$
No 2   (9% KH_2PO_4 + 91% H_2O) $-K_2SO_4$
No 3   (3% K_2SO_4 + 97% H_2O) $-KH_2PO_4$
No 4   (5% K_2SO_4 + 95% H_2O) $-KH_2PO_4$
No 5   (12% KH_2PO_4 + 88% H_2O) $-K_2SO_4$
No 6   (16% KH_2PO_4 + 84% H_2O) $-K_2SO_4$

Solubility isotherms at various temperatures.

sect No	KH_2PO_4 mass%	concn^a	mol/kg^b	K_2SO_4 mass%	concn^a	mol/kg^b	H_2O mass%	concn^a	solid^c phase
				temp. = 0°C.					
bin.	0	0	0	6.9	100	0.42	93.1	13238	A
1	3.8	30.43	0.30	5.6	69.57	0.35	90.6	10932	"
2	8.6	54.38	0.72	4.4	45.62	0.29	86.9	8461	"
	10.8	61.90	0.93	4.2	38.1	0.28	85.0	7488	A + B
3	11.3	73.2	0.96	2.6	26.8	0.17	86.1	8534	B
bin.	12.2	100	1.02	0	0	0	87.8	11077	"

(continued next page)

AUXILIARY INFORMATION

METHOD/APPARATUS/PROCEDURE:	SOURCE AND PURITY OF MATERIALS:
A visual polythermic method was used (1). The isotherms were constructed by inter-polation.	Chemically pure KH_2PO_4 and K_2SO_4 were recrystallized twice.

	ESTIMATED ERROR:
	No information is given.

	REFERENCES:
	1. Bel'tschev, F.V.; Bergman, A.G. *Zh. Prikl. Khim.* <u>1944</u>, *17*, 9.

COMPONENTS:	ORIGINAL MEASUREMENTS:
(1) Potassium dihydrogenphosphate; KH_2PO_4; [7778-77-0] (2) Dipotassium sulfate; K_2SO_4; [7778-80-5] (3) Water; H_2O; [7732-18-5]	Bel'tschev, F.V. *Trudy Beloruss. S.-Kh. Akad.* <u>1953</u>, *19*, 145-9.

EXPERIMENTAL VALUES cont'd:

The following sections were studied in the binary salt-water systems:

No 1 (4% KH_2PO_4 + 96% H_2O) $-K_2SO_4$
No 2 (9% KH_2PO_4 + 91% H_2O) $-K_2SO_4$
No 3 (3% K_2SO_4 + 97% H_2O) $-KH_2PO_4$
No 4 (5% K_2SO_4 + 95% H_2O) $-KH_2PO_4$
No 5 (12% KH_2PO_4 + 88% H_2O) $-K_2SO_4$
No 6 (16% KH_2PO_4 + 84% H_2O) $-K_2SO_4$

Solubility isotherms at various temperatures.

sect No	KH_2PO_4 mass%	concn [a]	mol/kg [b]	K_2SO_4 mass%	concn [a]	mol/kg [b]	H_2O mass%	concn [a]	solid phase [c]
				temp. = 10°C.					
bin.	0	0	0	8.4	100	0.52	91.6	10591	A
1	3.8	25.91	0.31	7.1	74.09	0.45	89.1	9157	"
2	8.4	47.62	0.21	5.8	52.38	0.38	85.8	7558	"
5	11.4	59.42	1.00	5.1	40.58	0.35	83.5	6526	"
	14.3	64.06	1.29	4.8	36.0	0.34	81.9	6060	A + B
4	13.4	67.12	1.19	4.3	32.88	0.29	82.3	6257	B
3	14.0	78.46	1.23	2.4	21.54	0.16	83.6	7138	"
bin.	14.9	100	1.28	0	0	0	85.1	8746	"
				temp. = 20°C.					
bin.	0	0	0	9.8	100	0.62	90.2	8939	A
1	3.7	22.58	0.30	8.5	77.42	0.55	87.8	7859	"
2	8.3	42.25	0.72	7.2	57.75	0.48	84.5	6605	"
5	11.2	52.56	0.99	6.5	47.44	0.45	82.3	5855	"
	16.0	66.66	1.49	5.2	33.34	0.37	78.8	5028	A + B
4	16.3	71.08	1.50	4.2	28.92	0.30	79.5	5316	B
5	17.0	81.58	1.54	2.4	18.42	0.17	80.6	5813	"
bin.	17.8	100	1.59	0	0	0	82.2	7018	"
				temp. = 30°C.					
bin.	0	0	0	11.4	100	0.73	88.6	7566	A
1	3.6	18.57	0.30	10.0	81.43	0.66	86.47	6850	"
2	8.2	38.46	0.72	8.4	61.54	0.57	83.4	5934	"
5	11.1	48.19	1.00	7.6	51.81	0.53	81.3	5436	"
6	15.0	59.78	1.40	6.4	40.22	0.40	78.6	4742	"
	18.6	68.0	1.82	6.6	32.0	0.50	75.8	4207	A + B
4	19.1	75.26	1.82	4.0	24.74	0.29	76.9	4589	B
5	19.8	85.58	1.86	2.2	14.12	0.16	78.0	5094	"
bin.	21	100	1.95	0	0	0	79	5693	"

[a] The concentration unit is: mol/100 mol solute.

[b] These values were calculated by the compiler.

[c] The solid phases are: A = K_2SO_4; B = KH_2PO_4.

(continued next page)

COMPONENTS:	ORIGINAL MEASUREMENTS:
(1) Potassium dihydrogenphosphate; KH_2PO_4; [7778-77-0] (2) Dipotassium sulfate; K_2SO_4; [7778-80-5] (3) Water; H_2O; [7732-18-5]	Bel'tschev, F.V. *Trudy Beloruss. S.-Kh. Akad.* <u>1953</u>, *19*, 145-9.

EXPERIMENTAL VALUES cont'd:

Composition and crystallization temperature of the monovariant points.

sect No.	KH_2PO_4 mass%	concn[a]	mol/kg	K_2SO_4 mass%	concn[a]	mol/kg	H_2O mass%	concn[a]	$t/°C.$	solid phases[b]
bin.	11.4	100	0.94	0	0	0	88.6	11708	-2.4	A + B
3	10.7	67.24	0.90	2.6	32.76	0.17	86.7	8297	-2.6	A + B
	10.3	62.29	0.88	4.2	38.71	0.28	85.5	7655	-3.1	A + B + C
2	8.6	55.35	0.72	4.3	44.64	0.28	87.1	8626	-2.7	B + C
1	3.8	31.11	0.30	5.4	68.89	0.34	90.8	11198	-1.7	B + C
bin.	0	0	0	6.6	100	0.40	93.4	13531	-1.5	B + C

[a] The concentration unit is: mol/100 mol of solute.

[b] The solid phases are: A = KH_2PO_4; B = ice; C = K_2SO_4.

COMPONENTS:	ORIGINAL MEASUREMENTS:
(1) Potassium dihydrogenphosphate; KH_2PO_4; [7778-77-0] (2) Potassium chloride; KCl; [7747-40-7] (3) Water; H_2O; [7732-18-5]	Brunisholz, G.; Bodmer, M. *Helv. Chim. Acta* <u>1963</u>, *46*, 288, 2566-74.

VARIABLES:	PREPARED BY:
Temperature and composition.	J. Eysseltová

EXPERIMENTAL VALUES:

Solubility isotherms in the KH_2PO_4-KCl-H_2O system.

K^+	Cl^-	H_2O		$KH_2PO_4{}^b$		KCl^b		solid $_c$
ion%	ion%	conc^a	mass%	mass%	mol/kg	mass%	mol/kg	phase^c
				temp. = 0°C.				
93.84	90.73	1358	77.28	1.32	0.12	21.38	3.71	A
88.55	82.76	1271	76.69	2.62	0.25	20.68	2.62	A + B
77.07	65.48	1627	81.95	4.38	0.39	13.66	2.23	B
62.12	42.81	1888	85.45	6.52	0.58	8.02	1.25	"
61.23	41.55	1901	85.61	6.63	0.56	7.75	1.21	"
48.13	21.91	1919	86.96	8.91	0.75	4.11	0.63	"
47.66	20.99	1925	87.06	9.00	0.76	3.93	0.60	"
33.33	0	1778	87.58	12.41	1.04	0.00	0.00	"
				temp. = 25°C.				
93.02	89.62	1057	72.67	1.79	0.18	25.52	4.71	A
87.18	80.71	972	71.73	3.58	0.36	24.67	4.61	A + B
75.03	62.02	1158	76.65	6.33	0.60	17.00	2.97	B
60.00	40.00	1248	79.74	9.66	0.89	10.58	1.78	"
46.23	18.86	1177	80.63	14.01	1.27	5.35	0.89	"
33.33	0	1012	80.06	19.93	1.82	0.00	0.00	"
				temp. = 50°C.				
100	100	967.4	70.01	0.00	0.00	29.98	5.74	A
84.91	77.30	783.1	67.47	4.92	0.53	27.59	5.48	A + B
69.54	54.28	885.7	72.25	9.40	0.95	18.34	3.40	B
51.26	26.93	834.0	73.82	16.30	1.62	9.87	1.79	"
33.00	0	624.4	71.30	28.70	2.96	0.00	0.00	"

(continued next page)

AUXILIARY INFORMATION

METHOD/APPARATUS/PROCEDURE:	SOURCE AND PURITY OF MATERIALS:
A previously described method was used (1). The analytical procedures were the following: chloride was titrated potentiometrically with $AgNO_3$; $H_2PO_4^-$ was converted to H_3PO_4 by ion exchange and then titrated acidimetrically using chlorophenol red as indicator; potassium was determined gravimetrically as $KClO_4$ or as the tetraphenylborate; water was determined by difference.	No information is given.
	ESTIMATED ERROR: No information is given.
	REFERENCES: 1. Flatt, R. *Chimia* <u>1962</u>, *6*, 62.

COMPONENTS:	ORIGINAL MEASUREMENTS:
(1) Potassium dihydrogenphosphate; KH_2PO_4; [7778-77-0] (2) Potassium chloride; KCl; [7747-40-7] (3) Water; H_2O; [7732-18-5]	Brunisholz, G.; Bodmer, M. *Helv. Chim. Acta* <u>1963</u>, *46*, 288, 2566-74.

EXPERIMENTAL VALUES cont'd:

Solubility isotherma in the KH_2PO_4-KCl-H_2O system.

K^+	Cl^-	H_2O		KH_2PO_4[b]		KCl[b]		solid
ion%	ion%	conc[a]	mass%	mass%	mol/kg	mass%	mol/kg	phase[c]
				temp. = 75°C.				
91.90	87.85	753.0	65.61	2.66	0.29	31.71	6.48	A
87.12	80.67	704.9	64.80	4.47	0.50	30.71	6.35	"
80.43	70.61	629.5	63.19	7.43	0.86	29.36	6.23	A + B
70.92	56.43	659.5	65.74	10.94	1.22	23.30	4.75	B
55.65	33.40	615.0	66.76	18.22	2.00	15.01	3.01	"
46.30	19.56	563.3	66.50	23.93	2.64	9.56	1.92	"
33.00	0	400.7	61.38	38.61	4.62	0.00	0.00	"

[a]The concentration unit is: mol H_2O/100 g equiv of the salts.

[b]These values were calculated by the compiler.

[c]The solid phases are: A = KCl; B = KH_2PO_4.

COMPONENTS:	ORIGINAL MEASUREMENTS:
(1) Potassium dihydrogenphosphate; KH_2PO_4; [7778-77-0] (2) Potassium chloride; KCl; [7747-40-7] (3) Water; H_2O; [7732-18-5]	Filipescu, L. *Rev. Chim. (Bucharest)* <u>1971</u>, *22*, 533-40.

VARIABLES:	PREPARED BY:
Composition and temperature.	J. Eysseltová

EXPERIMENTAL VALUES:

Solubility isotherms in the KH_2PO_4–KCl–H_2O system.

$t/°C$	d $g\,cm^{-3}$	PO_4^{3-} $concn^a$	Cl^- $concn^a$	H_2O M^b	$KH_2PO_4^c$ mass%	mol/kg	KCl^c mass%	mol/kg	solid phase^d
20	1.175	0.0000	0.4591	1210.0	0.00	0.00	25.50	4.59	A
20	-----	0.0301	0.4526	1149.6	1.01	0.10	24.97	4.52	"
20	-----	0.0691	0.4443	1082.0	2.30	0.23	24.31	4.44	"
20	1.170	0.1009	0.4358	1035.1	3.33	0.33	23.70	4.35	A + B
20	-----	0.1387	0.4256	1106.6	4.55	0.46	22.99	4.25	B
20	-----	0.1753	0.2789	1223.1	6.17	0.58	16.15	2.78	"
20	-----	0.2275	0.2021	1293.2	8.23	0.75	12.01	2.02	"
20	-----	0.2774	0.1472	1308.3	10.18	0.92	8.88	1.47	"
20	-----	0.3443	0.0898	1278.3	12.77	1.14	5.47	0.89	"
20	-----	0.4186	0.0380	1216.6	15.58	1.39	2.32	0.38	"
20	1.131	0.4855	0.0000	1144.2	18.04	0.01	0.00	0.00	"
40	1.187	0.0000	0.5399	1028.9	0.00	0.00	28.70	5.39	A
40	-----	0.0214	0.5366	994.5	0.68	0.07	28.37	5.36	"
40	-----	0.0872	0.5223	911.4	2.76	0.29	27.25	5.22	"
40	1.185	0.1370	0.5126	855.3	4.30	0.45	26.46	5.12	A + B
40	-----	0.2197	0.3757	933.0	7.22	0.73	20.30	3.75	B
40	-----	0.3131	0.2701	952.6	10.57	1.04	14.99	2.70	"
40	-----	0.4438	0.1644	913.3	15.20	1.47	9.25	1.64	"
40	-----	0.6008	0.0655	833.2	20.62	2.00	3.69	0.65	"
40	-----	0.7348	0.0000	756.1	25.00	2.44	0.00	0.00	"

(continued next page)

AUXILIARY INFORMATION

METHOD/APPARATUS/PROCEDURE:	SOURCE AND PURITY OF MATERIALS:
The mixtures were equilibrated isothermally for 5 hours while being stirred by a stream of inert gas. The potassium content was determined by using a flame photometer, the dihydrogenphosphate was determined acidimetrically using thymolphthalein as indicator, and chloride was determined mercurimetrically.	The KCl and the KH_2PO_4 were recrystallized three times before being used.

ESTIMATED ERROR:
The temperature at 20 and 40°C were controlled to within ± 0.05°C while the temperatures at 60 and 80°C were controlled to within ± 0.1°C.
REFERENCES:

COMPONENTS:	ORIGINAL MEASUREMENTS:
(1) Potassium dihydrogenphosphate; KH_2PO_4; [7778-77-0]	Filipescu, L.
(2) Potassium chloride; KCl; [7747-40-7]	Rev. Chim. (Bucharest) <u>1971</u>, 22, 533-40.
(3) Water; H_2O; [7732-18-5]	

EXPERIMENTAL VALUES cont'd:

Solubility isotherms in the KH_2PO_4-KCl-H_2O system.

$t/°C$	d g cm^{-3}	PO_4^{3-} concn[a]	Cl^- concn[a]	H_2O M[b]	KH_2PO_4[c] mass%	mol/kg	KCl[c] mass%	mol/kg	solid phase[d]
60	1.201	0.0000	0.6151	903.1	0.00	0.00	31.44	6.15	A
60	-----	0.0643	0.6059	828.9	1.96	0.21	30.50	6.05	"
60	-----	0.1539	0.5908	746.0	4.62	0.51	29.16	5.90	"
60	1.204	0.1931	0.5832	715.6	5.75	0.64	28.56	5.83	A + B
60	-----	0.2604	0.4695	761.1	8.04	0.86	23.84	4.69	B
60	-----	0.3771	0.3342	781.0	12.04	1.25	17.54	3.34	"
60	-----	0.5116	0.2388	740.3	16.45	1.70	12.62	2.38	"
60	-----	0.7121	0.1424	650.1	22.60	2.37	7.42	1.42	"
60	1.240	1.0554	0.0000	526.4	32.37	3.51	0.00	0.00	"
80	1.214	0.0000	0.6836	812.6	0.00	0.00	33.76	6.83	A
80	-----	0.0888	0.6731	729.1	2.61	0.29	32.54	6.73	"
80	-----	0.1970	0.6553	652.6	5.66	0.65	30.96	6.55	"
80	1.230	0.2870	0.6370	601.2	8.11	0.95	29.58	6.37	A + B
80	-----	0.4256	0.4604	627.0	12.56	1.41	22.34	4.60	B
80	-----	0.6005	0.3327	595.3	17.91	2.00	16.31	3.32	"
80	-----	0.9132	0.1871	504.9	26.66	3.04	8.97	1.87	"
80	-----	1.2355	0.0832	421.3	34.54	4.11	3.82	0.83	"
80	1.309	1.5162	0.0000	366.4	40.75	5.05	0.00	0.00	"

[a] The concentration unit is: equiv/100 g water.

[b] The concentration unit is: mol/100 equiv of salts.

[c] These values were calculated by the compiler.

[d] The solid phases are: A = KCl; B = KH_2PO_4.

COMPONENTS:	ORIGINAL MEASUREMENTS:
(1) Potassium dihydrogenphosphate; KH_2PO_4; [7778-77-0] (2) Ammonium dihydrogenphosphate; $NH_4H_2PO_4$; [7722-76-1] (3) Water; H_2O; [7732-18-5]	Bergman, A.G.; Gladkovskaya, A.A.; Galushkina, R.A. *Zh. Neorg. Khim.* 1972, *17*, 2055-6.
VARIABLES: Temperature and composition.	PREPARED BY: J. Eysseltová

EXPERIMENTAL VALUES:

Original mixture	Component added	conc.[a]	$t/°C.$	solid phases[b]
12.5% KH_2PO_4 + 87.5% H_2O	$NH_4H_2PO_4$	11.5	+6.0	A + B
15.0% KH_2PO_4 + 85.0% H_2O	"	11.6	+14.2	"
" " " "	"	17.3	+17.6	"
71.72% $NH_4H_2PO_4$ + 28.28% KH_2PO_4	H_2O	20.5	-4.5	B + C
" " " "	"	32.6	+25.0	A + B
45.81% $NH_4H_2PO_4$ + 54.19% KH_2PO_4	"	18.2	-4.0	A + C
" " "	"	24.5	+13.6	A + B
" " "	"	25.0	+14.0	"
25.0% $NH_4H_2PO_4$ + 75.0% KH_2PO_4	"	14.3	-3.6	A + C
10.0% $NH_4H_2PO_4$ + 90.0% H_2O	KH_2PO_4	10.3	-4.0	"

[a] This is the mass% of the component added.

[b] The solid phases are: A = β-solid soln; B = α-$NH_4H_2PO_4$; C = ice.

Compiler's comment: It is not possible to construct a legitimate phase diagram on the basis of the data that are given. The conc. of component added has the meaning given to it only if the added component is a salt. When water is the added component, the relation w_{H_2O} = 100 - conc. of added component is valid. With this assumption the following compositions of points lying on the eutectic curve were calculated.

(continued next page)

AUXILIARY INFORMATION

METHOD/APPARATUS/PROCEDURE:	SOURCE AND PURITY OF MATERIALS:
The only information given is that a visually polythermic method (1) was used.	Chemically pure KH_2PO_4 and $NH_4H_2PO_4$ were recrystallized and dried before being used. Bidistilled water was used.
	ESTIMATED ERROR: No information is given.
	REFERENCES: 1. Bergman, A.G.; Luzhnaya, N.P. *Fiziko-Khimicheskie Osnovy Izucheniya i Ispol'zovaniya Solyanykh Mestorozhdeniy Khlorid-sul'fatnogo Tipa*, Moscow, IAN SSSR, 1951.

COMPONENTS:	ORIGINAL MEASUREMENTS:
(1) Potassium dihydrogenphosphate; KH_2PO_4; [7778-77-0]	Bergman, A.G.; Gladkovskaya, A.A.; Galushkina, R.A.
(2) Ammonium dihydrogenphosphate; $NH_4H_2PO_4$; [7722-76-1]	*Zh. Neorg. Khim.* <u>1972</u>, *17*, 2055-6.
(3) Water; H_2O; [7732-18-5]	

EXPERIMENTAL VALUES cont'd:

KH_2PO_4		$NH_4H_2PO_4$		H_2O		
mass%	mol/kg	mass%	mol/kg	mass%	$t/°C.$	solid phases[a]
11.21	1.06	11.50	1.29	77.28	+6.0	A + B
13.44	1.32	11.60	1.34	74.95	+14.2	"
12.78	1.34	17.30	2.15	69.91	+17.6	"
5.80	0.53	14.70	1.60	79.50	-4.5	B + C
9.22	1.00	23.38	3.01	67.40	+25.0	A + B
9.86	0.88	8.33	0.88	81.80	-4.0	A + C
13.28	1.29	11.22	1.29	75.50	+13.6	A + B
13.55	1.32	11.45	1.32	75.00	+14.0	"
7.72	0.63	2.58	0.24	89.70	-3.6	A + C
10.30	0.94	9.06	0.98	80.63	-4.0	"

The composition of the eutectic point is:

8.50	0.07	11.40	0.53	80.10	-4.5	A + B + C

[a]The solid phases are the same as those given in footnote b above.

COMPONENTS:	ORIGINAL MEASUREMENTS:
(1) Potassium dihydrogenphosphate; KH_2PO_4; [7778-77-0]	Shenkin, Ya.S.; Ruchnova, S.A.; Rodionova, N.A.
(2) Ammonium dihydrogenphosphate; $NH_4H_2PO_4$; [7722-76-1]	*Zh. Neorg. Khim.* 1972, *17*, 3368-9.
(3) Water; H_2O; [7732-18-5]	

VARIABLES:	PREPARED BY:
Composition and temperature at atmospheric pressure.	J. Eysseltová

EXPERIMENTAL VALUES:

Composition and boiling points of saturated solutions
in the $NH_4H_2PO_4$-KH_2PO_4-H_2O system.

KH_2PO_4		$NH_4H_2PO_4$		H_2O	
mass%	mol/kg	mass%	mol/kg	mass%	b.p./°C
52.70	8.18	0	0	47.30	105.1
51.78	8.46	3.28	0.63	44.94	105.3
49.40	8.30	6.90	1.37	43.70	106.2
47.46	8.32	10.67	2.21	41.87	105.7
44.58	8.20	15.52	3.38	39.90	106.9
43.60	8.25	17.57	3.93	38.83	107.1
42.60	8.30	19.70	4.54	37.70	108.0
38.78	8.07	25.94	6.39	35.28	108.9
38.44	8.41	28.00	7.25	33.56	108.6
37.23	8.19	29.39	7.65	33.38	106.6
35.31	8.12	32.76	8.91	31.93	108.8
35.08	7.75	31.69	8.28	33.23	108.8
32.64	7.31	34.56	9.15	32.80	106.4
30.57	6.89	36.86	9.83	32.57	106.9
28.15	6.54	40.23	11.05	31.62	109.7
27.92	6.79	41.87	12.04	30.21	110.6
27.09	6.62	42.86	12.39	30.05	109.8
25.36	6.16	44.42	12.77	30.22	110.3
24.79	6.09	45.32	13.17	29.89	109.5
21.91	5.50	48.85	14.52	29.24	109.4
20.24	4.92	49.59	14.28	30.17	111.4

(continued next page)

AUXILIARY INFORMATION

METHOD/APPARATUS/PROCEDURE:	SOURCE AND PURITY OF MATERIALS:
The method used to determine the solubility has been described earlier (1).	Chemically pure KH_2PO_4 and $NH_4H_2PO_4$ were used.
	ESTIMATED ERROR:
	No information is given.
	REFERENCES:
	1. Shenkin, Ya.S.; Ruchnova, S.A.; Shenkina, A.P. *Zh. Neorg. Khim.* 1968, *13*, 256.

COMPONENTS:	ORIGINAL MEASUREMENTS:
(1) Potassium dihydrogenphosphate; KH_2PO_4; [7778-77-0]	Shenkin, Ya.S.; Ruchnova, S.A.; Rodionova, N.A.
(2) Ammonium dihydrogenphosphate; $NH_4H_2PO_4$; [7722-76-1]	*Zh. Neorg. Khim.* 1972, *17*, 3368-9.
(3) Water; H_2O; [7732-18-5]	

EXPERIMENTAL VALUES cont'd:

Composition and boiling points of saturated solutions
in the $NH_4H_2PO_4$-KH_2PO_4-H_2O system.

KH_2PO_4		$NH_4H_2PO_4$		H_2O	
mass%	mol/kg	mass%	mol/kg	mass%	b.p./°C.
19.07	5.14	53.69	17.13	27.24	110.9
17.01	4.35	54.27	16.42	28.72	112.0
14.82	3.72	55.91	16.60	29.27	109.6
14.71	3.76	56.57	17.12	28.72	110.8
8.26	2.05	62.15	18.25	29.59	110.4
6.83	1.80	65.35	20.41	27.82	109.9
0	0	68.30	18.72	31.70	110.5

COMPONENTS:	ORIGINAL MEASUREMENTS:
(1) Potassium dihydrogenphosphate; KH_2PO_4; [7778-77-0] (2) Potassium hydrogenselenate; $KHSeO_4$; [25105-33-3] (3) Water; H_2O; [7732-18-5]	Zbořilová, L.; Krejčí, J. *Scripta Fac. Sci. Nat. UJEP Brunensis, Chemie 1* <u>1972</u>, 77-80.
VARIABLES:	PREPARED BY:
Composition at 20°C.	J. Eysseltová

EXPERIMENTAL VALUES:

Composition of the phases in the KH_2PO_4-$KHSeO_4$-H_2O system at 20°C.

saturated solution					solid phase			
KH_2PO_4		$KHSeO_4$		H_2O	Se	P	$KHSeO_4$	KH_2PO_4
mass%	mol/kg^a	mass%	mol/kg^a	mass%	mass%	mass%	mass%	mass%
0	0	43.31	4.16	56.69				
1.51	0.18	38.88	3.56	59.50	35.40	4.20	80.93	19.10
3.21	0.38	35.96	3.22	60.86	34.70	4.46	80.00	19.60
4.29	0.51	34.64	3.09	61.03	24.75	8.90	56.98	40.85
5.06	0.55	28.28	2.31	66.66	25.30	9.40	58.20	41.80
5.37	0.55	23.49	1.80	71.18	24.40	9.47	56.85	41.90
8.18	0.84	20.87	1.60	70.94	24.80	10.08	57.80	41.60
7.64	0.75	17.77	1.30	74.57	8.06	18.46	18.50	81.50
8.15	0.75	12.94	0.89	78.89	1.18	21.80	2.72	96.80
8.91	0.75	4.46	0.28	86.53	0.33	22.40	0.76	98.40
18.50	1.66	0	0	81.50				

calculated for $KH_2PO_4 \cdot KHSeO_4$ 57.36   42.67

calculated for $3KH_2PO_4 \cdot KHSeO_4$ 80.15   19.85

^{a}The mol/kg H_2O values were calculated by the compiler.

AUXILIARY INFORMATION

METHOD/APPARATUS/PROCEDURE:	SOURCE AND PURITY OF MATERIALS:
Saturated solutions containing $KHSeO_4$ and KH_2PO_4 in molar ratios of 9:1 to 1:9 were prepared at higher temperatures and equilibrated in a thermostat for several hours. Solid and liquid phases were analyzed. Selenium was determined iodometrically (2) and phosphorus was determined colorimetrically (3, 4).	$KHSeO_4$ was synthesized by the reaction of K_2SeO_4 with H_2SeO_4 (1). No other details are given.

	ESTIMATED ERROR:
	No information is given.
	REFERENCES:
	1. Dostál, K.; Krejčí, J. *Z. Anorg. Allg. Chem.* <u>1958</u>, *296*, 29. 2. Blanka, B; et al. *Coll. Czech. Chem. Soc.* <u>1963</u>, *28*, 3424. 3. Bernhart. D.N.; Wreath, A.R. *Anal. Chem.* <u>1955</u>, *27*, 440. 4. Netherton, L.E.; Wreath, A.R. *Anal. Chem.* <u>1955</u>, *27*, 860.

COMPONENTS:	ORIGINAL MEASUREMENTS:
(1) Potassium dihydrogenphosphate; KH_2PO_4; [7778-77-0] (2) Dipotassium sulfate; K_2SO_4; [7778-80-5] (3) Water; H_2O; [7732-18-5]	Gladkovskaya, A.A.; Bergman, A.G. *Tr. Kuban. S.-Kh. In-ta* 1975, *102*, 130, 31-4.
VARIABLES: Composition and temperature.	PREPARED BY: J. Eysseltová

EXPERIMENTAL VALUES:

original mixture	added	A^a	$t/°C.$	solid phases
10% KH_2PO_4 + 90% K_2SO_4	H_2O	8.6	+7.2	$K_2SO_4 \cdot H_2O$ + K_2SO_4
" " " "	"	7.0	-1.2	$K_2SO_4 \cdot H_2O$ + ice
20% KH_2PO_4 + 80% K_2SO_4	"	8.6	+3.0	K_2SO_4 + $K_2SO_4 \cdot H_2O$
" " " "	"	7.5	-1.2	ice + $K_2SO_4 \cdot H_2O$
43.85% KH_2PO_4 + 56.15% K_2SO_4	"	8.6	-2.0	ice + K_2SO_4
16% KH_2PO_4 + 84% H_2O	K_2SO_4	7.0	+15.6	KH_2PO_4 + K_2SO_4

[a] This is the mass% of the component added.

COMMENT: It is impossible to construct a valid phase diagram on the basis of the above data. The compiler's opinion is that the value A had its given meaning only if the component added was a salt. In the case of water, the relation w_{H_2O} = 100 - A is valid. On the basis of this assumption the compiler has calculated the following values.

KH_2PO_4		K_2SO_4		H_2O		
mass%	mol/kg	mass%	mol/kg	mass%	$t/°C.$	solid phases[a]
0.86	0.06	7.74	0.48	91.4	+7.2	A + B
0.70	0.05	6.30	0.38	93.0	-1.2	A + C
1.72	0.13	6.88	0.43	91.4	+3.0	A + B
1.50	0.11	6.00	0.37	92.5	-1.2	A + C
3.77	0.30	4.83	0.30	91.4	-2.0	B + C
14.95	1.40	7.00	0.51	78.05	+15.6	B + D

The solid phases are: A = $K_2SO_4 \cdot H_2O$; B = K_2SO_4; C = ice; D = KH_2PO_4.

(continued next page)

AUXILIARY INFORMATION

METHOD/APPARATUS/PROCEDURE:	SOURCE AND PURITY OF MATERIALS:
A visual polythermic method was used (1). Solid carbon dioxide was used as the cooling agent.	No information is given.
	ESTIMATED ERROR: No details are given.
	REFERENCES: 1. Bergman, A.G.; Luzhnaya, N.P. *Fiziko-khimicheskie Osnovy Izuchenija i Ispol'zovanija Soljanykh Mestorozhdenii Khlorid-sul'fatnogo Tipa*, Moscow, IAN SSSR, 1951.

COMPONENTS:	ORIGINAL MEASUREMENTS:
(1) Potassium dihydrogenphosphate; KH_2PO_4; [7778-77-0]	Gladkovskaya, A.A.; Bergman, A.G.
(2) Dipotassium sulfate; K_2SO_4; [7778-80-5]	*Tr. Kuban. S.-Kh. In-ta* <u>1975</u>, *102*, 130, 31-4.
(3) Water; H_2O; [7732-18-5]	

EXPERIMENTAL VALUES cont'd:

The authors give the following triple points:

KH_2PO_4		K_2SO_4		H_2O		
mass%	mol/kg[a]	mass%	mol/kg[a]	mass%	t/°C.	solid phases
10.3	0.07	4.2	0.01	88.5	+3.1	ice + KH_2PO_4 + K_2SO_4
2.5	0.01	6.2	0.02	91.3	+1.2	ice + K_2SO_4 + $K_2SO_4 \cdot H_2O$

[a]These values were calculated by the compiler.

COMPONENTS:	ORIGINAL MEASUREMENTS:
(1) Potassium dihydrogenphosphate; KH_2PO_4; [7778-77-0] (2) Potassium chloride; KCl; [7747-40-7] (3) Water; H_2O; [7732-18-5]	Mráz, R.; Srb, V.; Tichý, S.; Vosolsobě, J. *Chem. Prům.* 1976, *26*, 511-4.

VARIABLES:	PREPARED BY:
Composition and temperature.	J. Eysseltová

EXPERIMENTAL VALUES:

Solubility isotherms in the KH_2PO_4-KCl-H_2O system.

KH_2PO_4		KCl		H_2O	
mass%	mol/kg[a]	mass%	mol/kg[a]	mass%[a]	solid phases
			temp. = 25°C.		
20.3	1.87	0	0	79.7	KH_2PO_4
20.8	1.93	0	0	79.2	"
17.6	1.61	2.2	0.37	80.2	"
16.6	1.55	4.8	0.82	78.6	"
13.5	1.21	6.0	1.00	80.7	"
13.3	1.25	8.3	1.42	78.4	"
11.0	1.05	12.0	2.09	77.0	"
8.7	0.83	14.0	2.43	77.3	"
7.5	0.71	15.4	2.68	77.1	"
7.0	0.69	18.6	3.36	74.4	"
7.1	0.72	20.2	3.73	72.7	"
5.7	0.57	20.4	3.70	73.9	"
5.5	0.57	23.4	4.41	71.1	"
5.0	0.52	23.8	4.48	71.2	KH_2PO_4 + KCl
5.6	0.59	24.9	4.80	69.5	KCl
3.0	0.31	25.3	4.73	71.7	"
1.5	0.15	25.5	4.69	73.0	"
0	0	26.6	4.86	73.4	"
0	0	26.4	4.81	73.6	"

(continued next page)

AUXILIARY INFORMATION

METHOD/APPARATUS/PROCEDURE:	SOURCE AND PURITY OF MATERIALS:
Solutions were saturated at a temperature 5 K higher than that of the respective isotherm. The samples were equilibrated by stirring for 4 hours. The mixtures were then allowed to stand for 1 hour before samples were taken for analyses. Chlorides were precipitated by adding excess $AgNO_3$ and then back-titrating the excess $AgNO_3$ with rhodanine. The dihydrogenphosphate ions were precipitated by adding excess bismuth nitrate and the excess bismuth was back-titrated with Komplexon III.	No information is given.
	ESTIMATED ERROR: The temperature was controlled to within ± 0.2 K. The accuracy of the phosphorus analysis was at least ± 3%.
	REFERENCES:

COMPONENTS:	ORIGINAL MEASUREMENTS:
(1) Potassium dihydrogenphosphate; KH_2PO_4; [7778-77-0] (2) Potassium chloride; KCl; [7747-40-7] (3) Water; H_2O; [7732-18-5]	Mráz, R.; Srb, V.; Tichý, S.; Vosolsobě, J. *Chem. Prům.* 1976, *26*, 511-4.

EXPERIMENTAL VALUES cont'd:

Solubility isotherms in the KH_2PO_4-KCl-H_2O system.

KH_2PO_4		KCl		H_2O	
mass%	mol/kg[a]	mass%	mol/kg[a]	mass%[a]	solid phases
temp. = 50°C.					
29.0	3.00	0	0	71.00	KH_2PO_4
22.0	2.17	3.5	0.63	74.5	"
14.9	1.42	8.0	1.40	77.1	"
12.4	1.20	11.7	2.07	75.9	"
12.5	1.24	13.6	2.47	73.9	"
9.7	0.95	15.6	2.80	74.7	"
6.5	0.66	21.4	3.98	72.1	"
5.0	0.53	25.5	4.92	69.5	"
5.3	0.58	27.0	5.34	67.7	KH_2PO_4 + KCl
0	0	29.6	5.64	70.4	KCl
temp. = 75°C.					
39.2	4.73	0	0	60.8	KH_2PO_4
34.7	4.00	1.7	0.40	63.6	"
31.5	3.51	2.6	0.53	65.9	"
27.3	3.04	6.7	1.36	66.0	"
25.0	2.83	10.1	2.09	64.9	"
20.1	2.22	13.5	2.72	66.4	"
16.1	1.78	17.4	3.51	66.5	"
16.4	1.89	20.0	4.22	63.6	"
12.2	1.43	25.0	5.33	62.8	"
9.1	1.11	30.6	6.81	60.3	KH_2PO_4 + KCl
5.0	0.59	32.5	6.97	62.5	KCl
2.9	0.33	31.8	6.53	65.3	"
0	0	32.2	6.66	66.8	"

[a]These values were calculated by the compiler.

COMPONENTS:	ORIGINAL MEASUREMENTS:
(1) Potassium dihydrogenphosphate; KH_2PO_4; [7778-77-0] (2) Potassium nitrate; KNO_3; [7757-79-1] (3) Water; H_2O; [7732-18-5]	Shenkin, Ya.S.; Gorozhankin, E.V. Zh. Neorg. Khim. <u>1976</u>, 21, 2293-5.

VARIABLES:	PREPARED BY:
Composition and temperature at atmospheric pressure.	J. Eysseltová

EXPERIMENTAL VALUES:

Composition and boiling points in the KH_2PO_4-KNO_3-H_2O system.

KH_2PO_4		KNO_3		H_2O				solid[c]
mass%	mol/kg[a]	mass%	mol/kg[a]	mass%[a]	b.p./°C	-lg N[b]		phase[c]
52.7	8.18	0	0	47.3	105.1			A
47.3	6.96	2.8	0.555	49.9	105.1	0.0564		"
46.2	6.78	3.7	0.730	50.1	105.1	0.0495		"
42.6	6.20	6.9	1.35	50.5	105.5	0.0560		"
41.4	6.08	8.6	1.70	50.0	105.5	0.0570		"
37.8	5.63	12.9	2.59	49.3	106.0	0.0603		"
35.5	5.32	15.5	3.13	49.0	106.4	0.0610		"
32.1	4.79	18.7	3.76	49.2	106.9	0.0623		"
32.5	4.91	18.9	3.85	48.6	106.5	0.0618		"
30.6	4.72	21.8	4.53	47.6	106.9	0.0623		"
29.8	5.10	27.3	6.29	42.9	107.2	0.0717		A + B
26.4	4.47	30.2	6.88	43.4	107.7	0.0826		B
24.5	4.10	31.6	7.12	43.9	107.9	0.0736		"
23.7	3.93	32.0	7.14	44.3	106.7	0.0801		"
22.2	3.61	32.7	7.18	45.1	107.2	0.0781		"
21.2	3.57	35.2	7.99	43.6	107.2			"
15.7	2.86	43.9	10.75	40.4	108.7	0.0880		"
15.0	2.83	46.0	11.67	39.0	109.2	0.1016		"
14.8	2.86	47.0	12.17	38.2	109.2	0.1016		"
12.3	2.45	51.1	13.81	36.6	109.9	0.1077		"
11.7	2.36	51.8	14.04	36.5	110.0	0.1125		"
10.5	2.20	54.5	15.40	35.0	110.2	0.1134		"
8.1	1.94	61.2	19.71	30.7	112.8	0.1258		"
7.7	2.00	64.0	22.37	28.3	115.1	0.1438		"

(continued next page)

AUXILIARY INFORMATION

METHOD/APPARATUS/PROCEDURE:	SOURCE AND PURITY OF MATERIALS:
The method is the same as that described earlier (1). Phosphorus was determined gravimetrically as $Mg_2P_2O_7$. Nitrate ion was reduced with Deward alloy and the NH_3 was distilled, but no further information is given.	Chemically pure salts were recrystallized before being used.
	ESTIMATED ERROR:
	No information is given.
	REFERENCES:
	1. Shenkin, Ya.S.; Rushnova, S.A.; Shenkina, A.P. Zh. Neorg. Khim. <u>1968</u>, 13, 256.

COMPONENTS:	ORIGINAL MEASUREMENTS:
(1) Potassium dihydrogenphosphate; KH_2PO_4; [7778-77-0]	Shenkin, Ya.S; Gorozhankin, E.V.
(2) Potassium nitrate; KNO_3; [7757-79-1]	*Zh. Neorg. Khim.* 1976, *21*, 2293-5.
(3) Water; H_2O; [7732-18-5]	

EXPERIMENTAL VALUES cont'd:

Composition and boiling points in the KH_2PO_4-KNO_3-H_2O system.

KH_2PO_4		KNO_3		H_2O			solid
mass%	mol/kg[a]	mass%	mol/kg[a]	mass%[a]	b.p./°C	-lg N[b]	phase[c]
5.9	1.64	67.6	25.23	26.5	115.6	0.1592	B
4.1	1.10	68.4	24.60	27.5	116.2	0.1772	"
4.0	1.21	71.7	29.19	24.3	118.0	0.1774	"
3.3	1.02	73.0	30.47	23.7	118.0	0.1910	"
0	0	74.6	29.05	25.4	114.5	0.1960	"

[a] These values were calculated by the compiler.

[b] N is the mol fraction of water in the system.

[c] The solid phases are: A = KH_2PO_4; B = KNO_3.

According to the authors, the "temperature depression" (not defined) is a nearly linear function of -lg N.

COMPONENTS:	ORIGINAL MEASUREMENTS:
(1) Potassium dihydrogenphosphate; KH_2PO_4; [7778-77-0] (2) Potassium chloride; KCl; [7747-40-7] (3) Water; H_2O; [7732-18-5]	Solov'ev, A.P.; Balashova, E.F.; Verendyakina, N.A.; Zyuzina, L.F. *Vzaymodeystvie Khloridov Kaliya, Magniya, Amoniyas ich Nitratami i Fosfatami* <u>1977</u>, 3-11.

VARIABLES:	PREPARED BY:
Composition at 25°C.	J. Eysseltová

EXPERIMENTAL VALUES:

Solubility in the KH_2PO_4-KCl-H_2O system at 25°C.

KH_2PO_4		KCl		H_2O		
mass%	mol/kg[a]	mass%	mol/kg[a]	mass%[a]	refr. index	solid phase
----	0.000	26.30	4.786	73.70	1.3714	KCl
2.25	0.226	24.75	4.547	73.00	1.3711	"
4.47	0.458	23.90	4.475	71.63	1.3710	KCl + KH_2PO_4
4.90	0.485	20.98	3.796	74.12	1.3669	KH_2PO_4
6.50	0.636	18.42	3.290	75.08	1.3643	"
7.60	0.726	15.52	2.707	76.88	1.3611	"
10.92	1.047	12.51	2.191	76.57	1.3591	"
12.59	1.187	9.48	1.631	77.93	1.3571	"
14.65	1.354	5.90	0.995	79.45	1.3566	"
20.30	1.871	-----	0.000	79.70	1.3550	"

[a]These values were calculated by the compiler.

AUXILIARY INFORMATION

METHOD/APPARATUS/PROCEDURE:	SOURCE AND PURITY OF MATERIALS:
Equilibrium was reached isothermally during the course of 1 to 3 days. The chloride content was determined by the Volhard method, the dihydrogenphosphate was precipitated as NH_4MgPO_4, the excess of Mg was titrated compleximetrically. The index of refraction was measured with an IRF-22 refractometer.	The salts were either reagent grade or chemically pure. They were recrystallized twice before being used.
	ESTIMATED ERROR: The temperature was controlled to within ± 0.1°C. No other information is given.
	REFERENCES:

COMPONENTS:	ORIGINAL MEASUREMENTS:
(1) Potassium dihydrogenphosphate; KH_2PO_4; [7778-77-0] (2) Ammonium dihydrogenphosphate; $NH_4H_2PO_4$; [7722-76-1] (3) Water; H_2O; [7732-18-5]	Solov'ev, A.P.; Balashova, E.F.; Verendyakina, N.A.; Zyuzina, L.F. *Vzaymodeystvie Khloridov Kaliya, Magniya, Amoniya s ich Nitratami i Fosfatami* <u>1977</u>, 3-11.

VARIABLES:	PREPARED BY:
Composition at 25°C.	J. Eysseltová

EXPERIMENTAL VALUES:

Composition of saturated solutions in the KH_2PO_4-$NH_4H_2PO_4$-H_2O system at 25°C.

KH_2PO_4		$NH_4H_2PO_4$		H_2O		
mass%	mol/kg[a]	mass%	mol/kg[a]	mass%[a]	refr. index	solid phases
21.60	2.024	----	0.00	78.40	1.3550	KH_2PO_4
20.80	2.034	4.08	0.472	75.12	1.3630	solid soln
19.24	1.890	5.98	0.695	74.78	1.3665	" "
16.10	1.698	14.25	1.778	69.65	1.3720	" "
13.60	1.478	18.80	2.417	67.60	1.3742	" "
11.37	1.264	22.57	2.969	66.06	1.3750	" "
9.24	0.995	24.20	2.877	66.56	1.3760	" "
7.00	0.771	26.30	3.427	66.70	1.3770	" "
0.30	0.030	28.05	3.403	71.65	1.3780	" "
-----	0.000	29.30	3.613	70.70	1.3780	$NH_4H_2PO_4$

[a]These values were calculated by the compiler.

AUXILIARY INFORMATION

METHOD/APPARATUS/PROCEDURE:	SOURCE AND PURITY OF MATERIALS:
The mixtures were equilibrated for 1-3 days in a thermostat. The ammonium ion content was determined by the Kjeldahl method. The $H_2PO_4^-$ was precipitated as NH_4MgPO_4, and the excess magnesium was titrated complex-imetrically. The refractive index was measured with a IRF-22 refractometer.	The salts were reagent grade or chemically pure and were recrystallized before being used.

	ESTIMATED ERROR:
	The temperature was controlled to within ± 0.1 K. No other information is given.

	REFERENCES:

COMPONENTS:	ORIGINAL MEASUREMENTS:
(1) Potassium dihydrogenphosphate; KH_2PO_4; [7778-77-0]	Beremzhanov, B.A.; Voronina, L.V.; Savich, R.F.
(2) Potassium borate, KBO_2; [13709-94-9]	_Khim. Khim. Tekhnol._ (Alma Ata) <u>1978</u>, 173-8.
(3) Water; H_2O; [7732-18-5]	

VARIABLES:	PREPARED BY:
Composition at 25 and 50°C.	J. Eysseltová

EXPERIMENTAL VALUES:
Composition of saturated solutions in the KH_2PO_4-KBO_2-H_2O system.

KH_2PO_4			KBO_2			refr.		solid[b]
mass%	mol%	mol/kg[a]	mass%	mol%	mol/kg[a]	index	pH	phase
				temp. = 25°C.				
---- [a]	----	----	0.367[a]	0.081	0.450	1.441	13.95	A
89.5 [a]	53.12	62.9	-----	-----	-----	1.360	7	B
66.05	20.34	14.30	0.023	0.012	0.0083	1.410	7.23	"
41.47	8.74	5.21	0.054	0.017	0.0113	1.399	6.98	"
39.17	7.67	4.74	0.055	0.018	0.0110	1.396	6.53	"
38.40	7.56	4.58	0.056	0.018	0.0111	1.394	6.32	"
35.52	6.73	4.05	0.057	0.018	0.0108	1.390	6.24	"
29.18	5.07	3.03	0.094	0.026	0.0162	1.385	5.97	"
27.65	4.75	2.81	0.129	0.035	0.0218	1.374	5.81	"
27.26	4.72	2.76	0.145	0.040	0.0244	1.371	5.74	"
26.50	4.47	2.66	0.164	0.047	0.0273	1.366	5.59	"
25.73	4.19	2.55	0.176	0.048	0.0290	1.361	5.31	"
24.96	4.13	2.45	0.201	0.055	0.0328	1.360	5.06	"
20.16	3.28	1.86	0.293	0.076	0.0450	1.368	4.32	"
17.28	2.56	1.54	0.374	0.096	0.0554	1.356	4.64	A + B
13.44	1.84	1.15	0.328	0.076	0.0464	1.367	4.91	A
11.52	1.61	0.96	0.304	0.071	0.0421	1.386	6.82	"
10.37	1.39	0.85	0.300	0.071	0.0410	1.416	8.05	"
9.60	1.37	0.78	0.323	0.076	0.0438	1.425	11.24	"
9.22	1.31	0.75	0.304	0.072	0.0410	1.430	12.07	"
6.91	0.96	0.55	0.339	0.078	0.0446	1.435	13.05	"
4.61	0.56	0.36	0.351	0.079	0.0451	1.440	13.60	"
0.77	0.09	0.06	0.363	0.080	0.0449	1.441	13.85	"

(continued next page)

AUXILIARY INFORMATION

METHOD/APPARATUS/PROCEDURE:	SOURCE AND PURITY OF MATERIALS:
No details are given other than that a solubility method was used.	No information is given.
	ESTIMATED ERROR:
	No information is given.
	REFERENCES:

COMPONENTS:	ORIGINAL MEASUREMENTS:
(1) Potassium dihydrogenphosphate; KH_2PO_4; [7778-77-0]	Beremzhanov, B.A.; Voronina, L.V.; Savich, R.F.
(2) Potassium borate; KBO_2; [13709-94-9]	Khim. Khim. Tekhnol. (Alma Ata) 1978, 173-8.
(3) Water; H_2O; [7732-18-5]	

EXPERIMENTAL VALUES cont'd:

Composition of saturated solutions in the KH_2PO_4-KBO_2-H_2O system.

KH_2PO_4			KBO_2			refr.		solid[b]
mass%	mol%	mol/kg[a]	mass%	mol%	mol/kg[a]	index	pH	phase

temp. = 50°C.

mass%	mol%	mol/kg[a]	mass%	mol%	mol/kg[a]	index	pH	phase
----	----	-----	0.409[a]	0.090	0.0502	1.445	14.0	A
95.3[a]	72.91	149.8	-----	-----	-----	1.365	7.20	B
55.30	13.94	9.10	0.051	0.020	0.0139	1.360	6.21	"
47.62	10.73	6.69	0.059	0.021	0.0138	1.364	6.03	"
43.78	9.30	5.73	0.059	0.017	0.0128	1.366	6.00	"
40.32	8.06	4.97	0.059	0.016	0.0121	1.368	5.98	"
36.48	7.12	4.22	0.063	0.021	0.0121	1.371	5.95	"
30.34	5.39	3.20	0.070	0.018	0.0123	1.374	5.90	"
29.95	5.36	3.14	0.073	0.022	0.0127	1.376	5.89	"
28.18	5.07	2.89	0.075	0.022	0.0128	1.379	5.87	"
27.65	4.73	2.81	0.103	0.031	0.0174	1.381	5.85	"
20.54	3.29	1.90	0.090	0.024	0.0139	1.384	5.82	A + B
13.06	1.83	1.10	0.059	0.014	0.0083	1.386	8.96	A
10.37	1.38	0.85	0.054	0.012	0.0074	1.388	9.88	"
8.45	1.17	0.68	0.070	0.015	0.0093	1.390	10.61	"
8.06	0.97	0.64	0.082	0.019	0.0109	1.394	11.15	"
5.38	0.74	0.42	0.117	0.026	0.0151	1.398	12.08	"
3.46	0.37	0.26	0.178	0.039	0.0225	1.405	12.46	"
2.69	0.36	0.20	0.199	0.044	0.0250	1.423	12.95	"
1.54	0.18	0.12	0.288	0.064	0.0358	1.432	13.20	"

[a]These values were calculated by the compiler.

[b]The solid phases are: A = KBO_2; B = KH_2PO_4.

COMPONENTS:	ORIGINAL MEASUREMENTS:
(1) Potassium dihydrogenphosphate; KH_2PO_4; [7778-77-0] (2) Potassium chloride; KCl; [7747-40-7] (3) Water; H_2O; [7732-18-5]	Khallieva, Sh.D. *Izv. AN Turkm. SSR. ser. khim.* <u>1978</u>, 3, 125-6.

VARIABLES:	PREPARED BY:
Composition at 40°C.	J. Eysseltová

EXPERIMENTAL VALUES:

Solubility isotherm for the KH_2PO_4-KCl-H_2O system at 40°C.

KH_2PO_4		KCl		H_2O	
mass%	mol/kg^a	mass%	mol/kg^a	mass%	solid phase
27.15	2.738	----	----	72.85	KH_2PO_4
17.12	1.658	7.04	1.244	75.84	"
12.82	1.243	11.46	2.029	75.72	"
9.73	0.929	13.39	2.335	76.88	"
7.29	0.741	20.47	3.800	72.24	"
4.21	0.443	25.97	4.988	69.82	KH_2PO_4 + KCl
3.945	0.416	26.48	5.104	68.575	"
----	----	28.01	5.218	71.99	KCl

^{a}The mol/kg H_2O values were calculated by the compiler.

AUXILIARY INFORMATION

METHOD/APPARATUS/PROCEDURE:	SOURCE AND PURITY OF MATERIALS:
The isothermal method was used. Equilibrium was checked by repeated analysis. Standard analytical methods were used to determine the amount of chloride, potassium and dihydrogenphosphate ions.	Reagent grade salts were used.

	ESTIMATED ERROR:
	The deviation from 40°C was no greater than ± 0.5 K. No other information is given.
	REFERENCES:

COMPONENTS:	ORIGINAL MEASUREMENTS:
(1) Potassium dihydrogenphosphate; KH_2PO_4; [7778-77-0] (2) Potassium nitrate; KNO_3; [7757-79-1] (3) Water; H_2O; [7732-18-5]	Girich, T.E.; Gulyamov, Yu.M.; Ganz, S.N. Zh. Neorg. Khim. 1979, 24, 1084-6.

VARIABLES:	PREPARED BY:
Composition at 25 and 50°C.	J. Eysseltová

EXPERIMENTAL VALUES:

Composition of the isotherms in the KH_2PO_4-KNO_3-H_2O system.

temp. = 25°C.

	mass percent			concn.[a]		indices[b]		mol/kg[c]		solid[d]
no	KH_2PO_4	KNO_3	H_2O	KH_2PO_4	KNO_3	KH_2PO_4	H_2O	KH_2PO_4	KNO_3	phase
1	20.49	0	79.51	33.94	0	100.0	2945.9	1.89	0	A
2	16.93	4.13	78.94	28.38	9.32	75.27	2651.5	1.57	0.52	"
3	14.96	7.45	77.59	25.52	17.12	59.85	2345.2	1.42	0.95	"
4	13.50	10.75	75.75	23.57	25.28	48.25	2046.7	1.31	1.40	"
5	12.23	13.95	73.82	21.92	33.67	39.43	1798.7	1.22	1.87	"
6	11.35	16.78	71.87	20.89	41.60	33.43	1600.4	1.16	2.31	"
7	11.00	18.46	70.54	20.62	46.64	30.65	1486.7	1.15	2.59	"
8	10.17	20.79	69.04	19.50	53.66	26.66	1366.7	1.15	3.00	"
9	10.15	22.11	67.74	19.82	58.17	25.42	1282.2	1.10	3.23	A + B
10	10.03	22.45	67.52	19.64	59.23	24.91	1267.7	1.09	3.29	"
11	7.07	23.79	69.14	13.51	61.31	18.06	1336.5	0.75	3.40	B
12	5.11	24.98	69.91	9.65	63.67	13.17	1363.7	0.54	3.53	"
13	0	27.49	72.51	0	67.57	0	1479.9	0	3.75	"

temp. = 50°C.

14	29.15	0	70.85	54.45	0	100.0	1836.7	3.02	0	A
15	25.63	4.21	70.16	49.75	10.98	81.91	1646.5	2.68	0.593	"
16	20.82	11.68	67.50	40.80	30.82	56.96	1396.1	2.27	1.71	"
17	19.95	13.11	66.94	39.44	34.89	53.06	1395.0	2.19	1.94	"
18	18.08	15.42	66.50	35.97	41.31	46.55	1294.0	2.00	2.29	"
19	14.73	22.01	63.26	30.76	61.90	33.20	1077.4	1.71	3.44	"
20	14.01	23.35	62.24	29.59	66.41	30.83	1041.6	1.65	3.77	"
21	12.32	26.37	61.31	26.53	76.56	25.74	939.3	1.48	4.25	"

(continued next page)

AUXILIARY INFORMATION

METHOD/APPARATUS/PROCEDURE:	SOURCE AND PURITY OF MATERIALS:
The mixtures were heated to a temperature 5-10 K above that desired and then placed in a thermostat at 25°C for 8 hours or at 50°C for 6 hours. The phosphate and nitrate ion contents were determined photocolori-metrically (2,3). The measurement of the physical properties has been described elsewhere (1).	A special purity grade of salts was used and the salts were recrystallized before being used.

ESTIMATED ERROR:

The temperature was controlled to within ± 0.1 K. No other details are given.

REFERENCES:
1. Protsenko, P.I.; Andreeva, T.A. Zh. Neorg. Khim. 1964, 9, 1441.
2. Moizhes, I.B. Rukovodstvo po Analizu v Proizvodstve Fosfora, Fosfornoy Kisloty i Udobroniy, Khimiya, Leningrad 1973.

(continued next page)

COMPONENTS:	ORIGINAL MEASUREMENTS:
(1) Potassium dihydrogenphosphate; KH_2PO_4; [7778-77-0] (2) Potassium nitrate; KNO_3; [7757-79-1] (3) Water; H_2O; [7732-18-5]	Girich, T.E.; Gulyamov, Yu.M.; Ganz, S.N. *Zh. Neorg. Khim.* <u>1979</u>, *24*, 1084-6.

EXPERIMENTAL VALUES cont'd:

Composition of the isotherms in the KH_2PO_4-KNO_3-H_2O system.

no	mass percent			concn.[a]		indices[b]		mol/kg[c]		solid[d] phase
	KH_2PO_4	KNO_3	H_2O	KH_2PO_4	KNO_3	KH_2PO_4	H_2O	KH_2PO_4	KNO_3	
					temp. = 50°C.					
22	10.70	28.90	60.40	23.39	85.14	21.55	920.1	1.30	4.73	A
23	7.84	36.80	55.36	18.72	115.86	13.92	742.8	1.04	6.58	"
24	7.54	38.90	53.56	18.61	129.42	12.58	675.5	1.03	7.18	"
25	7.44	39.41	53.15	18.52	132.11	12.30	663.7	0.96	7.34	A + B
26	6.98	39.63	53.39	17.30	132.29	11.56	668.5	1.03	7.33	"
27	5.60	41.06	53.34	13.86	137.18	9.18	662.0	0.77	7.61	B
28	5.16	41.48	53.36	12.79	138.51	8.45	661.0	0.71	7.69	"
29	4.50	42.28	53.22	11.16	141.82	7.31	654.7	0.62	7.86	"
30	0	46.25	53.75	0	153.35	0	652.1	0	8.51	"

[a] The concentration unit is: mol/1000 mol water.

[b] The concentration unit is: mol/100 mol of solute.

[c] These mol/kg H_2O values were calculated by the compiler.

[d] The solid phases are: A = KH_2PO_4; B = KNO_3.

The physical properties of the above solutions are given below.

no	viscosity/cP	density/g cm^{-3}	κ/S cm^{-1}
1	1.722	1.156	0.100
2	1.325	1.160	0.110
3	1.281	1.169	0.115
4	1.285	1.184	0.120
5	1.277	1.189	0.130
6	1.276	1.203	0.135
7	1.278	1.212	0.140
8	1.280	1.222	0.143
9	1.299	1.230	0.145
10	1.299	1.231	0.145
11	1.237	1.220	0.147
12	1.189	1.212	0.150
13	1.051	1.197	0.155
14	1.487	1.225	1.05
15	1.301	1.234	1.10
16	1.119	1.236	1.15
17	1.117	1.243	1.15
18	1.116	1.247	1.25
19	1.115	1.264	1.40
20	1.115	1.271	1.45
21	1.115	1.287	1.50
22	1.116	1.302	1.60
23	1.117	1.329	1.90
24	1.118	1.335	2.05
25	1.119	1.340	2.15
26	1.119	1.339	2.15
27	1.117	1.332	2.10
28	1.112	1.331	2.05
29	1.112	1.330	1.99
30	1.050	1.325	1.95

REFERENCES cont'd:

3. Lure, Yu.Yu.; Rybnikova, A.I.
   Khimicheskii Analiz Proizvodstvennykh Stochnykh Vod, Izd. Khimiya, Moscow <u>1974</u>.

COMPONENTS:	ORIGINAL MEASUREMENTS:
(1) Potassium dihydrogenphosphate; KH_2PO_4; [7778-77-0] (2) Potassium chloride; KCl; [7747-40-7] (3) Ammonium dihydrogenphosphate; $NH_4H_2PO_4$; [7722-76-1] (4) Ammonium chloride; NH_4Cl; [12125-02-9] (5) Water; H_2O; [7732-18-5]	Askenasy, F.; Nessler, F. Z. Anorg. Chem. <u>1930</u>, 189, 305-28.

VARIABLES:	PREPARED BY:
Composition at 0°C.	J. Eysseltová

EXPERIMENTAL VALUES:

Points of simultaneous crystallization of several solid phases in the K^+, NH_4^+ || Cl^-, $H_2PO_4^-$ –H_2O system at 0°C.

mol/100 mol solute

$d/g\ cm^{-3}$	$H_2PO_4^-$	Cl^-	K^+	NH_4^+	H_2O	solid phases[a]
1.1100	7.6	92.4	8.0	92.0	925	"NH_4Cl" + "$NH_4H_2PO_4$"
1.1134	7.7	92.3	12.1	87.9	905	" " "
1.1300	8.1	91.9	17.9	82.1	888	" " "
1.1335	8.5	91.5	24.4	75.6	850	" " "
1.1385	8.7	91.3	28.7	71.3	810	ternary eutectic point
1.1504	8.3	91.7	29.8	70.2	865	"KCl" + "KH_2PO_4"
1.1514	8.3	91.7	32.7	67.3	887	" " "
1.1657	7.5	92.5	66.6	33.4	1135	" " "
1.1695	7.7	92.3	85.8	14.2	1310	" " "
1.1740	7.3	92.7	90.0	10.0	1380	" " "
1.1272	3.6	96.4	28.2	71.8	845	"KCl" + "NH_4Cl"

[a] A formula in quotation marks refers to a solid solution rich in that component.

(continued next page)

AUXILIARY INFORMATION

METHOD/APPARATUS/PROCEDURE:	SOURCE AND PURITY OF MATERIALS:
Binary eutonic solutions were prepared on the basis of a preliminary investigation of the boundary ternary systems. Samples on the curves for simultaneous crystallization of 2 salts were then prepared by adding a third component. The mixtures were shaken in a thermostat for 2-4 days. The solid phase was isolated by centrifuging. The analytical methods are not described.	No information is given.
	ESTIMATED ERROR:
	The temperature was controlled to within ± 0.1 K. No other information is given.
	REFERENCES:

COMPONENTS:	ORIGINAL MEASUREMENTS:
(1) Potassium dihydrogenphosphate; KH_2PO_4; [7778-77-0]	Askenasy, F.; Nessler, F.
(2) Potassium chloride; KCl; [7747-40-7]	Z. Anorg. Chem. 1930, 189, 305-28.
(3) Ammonium dihydrogenphosphate; $NH_4H_2PO_4$; [7722-76-1]	
(4) Ammonium chloride; NH_4Cl; [12125-02-9]	
(5) Water; H_2O; [7732-18-5]	

EXPERIMENTAL VALUES, cont'd:

Distribution of K^+ and NH_4^+ in the solid and liquid phases of some of the saturated solutions in the $(K,NH_4)H_2PO_4$ system.

liquid phase mol/100 mol of solute					solid phase mol%	
$H_2PO_4^-$	Cl^-	K^+	NH_4^+	H_2O	K^+	NH_4^+
9.8	90.2	57.8	42.2	1090	85.5	14.5
45.9	54.1	49.2	50.8	1690	78.9	21.1
33.8	66.2	45.5	54.5	1680	71.7	28.3
27.4	72.6	42.7	57.3	1450	66.1	33.9
27.2	72.8	43.3	56.7	1565	64.8	35.2
19.7	80.3	40.9	59.1	1135	62.2	37.8
13.1	86.9	38.9	61.1	1130	62.2	37.8
26.7	73.3	44.2	55.8	1555	56.4	45.6
24.5	75.5	36.9	63.1	1665	55.7	44.3
9.8	90.2	26.6	73.4	925	31.2	68.8
18.3	81.7	28.7	71.3	1235	29.2	70.8

The compiler has recalculated the data to give the following values:

$H_2PO_4^-$		Cl^-		K^+		NH_4^+		H_2O	solid phase[a]
mass%	mol/kg	mass%	mol/kg	mass%	mol/kg	mass%	mol/kg	mass%	
3.25	0.45	14.47	5.54	1.38	0.48	7.32	5.52	73.55	A + B
3.33	0.47	14.63	5.66	2.11	0.74	7.08	5.39	72.82	"
3.53	0.50	14.67	5.74	3.15	1.11	6.66	5.13	71.96	"
3.80	0.55	14.96	5.98	4.39	1.59	6.28	4.94	70.54	"
4.00	0.59	15.36	6.26	5.32	1.96	6.10	4.89	69.19	C
3.64	0.53	14.74	5.88	5.28	1.91	5.73	4.50	70.58	D + E
3.57	0.51	14.44	5.74	5.67	2.04	5.39	4.21	70.91	"
2.63	0.36	11.86	4.52	9.41	3.25	2.17	1.63	73.90	"
2.39	0.32	10.48	3.91	10.74	3.63	0.82	0.60	75.54	"
2.17	0.29	10.10	3.73	10.81	3.62	0.55	0.40	76.34	"
1.63	0.23	15.99	6.33	5.15	1.85	6.05	4.72	71.15	D + F

[a] The solid phases are: A = "NH_4Cl"; B = "$NH_4H_2PO_4$"; C = ternary eutectic point; D = "KCl"; E = "KH_2PO_4"; F = "NH_4Cl". These have the same meaning as in the table on the preceding page.

3.54	0.49	11.94	4.59	8.43	2.94	2.84	2.15	73.23	G[b]
11.23	1.50	4.84	1.77	4.85	1.61	2.31	1.66	76.65	"
8.48	1.11	6.07	2.18	4.60	1.50	2.54	1.80	78.28	"
7.80	1.04	7.56	2.78	4.90	1.63	3.03	2.19	76.68	"
7.30	0.96	7.15	2.58	4.68	1.53	2.83	2.01	78.02	"
6.86	0.96	10.22	3.93	5.74	2.00	3.82	2.89	73.34	"
4.65	0.64	11.28	4.27	5.56	1.91	4.03	3.00	74.46	"
7.21	0.95	7.23	2.61	4.81	1.57	2.80	1.99	77.93	"
6.31	0.81	7.12	2.51	3.83	1.23	3.02	2.10	79.69	"
4.10	0.58	13.81	5.41	4.49	1.59	5.71	4.40	71.88	"
6.05	0.82	9.88	3.67	3.82	1.29	4.38	3.20	75.84	"

[b] The solid phase, G. here refers to a precipitate designated as $(K,NH_4)H_2PO_4$.

COMPONENTS:	ORIGINAL MEASUREMENTS:
(1) Potassium dihydrogenphosphate; KH_2PO_4; [7778-77-0] (2) Calcium dihydrogenphosphate; $Ca(H_2PO_4)_2$; [10103-46-5] (3) Phosphoric acid; H_3PO_4; [7664-38-2] (4) Water; H_2O; [7732-18-5]	Flatt, R.; Brunisholz, G.; Bourgeois, J. *Helv. Chim. Acta* <u>1956</u>, *39*, 841-53.

VARIABLES:	PREPARED BY:
Composition at 25°C.	J. Eysseltová

EXPERIMENTAL VALUES:

Part 1. Composition of saturated solutions in the KH_2PO_4-$Ca(H_2PO_4)_2$ - H_3PO_4-H_2O system at 25°C.

soln. no.	eq% Ca^{2+}	eq% K^+	eq% H^+	mol H_2O/100 equiv of solute	solid phases[a]
1	2.0	12.6	85.4	68.5	A + B + C
2	8.2	12.4	79.4	162.6	A + B + D
3	11.2	14.0	74.8	278.5	B + D + E
4	16.1	6.6	77.3	260.0	A + D + E
5	0.9	7.6	91.5	55.0	A + C
6	1.6	10.0	88.4	63.8	"
7	1.9	11.8	86.3	66.3	"
8	2.7	12.7	84.6	81.3	A + B
9	4.7	12.5	82.8	114.6	"
10	5.8	12.3	81.9	133.7	"
11	7.0	12.5	80.5	147.8	"
12	8.1	12.3	79.6	159.1	"
13	9.0	11.1	79.9	162.4	A + D
14	11.0	9.6	79.4	184.5	"
15	12.6	8.6	78.8	221.5	"
16	13.8	7.6	78.6	234.5	"
17	17.1	4.8	78.1	288.3	A + E
18	18.9	2.1	79.0	309.0	"
19	9.0	12.5	78.5	183.4	B + D
20	9.4	12.7	77.9	192.0	"
21	9.5	12.8	77.7	199.4	"
22	9.9	12.9	77.2	216.1	"
23	10.3	13.1	76.6	231.8	"
24	9.5	15.6	74.9	324.4	B + E

(continued next page)

AUXILIARY INFORMATION

METHOD/APPARATUS/PROCEDURE:	SOURCE AND PURITY OF MATERIALS:
No experimental details are given. The composition of the double salt was calculated on the basis of the analysis of the saturated solution and of the wet residue. A small amount of KNO_3 was added to the solid phase in order to determine the amount of mother liquid adsorbed on the surface of the solid phase.	No information is given.
	ESTIMATED ERROR: No information is given.
	REFERENCES:

COMPONENTS:	ORIGINAL MEASUREMENTS:
(1) Potassium dihydrogenphosphate; KH_2PO_4; [7778-77-0]	Flatt, R.; Brunisholz, G.; Bourgeois, J.
(2) Calcium dihydrogenphosphate; $Ca(H_2PO_4)_2$; [10103-46-5]	Helv. Chim. Acta 1956, 39, 841-53.
(3) Phosphoric acid; H_3PO_4; [7664-38-2]	
(4) Water; H_2O; [7732-18-5]	

EXPERIMENTAL VALUES cont'd:

Part 1. Composition of saturated solutions in the KH_2PO_4-$Ca(H_2PO_4)_2$ - H_3PO_4-H_2O system at 25°C.

soln. no.	eq% Ca^{2+}	eq% K^+	eq% H^+	mol H_2O/100 equiv of solute	solid phases[a]
25	7.2	19.6	73.2	481.2	B + E
26	4.8	24.0	71.2	681.4	"
27	3.0	27.4	69.6	812.3	"
28	12.6	11.7	75.7	278.1	D + E
29	13.8	9.8	76.4	277.9	"
30	2.6	4.1	93.3	99.4	A
31	6.5	3.9	89.6	154.2	"
32	7.6	8.9	83.5	159.5	"
33	9.8	10.1	80.1	185.9	"
34	10.8	9.1	80.1	200.2	"
35	10.7	3.7	85.6	208.8	"
36	12.4	7.9	79.7	221.2	"
37	16.9	3.5	79.6	290.2	"
38	5.9	13.1	81.0	160.0	B
39	5.2	15.0	79.8	242.4	"
40	2.7	19.3	78.0	397.8	"
41	1.8	26.3	71.9	722.0	"
42	9.1	11.8	79.1	169.1	D
43	8.5	12.4	79.1	169.5	"
44	8.6	12.4	79.0	173.7	"
45	10.3	10.1	79.6	186.5	"
46	10.7	9.8	79.5	191.9	"
47	10.0	11.9	78.1	193.2	"
48	10.1	12.2	77.7	196.3	"
49	10.3	11.2	78.5	198.5	"
50	12.0	9.3	78.7	215.1	"
51	11.1	12.1	76.8	234.0	"
52	13.7	8.0	78.3	237.6	"
53	14.0	8.5	77.5	246.3	"

[a] The solid phases are: A = $Ca(H_2PO_4)_2$; B = KH_2PO_4; C = $KH_5(PO_4)_2$; D = $Ca_9K_4H_{32}(PO_4)_{18} \cdot 10H_2O$; E = $CaHPO_4$.

Part 2. The compiler has calculated the following values from the data given in Part 1 above.

soln. no.	$Ca(H_2PO_4)_2$ mass%	$Ca(H_2PO_4)_2$ mol/kg	KH_2PO_4 mass%	KH_2PO_4 mol/kg	H_3PO_4 mass%	H_3PO_4 mol/kg	H_2O mass%
1	8.91	1.62	32.65	10.22	34.96	15.19	23.48
2	24.67	2.80	21.69	4.23	16.04	4.35	37.61
3	25.36	2.23	18.43	2.79	7.71	1.62	48.49
4	36.28	3.44	8.65	1.41	10.03	2.27	45.05
5	4.51	0.91	22.16	7.68	52.13	25.08	21.20
6	7.45	1.39	27.14	8.71	42.48	18.92	22.90
7	8.60	1.59	31.04	9.89	37.23	16.45	23.09
8	11.32	1.84	30.97	8.68	31.49	12.25	26.22
9	17.07	2.28	26.40	6.06	24.53	7.82	32.00
10	19.59	2.41	24.15	5.11	21.54	6.33	34.72
11	22.27	2.63	23.13	4.70	18.43	5.20	36.17
12	24.62	2.83	21.74	4.29	16.46	4.52	37.18
13	26.88	3.08	19.28	3.80	16.55	4.53	37.30
14	30.47	3.31	15.46	2.89	14.77	3.83	39.30
15	31.73	3.16	12.59	2.16	12.79	3.04	42.89
16	33.46	3.27	10.71	1.80	12.11	2.83	43.72
17	36.50	3.30	5.96	0.92	10.22	2.20	47.32
18	38.54	3.40	2.49	0.38	10.53	2.22	47.32
19	25.48	2.73	20.57	3.79	14.02	3.58	39.92
20	25.93	2.72	20.37	3.67	12.97	3.25	40.73
21	25.75	2.65	20.17	3.57	12.52	3.07	41.56
22	25.76	2.54	19.52	3.31	11.48	2.71	43.24

(continued next page)

COMPONENTS:	ORIGINAL MEASUREMENTS
(1) Potassium dihydrogenphosphate; KH_2PO_4; [7778-77-0]	Flatt, R.; Brunisholz, G.; Bourgeois, J.
(2) Calcium dihydrogenphosphate; $Ca(H_2PO_4)_2$; [10103-46-5]	Helv. Chim. Acta 1956, 39, 841-53
(3) Phosphoric acid; H_3PO_4; [7664-38-2]	
(4) Water; H_2O; [7732-18-5]	

EXPERIMENTAL VALUES cont'd:

Part 2. The compiler has calculated the following values from the data given in Part 1 above.

	$Ca(H_2PO_4)_2$		KH_2PO_4		H_3PO_4		H_2O
soln. no.	mass%	mol/kg	mass%	mol/kg	mass%	mol/kg	mass%
23	25.81	2.47	19.09	3.14	10.42	2.38	44.67
24	20.23	1.63	19.31	2.67	7.34	1.41	53.12
25	12.34	0.83	19.54	2.26	4.69	0.75	63.42
26	6.57	0.39	19.10	1.96	2.60	0.37	71.73
27	3.63	0.20	19.28	1.87	1.49	0.20	75.60
28	28.27	2.52	15.26	2.34	8.48	1.80	47.98
29	30.70	2.76	12.68	1.60	9.07	1.94	47.55
30	10.93	1.45	10.02	2.29	46.89	14.88	32.14
31	21.50	2.34	7.50	1.40	31.76	8.26	39.23
32	23.68	2.65	16.13	3.10	21.96	5.86	38.22
33	27.53	2.93	16.50	3.02	15.80	4.01	40.16
34	29.10	3.00	14.26	2.52	15.16	3.73	41.48
35	29.05	2.85	5.84	0.98	21.52	5.04	43.59
36	31.42	3.11	11.64	1.98	13.83	3.27	43.11
37	36.2	3.23	4.36	0.67	11.60	2.48	47.82
38	18.54	2.05	23.94	4.55	18.86	4.98	38.66
39	13.66	1.19	22.92	3.44	14.45	3.01	48.98
40	5.48	0.38	22.78	2.70	9.63	1.58	62.10
41	2.41	0.14	20.44	2.02	2.93	0.40	74.22
42	26.63	2.99	20.08	3.88	15.23	4.08	38.06
43	25.04	2.79	21.24	4.06	15.33	4.08	38.39
44	25.05	2.75	21.00	3.96	15.04	3.94	38.91
45	28.67	3.07	16.34	3.01	15.07	3.85	39.92
46	29.29	3.10	15.60	2.84	14.71	3.72	40.40
47	27.35	2.88	18.92	3.42	13.09	3.29	40.63
48	27.34	2.86	19.22	3.45	12.515	3.12	40.90
49	27.82	2.88	17.58	3.13	13.38	3.31	41.22
50	30.78	3.10	13.87	2.40	12.92	3.11	42.43
51	27.49	2.64	17.43	2.87	10.51	2.40	44.57
52	33.02	3.20	11.21	1.87	11.74	2.72	44.03
53	33.00	3.16	11.65	1.92	10.69	2.44	44.65

COMPONENTS:	ORIGINAL MEASUREMENTS:
(1) Potassium dihydrogenphosphate; KH_2PO_4; [7778-77-0] (2) Potassium chloride; KCl; [7747-40-7] (3) Ammonium dihydrogenphosphate; $NH_4H_2PO_4$; [7722-76-1] (4) Ammonium chloride; NH_4Cl; [12125-02-9] (5) Water; H_2O; [7732-18-5]	Iovi, A.; Haiduc, C. *Rev. Roum. Chim.* <u>1971</u>, *16*, 1743-7.

VARIABLES:	PREPARED BY:
Temperature. Equimolar mixtures of KCl + $NH_4H_2PO_4$ or NH_4Cl and KH_2PO_4 are used.	J. Eysseltová

EXPERIMENTAL VALUES:

Molar solubility in equimolar mixtures in the K^+, $NH_4^+||Cl^-$, $H_2PO_4^-$-H_2O system.

		(A) KCl + $NH_4H_2PO_4$				(B) NH_4Cl + KH_2PO_4				
Nr.	$t/°C.$	$H_2PO_4^-$	Cl^-	K^+	NH_4^+	$H_2PO_4^-$	Cl^-	K^+	NH_4^+	$d/g\ cm^{-3}$
1	20	0.60	4.77	2.18	3.19	0.60	4.77	2.18	3.19	1.1585
2	25	0.70	4.76	2.18	3.28	0.70	4.72	2.18	3.24	1.1640
3	40	1.00	4.49	2.31	3.18	1.00	4.39	2.31	3.08	1.1690
4	60	1.50	4.36	2.44	3.42	1.50	4.36	2.44	3.42	1.1880
5	75	2.20	4.36	2.82	3.74	2.20	4.26	2.82	3.64	1.2360
6	80	2.50	4.23	2.95	3.78	2.50	4.22	2.95	3.77	1.2479

The compiler has recalculated these values as follows:

	$H_2PO_4^-$		chloride		potassium		ammonium	
Nr	mass%	mol/kg	mass%	mol/kg	mass%	mol/kg	mass%	mol/kg
1(A,B)	5.02	0.76	14.60	6.04	7.35	2.76	4.95	4.03
2(A)	5.83	0.89	14.50	6.07	7.32	2.78	5.07	4.18
2(B)	5.83	0.89	14.38	6.01	7.32	2.77	5.01	4.11
3(A)	8.29	1.30	13.62	5.86	7.72	3.01	4.89	4.14
3(B)	8.29	1.29	13.32	5.69	7.72	2.99	4.74	3.99
4(A,B)	12.24	2.05	13.01	5.96	8.03	3.33	5.18	4.67
5(A)	17.26	3.18	12.51	6.31	8.92	4.08	5.44	5.40
5(B)	17.26	3.16	12.22	6.12	8.92	4.05	5.30	5.22
6(A)	19.43	3.72	12.02	6.29	9.24	4.39	5.45	5.61
6(B)	19.43	3.71	11.99	6.27	9.24	4.38	5.43	5.59

(continued next page)

AUXILIARY INFORMATION

METHOD/APPARATUS/PROCEDURE:	SOURCE AND PURITY OF MATERIALS:
The samples were equilibrated by stirring for 10 hours in a thermostat. The potassium content was determined with the use of a type C Zeiss Jena Model III flame photometer and gravimetrically as $KClO_4$, phosphate ion was determined by a compleximetric titration with $MgCl_2$ using Eriochrome Black T as indicator (1), chloride was determined by potentiometric titration with $AgNO_3$, and NH_3 was determined by the Kjeldahl method.	No information is given.
	ESTIMATED ERROR:
	No information is given.
	REFERENCES:
	1. Liteanu, C. *Chimie Analitica Cantitativa*, Ed. didactica si pedagogica, Bucuresti, <u>1964</u>, p. 508.

COMPONENTS:	ORIGINAL MEASUREMENTS:
(1) Potassium dihydrogenphosphate; KH_2PO_4; [7778-77-0]	Iovi, A.; Haiduc, C.
(2) Potassium chloride; KCl; [7747-40-7]	Rev. Roum. Chim. 1971, 16, 1743-7.
(3) Ammonium dihydrogenphosphate; $NH_4H_2PO_4$; [7722-76-1]	
(4) Ammonium chloride; NH_4Cl; [12125-02-9]	
(5) Water; H_2O; [7732-18-5]	

EXPERIMENTAL VALUES cont'd:

Molar solubility in equimolar mixtures in the K^+, $NH_4^+||Cl^-$, $H_2PO_4^- - H_2O$ system.

The following concentrations are expressed as: mol/100 mol of solute

Nr	$H_2PO_4^-$	chloride	potassium	ammonium	water
1(A,B)	11.17	88.83	40.64	59.36	818
2(A)	12.78	87.22	39.94	60.06	798
2(B)	12.89	87.19	40.26	59.74	807
3(A)	18.15	81.85	42.83	57.17	804
3(B)	18.48	81.52	42.78	57.22	796
4(A,B)	25.59	74.41	41.62	58.38	694
5(A)	33.50	66.50	43.03	56.97	586
5(B)	34.05	65.95	43.68	56.32	599
6(A)	37.16	62.84	43.90	56.10	556
6(B)	37.17	62.83	43.93	56.07	557

COMPONENTS:	ORIGINAL MEASUREMENTS:
(1) Potassium oxide; K_2O; [12136-45-7] (2) Ammonia; NH_3; [7664-42-7] (3) Phosphoric acid; H_3PO_4; [7664-38-2] (4) Diphosphoric acid: $H_4P_2O_7$; [2466-09-3] (5) Water; H_2O; [7732-18-5]	Frazier, A.W.; Dillard, E.F.; Thrasher, R.D.; Waerstad, K.R. J. Agʌ. Food Chem. 1973, 21, 700-4

VARIABLES:	PREPARED BY:
Composition at 298 K.	J. Eysseltová

EXPERIMENTAL VALUES:

Solubility in the $NH_3-K_2O-H_3PO_4-H_4P_2O_7-H_2O$ system at 25°C.

Soln. no.	pH	$(NH_4)_2O$ mass%	K_2O mass%	total P_2O_5 mass%	ortho P_2O_5 mass%	pyro P_2O_5 % of total	solid a phases
1	7.18	12.4	14.9	34.0	14.2	58	A,C,L
2	6.65	16.1	8.9	36.9	13.0	65	A,C,L
3	6.30	19.6	5.5	40.1	13.5	66	A,B,C,L
4	6.00	18.7	5.7	39.0	12.3	68	B,C,L
5	5.93	18.5	6.5	40.3	18.0	55	B,C,L,M
6	5.88	18.8	5.0	41.2	16.3	60	B,C,M
7	5.28	16.9	5.5	42.3	9.6	77	B,C,M
8	5.18	16.6	6.8	42.6	9.1	79	B,C,E,M
9	5.10	17.3	5.4	42.6	8.6	80	B,E,M
10	6.95	11.2	16.5	34.5	17.5	49	C,L,M
11	6.30	15.7	9.7	37.7	17.8	53	C,L,M
12	6.13	16.9	8.0	38.9	17.9	54	C,L,M
13	4.55	9.8	13.5	39.9	6.8	83	D,E,M
14	5.08	9.9	14.3	40.7	5.8	86	D,E,M
15	4.95	13.5	16.3	41.2	4.8	88	C,M
16	5.18	11.1	17.9	39.2	4.7	88	C,M
17	6.10	8.5	22.7	33.6	10.0	70	C,M
18	7.40	3.6	27.4	32.8	10.1	69	C,J,M
19	6.95	3.1	27.4	34.0	4.2	79	C,J,M
20	6.49	2.7	27.7	36.0	5.0	86	C,G,J,M
21	6.23	3.3	26.3	36.1	5.8	84	C,G,M
22	5.45	7.3	16.9	38.7	5.8	85	C,G,M

(continued next page)

AUXILIARY INFORMATION

METHOD/APPARATUS/PROCEDURE:	SOURCE AND PURITY OF MATERIALS:
For each mixture the pH was adjusted to a selected value. The most acidic solutions were prepared by the use of a cation exchange resin. The mixtures were equilibrated in a water bath for 4 weeks.	Reagent grade KH_2PO_4, $NH_4H_2PO_4$, KOH, and NH_4OH were used. The $K_2H_2P_2O_7$ and $(NH_4)_3P_2O_7 \cdot H_2O$ were recrystallized.
	ESTIMATED ERROR: No information is given.
	REFERENCES:

COMPONENTS:	ORIGINAL MEASUREMENTS:
(1) Potassium oxide; K_2O; [12136-45-7] (2) Ammonia; NH_3; [7664-42-7] (3) Phosphoric acid; H_3PO_4; [7664-38-2] (4) Diphosphoric acid; $H_4P_2O_7$; [2466-09-3] (5) Water; H_2O; [7732-18-5]	Frazier, A.W.; Dillard, E.F.; Thrasher, R.D.; Waerstad, K.R. J. Agr. Food Chem. 1973, 21, 700-4

EXPERIMENTAL VALUES cont'd:

Solubility in the $NH_3-K_2O-H_3PO_4-H_4P_2O_7-H_2O$ system at 25°C.

Soln no.	pH	$(NH_4)_2O$ mass%	K_2O mass%	total P_2O_5 mass%	ortho P_2O_5 mass%	pyro P_2O_5 % of total	solid phases[a]
23	5.11	9.3	16.2	40.1	1.7	88	C,G,M
24	5.10	9.9	15.9	41.0	6.7	84	C,F,G,M
25	4.75	8.5	12.4	39.9	4.9	88	F,G,M
26	4.0	---	6.9	10.2	10.2	0	M
27	6.1	---	17.7	18.1	18.1	0	M
28	7.9	---	35.0	29.2	29.2	0	M,N
29	10.1	---	34.1	25.2	25.2	0	N

[a]The solid phases are: A = $(NH_4)_4P_2O_7$; B = $(NH_4)_3HP_2O_7 \cdot H_2O$; C = $(NH_4,K)_3HP_2O_7 \cdot H_2O$,

mole ratio N/P 1.0 to 2.8; D = $(NH_4)_2H_2P_2O_7$; E = $(NH_4,K)_2H_2P_2O_7$, mole ratio N/P

is 1.9 to 3.0; F = $(NH_4,K)_2H_2P_2O_7 \cdot 0.5H_2O$, mole ratio N/P is 0.61 to 0.74;

G = $K_2H_2P_2O_7$; J = $K_3HP_2O_7 \cdot 3H_2O$; L = $(NH_4)_2HPO_4$; M = $(NH_4,K)H_2PO_4$; N = $K_2HPO_4 \cdot 3H_2O$.

COMMENT: The "mole ratio N/P" is probably a typographical error and should be
"mole ratio N/K"--compiler.

The compiler has recalculated the above values to give the following.

Soln no	$(NH_4)_2O$ mol/kg	K_2O mol/kg	total P_2O_5 mol/kg	ortho P_2O_5 mol/kg
1	6.15	4.09	6.19	2.58
2	8.11	2.48	6.82	2.40
3	10.82	1.68	8.12	2.73
4	9.81	1.65	7.51	2.37
5	10.24	1.99	8.18	3.65
6	10.32	1.52	8.29	3.28
7	9.20	1.65	8.44	1.92
8	9.38	2.12	8.83	1.89
9	9.58	1.65	8.65	1.75
10	5.69	4.63	6.43	3.26
11	8.17	2.79	7.20	3.40
12	8.98	2.35	7.57	3.48
13	5.11	3.89	7.64	1.30
14	5.42	4.32	8.17	1.16
15	8.94	5.97	10.01	1.17
16	6.70	5.98	8.68	1.04
17	4.64	6.85	6.72	2.00
18	1.91	8.04	6.38	1.97
19	1.68	8.19	6.75	0.83
20	1.54	8.75	7.55	1.05
21	1.85	8.14	7.41	1.19
22	3.78	4.84	7.35	1.10
23	5.19	5.00	8.21	0.35
24	5.73	5.08	8.70	1.42
25	4.16	3.36	7.17	0.88
26	----	0.88	0.87	0.87
27	----	2.93	1.97	1.97
28	----	10.38	5.74	5.74
29	----	9.64	4.09	4.09

COMPONENTS:	ORIGINAL MEASUREMENTS:
(1) Potassium dihydrogenphosphate; KH_2PO_4; [7778-77-0] (2) Potassium chloride; KCl; [7447-40-7] (3) Calcium dihydrogenphosphate; $Ca(H_2PO_4)_2$; [10103-46-5] (4) Calcium chloride; $CaCl_2$ [10043-52-4] (5) Water; H_2O; [7732-18-5]	Timoshenko, Yu.M.; Gilyazova, G.N. *Zh. Neorg. Khim.* <u>1981</u>, *26*, 1104-6.

VARIABLES:	PREPARED BY:
Composition at 25°C.	J. Eysseltová

EXPERIMENTAL VALUES:

Composition of saturated solutions at 25°C.

d g cm^{-3}	η/cP	$Ca(H_2PO_4)_2$ mass%	$CaCl_2$ mass%	KH_2PO_4 mass%	KCl mass%	mol/100 mol solute Ca^{2+}	$2H_2PO_4^-$	H_2O	solid phase[a]
1.1594	1.374	1.14	----	19.94	----	6.21	100	5609	A + B
1.1672	1.118	1.21	----	12.78	8.61	4.69	47.43	3911	"
1.1781	0.983	1.26	----	4.30	22.89	3.09	9.04	2273	A + B + C
1.1874	0.901	1.22	-----	-----	26.01	2.89	2.95	2248	A + C
1.3805	7.083	1.23	19.08	-----	16.17	62.04	1.82	1236	"
1.4473	9.256	1.14	44.97	-----	-----	100	1.18	729	A + D
1.4730	10.91	1.21	43.50	-----	3.82	93.86	1.23	684	A + C + D

[a] The solid phases are: A = $Ca(H_2PO_4)_2 \cdot H_2O$; B = KH_2PO_4; C = KCl; D = $CaCl_2 \cdot 6H_2O$.

COMMENT: The authors emphasize that in the $Ca(H_2PO_4)_2$-KH_2PO_4-H_2O and $Ca(H_2PO_4)_2$-$CaCl_2$-H_2O systems the solubility of each salt component is not influenced appreciably by the presence of the other one.

AUXILIARY INFORMATION

METHOD/APPARATUS/PROCEDURE:	SOURCE AND PURITY OF MATERIALS:
The solutions were equilibrated with excess solid phase for 3-7 days. Viscosity was measured with an Ostwald viscometer. The density was measured with a pycnometer. The chloride content was determined by the Volhard method, the $H_2PO_4^-$ content was determined alkalimetrically after ion exchange, calcium content was determined by a compleximetric titration. Where the amount of phosphate ion was small, it was determined gravimetrically as ammonium phosphomolybdate.	No information is given.
	ESTIMATED ERROR:
	No information is given.
	REFERENCES:

COMPONENTS:	ORIGINAL MEASUREMENTS:
(1) Potassium dihydrogenphosphate; KH_2PO_4; [7778-77-0] (2) Formamide; CH_3NO; [75-12-7] (3) Water; H_2O; [7732-18-5]	Becker, B. J. Chem. Eng. Data 1969, 14, 431-2.
VARIABLES: Composition at 25°C.	PREPARED BY: J. Eysseltová

EXPERIMENTAL VALUES:

Solubility in the KH_2PO_4-$HCONH_2$-H_2O system at 25°C.

Initial composition (mass%)			saturated solution (mass%)		
KH_2PO_4	$HCONH_2$	H_2O	KH_2PO_4	$HCONH_2$	H_2O
40.7	7.6	51.7	15.2	10.9	73.9
29.8	20.3	49.9	10.3	25.6	64.1
32.9	33.6	33.5	5.8	46.5	44.7
28.6	43.0	28.4	3.9	58.4	37.7
32.2	49.5	18.3	2.7	71.0	26.3
29.6	59.7	10.7	1.6	82.7	15.7
			1.6	85.1	13.3
20.0	74.0	6.0	0.9	91.4	7.5
			1.0	95.1	3.9
40	60	0	1.0	99.0	0

The only solid phase in equilibrium with the above saturated solutions was KH_2PO_4.

AUXILIARY INFORMATION

METHOD/APPARATUS/PROCEDURE:	SOURCE AND PURITY OF MATERIALS:
Excess salt was added to aqueous formamide, the mixture was shaken vigorously for at least 24 hours and then filtered through a fritted glass filter. The nitrogen content was determined by the Kjeldahl method, potassium was determined by the Perrin method and phosphorus was determined gravimetrically by the quinoline molybdate method.	Fresh reagent grade salts were used without further purification. Matheson, Coleman, Bell 99% formamide was used without further purification. It had a m.p. of 2.5 ± 0.1°C in good agreement with the literature value (1).
	ESTIMATED ERROR: No information is given.
	REFERENCES: 1. Smith, G.F. J. Chem. Soc. 1931, 3527.

COMPONENTS:	ORIGINAL MEASUREMENTS:
(1) Potassium dihydrogenphosphate; KH_2PO_4; [7778-77-0] (2) Ethanamine,N,N,diethylphosphate 1:1 (triethylamine phosphate); $C_6H_{18}NO_4P$; [10138-93-9] (3) Water; H_2O; [7732-18-5]	Filipescu, L. *Rev Chim.* (*Bucharest*) <u>1971</u>, *22*, 533-40.

VARIABLES:	PREPARED BY:
Temperature and composition.	J. Eysseltová

EXPERIMENTAL VALUES:

Solubility isotherms in the $KH_2PO_4-((C_2H_5)_3N)H_3PO_4-H_2O$ system.

$t/°C.$	d g cm^{-3}	K^+ concn[b]	$(C_2H_5)_3N$ concn[b]	H_2O M[c]	KH_2PO_4[a] mass%	KH_2PO_4[a] mol/kg	$(C_2H_5)_3NH_3PO_4$[a] mass%	$(C_2H_5)_3NH_3PO_4$[a] mol/kg	solid phase[d]
20	-----	0.1124	0.0417	1201.1	4.49	1.12	7.32	0.41	A
20	-----	0.0765	0.0896	1113.9	2.86	0.76	14.71	0.89	"
20	-----	0.0602	0.1839	758.0	1.95	0.60	26.28	1.83	"
20	1.118	0.0664	0.3391	456.6	1.76	0.66	39.60	3.39	"
20	1.144	0.0798	0.5540	292.1	1.69	0.79	51.57	5.54	"
20	1.205	0.3719	3.4538	48.4	2.00	3.71	85.48	34.53	A + B
40	-----	0.1623	0.0746	781.1	6.02	1.62	12.15	0.74	A
40	-----	0.1045	0.2427	533.3	3.09	1.04	31.58	2.42	"
40	-----	0.1026	0.3534	406.9	2.65	1.02	40.21	3.53	"
40	1.1311	0.1096	0.4039	360.6	2.68	1.09	43.39	4.03	"
40	1.229	0.6778	4.6127	35.0	2.92	6.77	87.54	46.12	A + B
60	1.181	0.2559	0.0762	557.5	9.15	2.55	11.97	0.76	A
60	-----	0.2415	0.0894	559.5	8.50	2.41	13.83	0.89	"
60	1.138	0.1800	0.2251	457.0	5.33	1.80	29.30	2.25	"
60	-----	0.2434	1.1271	135.1	3.29	2.43	66.90	11.27	"
60	-----	0.9418	5.6949	27.9	3.34	9.41	88.82	56.94	A + B
80	1.221	0.3778	0.0963	390.5	12.57	3.77	14.07	0.96	A
80	1.197	0.3361	0.1694	366.3	10.23	3.36	22.64	1.69	"
80	-----	0.2459	0.5914	221.2	4.87	2.45	51.45	5.91	"

[a] These values were calculated by the compiler.
[b] The concentration unit is: equiv/100 g water.
[c] The concentration unit is: mol/100 equiv of solute.
[d] The solid phases are: A = KH_2PO_4; B = $((C_2H_5)_3)NH_3PO_4$.

AUXILIARY INFORMATION

METHOD/APPARATUS/PROCEDURE:	SOURCE AND PURITY OF MATERIALS:
The samples were equilibrated isothermally by stirring for 5 hours with a stream of inert gas. The potassium content was determined with a flame photometer, the $(C_2H_5)_3NH^+$ content was determined by the Kjeldahl method.	The KH_2PO_4 was recrystallized 3 times before use. The triethylamine phosphate was synthesized from phosphoric acid and triethylamine.

	ESTIMATED ERROR:
	The temperature was controlled to within ± 0.05°C at 20 and 40°C and to within ± 0.1°C at 60 and 80°C.

	REFERENCES:

COMPONENTS:	ORIGINAL MEASUREMENTS:
(1) Potassium dihydrogenphosphate; KH_2PO_4; [7778-77-0]	Bergman, A.G.; Gladkovskaya, A.A.; Galushkina, R.A.
(2) Ammonium dihydrogenphosphate; $NH_4H_2PO_4$; [7722-76-1]	*Zh. Neorg. Khim.* <u>1973</u>, *18*, 1978-80.
(3) Urea; CH_4N_2O; [57-13-6]	
(4) Water; H_2O; [7732-18-5]	

VARIABLES:	PREPARED BY:
Temperature and composition.	J. Eysseltová

EXPERIMENTAL VALUES:

Four sections through the system were investigated. The sections are:

No. 1 (45.81% $NH_4H_2PO_4$ + 54.19% KH_2PO_4) - urea - water
No. 2 (71.72% $NH_4H_2PO_4$ + 28.28% KH_2PO_4) - urea - water
No. 3 (65.70% $NH_4H_2PO_4$ + 34.30% urea) - KH_2PO_4 - H_2O
No. 4 (85.18% $NH_4H_2PO_4$ + 14.82% urea) - KH_2PO_4 - H_2O

Solubility data for saturated solutions in the urea-$NH_4H_2PO_4$-KH_2PO_4-H_2O system.

sect no	urea mass%	mol/kg[a]	$NH_4H_2PO_4$ mass%	mol/kg[a]	KH_2PO_4 mass%	mol/kg[a]	H_2O mass%	t/°C	solid phases[b]
	1.80	0.38	9.34	1.04	11.06	1.04	77.80	+6.4	A + B + C
	1.50	0.30	8.06	0.86	9.54	0.86	80.90	−4.1	A + B + D
1	12.60	2.84	6.23	0.73	7.37	0.73	73.8	−8.0	B + C + D
	40.00	13.48	4.86	0.85	5.74	0.85	49.4	+8.0	C + F + G
	32.80	3.41	4.20	0.62	5.00	0.63	58.00	−6.0	C + E + F
	28.40	8.50	8.98	1.40	10.62	1.40	55.60	−18.0	C + D + E
	1.20	0.28	21.30	2.67	8.40	0.89	69.10	+ 19.4	A + B + C
	1.40	0.29	14.34	1.58	5.66	0.52	78.60	−5.5	A + B + D
2	24.50	6.76	10.90	1.57	4.30	0.52	60.30	−8.8	B + C + E
	32.00	9.68	9.32	1.47	3.68	0.49	55.00	−7.0	C + E + F
	41.00	12.88	8.65	1.41	3.35	0.46	53.00	+9.6	C + F + G
	24.60	6.72	10.40	1.48	4.10	0.49	60.90	−16.00	B + D + E
3	11.73	3.17	22.47	3.17	4.20	0.50	61.60	+26.5	A + B + C
	6.52	1.45	12.48	1.45	6.50	0.64	74.50	−8.0	A + B + D

(continued next page)

AUXILIARY INFORMATION

METHOD/APPARATUS/PROCEDURE:	SOURCE AND PURITY OF MATERIALS:
A visual polythermic method was used (1). Solid carbon dioxide was used as the cooling agent.	Chemically pure salts and bidistilled water were used.

	ESTIMATED ERROR:
	No information is given.

	REFERENCES:
	1. Bergman, A.G.; Luzhnaya, N.P. *Fiziko-Khimicheskie Osnovy Izucheniya i Ispol'zovaniya Solyanykh Mestorozhdeniy Khlorid-sul'fatnogo Tipa*, Moscow, IAN SSSR, <u>1951</u>.

COMPONENTS:	ORIGINAL MEASUREMENTS:
(1) Potassium dihydrogenphosphate; KH_2PO_4; [7778-77-0]	Bergman, A.G.; Gladkovskaya, A.A.; Galushkina, R.A.
(2) Ammonium dihydrogenphosphate; $NH_4H_2PO_4$; [7722-76-1]	*Zh. Neorg. Khim.* <u>1973</u>, *18*, 1978-80.
(3) Urea; CH_4N_2O; [57-13-6]	
(4) Water; H_2O; [7732-18-5]	

EXPERIMENTAL VALUES cont'd:

Four sections through the system were investigated. The sections are:

No. 1 (45.81% $NH_4H_2PO_4$ + 54.19% KH_2PO_4) – urea – water
No. 2 (71.72% $NH_4H_2PO_4$ + 28.28% KH_2PO_4) – urea – water
No. 3 (65.70% $NH_4H_2PO_4$ + 34.30% urea) – KH_2PO_4 – H_2O
No. 4 (85.18% $NH_4H_2PO_4$ + 14.82% urea) – KH_2PO_4 – H_2O

Solubility data for saturated solutions in the urea–$NH_4H_2PO_4$–KH_2PO_4–H_2O system.

sect no	urea mass%	urea mol/kg[a]	$NH_4H_2PO_4$ mass%	$NH_4H_2PO_4$ mol/kg[a]	KH_2PO_4 mass%	KH_2PO_4 mol/kg[a]	H_2O mass%	$t/°C$	solid phases[b]
4	4.30	10.52	24.70	3.15	3.00	0.32	68.00	+17.0	A + B + C
	2.70	5.22	15.70	1.74	3.50	0.32	78.10	-5.2	A + B + D

[a] The mol/kg H_2O values were calculated by the compiler.

[b] The solid phases are: A = α-$NH_4H_2PO_4$; B = (α-NH_4,K)H_2PO_4; C = (β-NH_4,K)H_2PO_4;

D = ice; E = α-urea; F = β-urea; G = γ-urea.

COMPONENTS:	ORIGINAL MEASUREMENTS:
(1) Potassium dihydrogenphosphate; KH_2PO_4; [7778-77-0] (2) Formamide; CH_3NO; [75-12-7] (3) Water; H_2O; [7732-18-5]	Beglov, B.M.; Tukhtaev, S.; Yugai, M.R. *Zh. Neorg. Khim.* <u>1980</u>, 25, 2283-5.

VARIABLES:	PREPARED BY:
Temperature and composition.	J. Eysseltová

EXPERIMENTAL VALUES:

Solutions coexisting with two or three solid phases.

composition (mass%)

$HCONH_2$	KH_2PO_4	H_2O	$t/°C.$	solid phases
0	11.5	88.5	−2.5	ice + KH_2PO_4
11.0	8.9	80.1	−7.6	"
18.4	8.0	73.6	−11.5	"
37.7	5.7	56.6	−25.0	"
57.8	3.6	38.6	−43.5	"
65.4	0	34.6	−45.5	ice + formamide
64.2	3.1	32.7	−51.2	ice + formamide + KH_2PO_4
68.1	2.7	29.2	−44.5	KH_2PO_4 + formamide
88.9	1.0	10.1	−15.0	"
94.4	1.1	4.5	−8.5	"
95.7	1.1	3.2	−6.3	"
98.8	1.2	0	−2.0	"

Solubility isotherms in the temperature range of −40 to 50°C are given only in graphical form.

AUXILIARY INFORMATION

METHOD/APPARATUS/PROCEDURE:	SOURCE AND PURITY OF MATERIALS:
A visual polythermic method was used (1) but no details are given.	Chemically pure KH_2PO_4 was recrystallized before use. Pure formamide was dehydrated and distilled under vacuum at 80-82°C.
	ESTIMATED ERROR: No details are given.
	REFERENCES: 1. Bergman, A.G.; Luzhnaya, N.P. *Fiziko-Khimicheskie Osnovy Izucheniya i Ispol'zovaniya Solyanykh Mestorzhdeniy Khlorid-Sul'fatnogo Tipa*, Moscow, IAN SSSR, <u>1951</u>.

COMPONENTS:	ORIGINAL MEASUREMENTS:
(1) Potassium dihydrogenphosphate; KH_2PO_4; [7778-77-0] (2) Formamide; CH_3NO; [75-12-7] (3) Water; H_2O; [7732-18-5]	Yugai, M.R.; Tukhtaev, S.; Beglov, B.M. *Uzb. Khim. Zh.* <u>1981</u>, *6*, 15-8.

VARIABLES:	PREPARED BY:
Composition at 50°C.	J. Eysseltová

EXPERIMENTAL VALUES:

Composition of saturated solutions in the KH_2PO_4-$HCONH_2$-H_2O system at 50°C.

$HCONH_2$	KH_2PO_4	H_2O
mass%	mass%	mass%
0	29.42	70.58
3.68	25.82	70.50
12.80	18.67	68.53
23.59	13.83	81.33
37.44	10.31	52.25
51.07	7.46	41.47
61.35	5.26	33.39
77.42	3.58	19.00
91.29	2.73	5.98
98.16	1.84	0

AUXILIARY INFORMATION

METHOD/APPARATUS/PROCEDURE:	SOURCE AND PURITY OF MATERIALS:
Equilibrium was approached isothermally by stirring for 1-2 days. Nitrogen content was determined by the Kjeldahl method, phosphorus was determined gravimetrically, and potassium was determined as the tetraphenylborate. The composition of the solid phase was determined by the Schreinemakers' method.	Chemically pure or analytically pure KH_2PO_4 was used. The formamide was dried and distilled under vacuum at 80-82°C.

	ESTIMATED ERROR:
	The temperature was controlled to within 0.1 K.

	REFERENCES:

COMPONENTS:	EVALUATOR:
(1) Dipotassium hydrogenphosphate; K_2HPO_4; [7758-11-4] (2) Water; H_2O; [7732-18-5]	J. Eysseltová Charles University Prague, Czechoslovakia May, 1985

CRITICAL EVALUATION:

THE BINARY SYSTEM

The only study of this system has been made by Ravich (1). A few other solubility values have been reported as part of a study of a multicomponent system (6,7). In both these studies the values reported are about 1% lower than those reported by Ravich (1). Ravich reports the eutectic of this system to be 36.78 mass% (3.33 mol/kg) K_2HPO_4 at 259.8 K; the transition of the hexahydrate to the trihydrate occurs at 287.5 K (solution composition is not given); and the transition of the trihydrate to the anhydrous salt takes place at 319 to 324 K and 71.26 to 72.64 mass% K_2HPO_4. He also observed metastable solutions saturated with the hexahydrate and the trihydrate. The regions in which the various phases exist are rather narrow. Because of the lack of solubility data from other sources, the treatment of data described in chapter 3 could not be used. The system has a pronounced tendency to form supersaturated solutions (2-5).

MULTICOMPONENT SYSTEMS

Solubility measurements have been reported for several multicomponent systems.

1. The K_2HPO_4-KBO_2-H_2O system. Data have been reported for 298 and 323 K (8). The data cannot be evaluated but it should be noted that the values reported for the K_2HPO_4-H_2O system differ from the values of Ravich (1) by about +30%.
2. The K_2HPO_4-$CO(NH_2)_2$-H_2O system. Two unspecified ternary compounds have been reported for this system (9). A later study of this system (10) gave a more detailed description and mentioned K_2HPO_4 and the α-, β-, γ-, and δ- modifications of urea as solid phases.
3. The K_2HPO_4-K_2CO_3-H_2O system. Solubility measurements have been made over the temperature interval of 253 to 353 K (11). Later these same investigators published the solubility polytherm of the quaternary system K_2HPO_4-K_2CO_3-$CO(NH_2)_2$-H_2O (12).
4. The K_2HPO_4-KNO_3-H_2O system. Only the components and their hydrates were found as solid phases in this system (13).
5. The K_2HPO_4-KCl-H_2O system. Solubility values have been determined at 298, 323 and 348 K (6). The authors reported $2KCl \cdot K_2HPO_4 \cdot 5H_2O$ as a solid phase at 298 K. They also emphasized the tendency of all solutions existing in contact with a phosphate-containing solid to form supersaturated solutions.
6. The K_2HPO_4-$(NH_4)_2HPO_4$-H_2O system. No ternary compounds were observed in this system (7,9,14). In contrast to this, the compound $NaNH_4HPO_4$ is present in the Na_2HPO_4-$(NH_4)_2HPO_4$-H_2O system (15). An analogous compound exists in the Na_2HPO_4-K_2HPO_4-H_2O system (16). This system is discussed in chapter 5.

Data have also been published for the K_2HPO_4-$NH_4H_2PO_4$-$(NH_4)_2HPO_4$-H_2O system (17), but the paper contains many uncertainties which make it impossible to discuss and evaluate the data.

References

1. Ravich, M.I. *Izv. AN SSSR, Ser. Khim.* 1938, 141.
2. Ravich, M.I. *Kaliy* 1936, *10*, 33.
3. Ravich, M.I. *Izv. Akad. Nauk SSSR* 1938, 167.
4. Berg, A.G. *Izv. Akad. Nauk SSSR* 1933, 167.
5. Berg, A.G. *Izv. Akad. Nauk SSSR* 1938, 147.
6. Mráz. R.; Srb. V.; Tichý, S.; Vosolsobě, J. *Chem. Prům.* 1976, *26*, 511.
7. Sokolov, S.J. *Kaliy* 1937, *2*, 28.
8. Beremzhanov, B.A.; Voronina, L.V.; Savich, R.F. *Khim. Khim. Tekhnol. (Alma Ata)* 1978, 29.
9. Bergman, A.G.; Dzuev, A.D. *Uch. Zap. Kabardino-Balkan. Univ., Ser. Sel.'-Khoz. Khim.-Biol.* 1966, *29*, 40.
10. Bergman, A.G.; Velikanova, L.V. *Zh. Neorg. Khim.* 1968, *13*, 1158.
11. Bergman, A.G.; Velikanova, L.V. *Zh. Neorg. Khim* 1968, *13*, 557.
12. Velikanova, L.V.; Bergman, A.G. *Izv. Vysch. Ucheb. Zaved. Khim. Khim. Tekhnol.* 1974, *17*, 7, 1513.
13. Endovitskaya, M.R.; Vereshchagina, V.I. *Zh. Neorg. Khim.* 1972, 17, 877.

(continued next page)

COMPONENTS:	EVALUATOR:
(1) Dipotassium hydrogenphosphate; K_2HPO_4; [7758-11-4] (2) Water; H_2O; [7732-18-5]	J. Eysseltová Charles University Prague, Czechoslovakia May, 1985

CRITICAL EVALUATION: (cont'd)

14. Bergman, A.G.; Dzuev, A.D.; Opredelnikova, L.V.; *Zh. Prikl. Khim.* 1967, *40*, 1838.
15. Platford, R.F. *J. Chem. Eng. Data* 1974, *19*, 166.
16. Ravich, M.I.; Popova, Z.V. *Izv. Akad. Nauk SSSR, Ser. Khim.* 1942, 268.
17. Torochestnikov, N.S.; Rodionova, T.M.; Kirsanova, L.D. *VINITI* 1979, 2909.

COMPONENTS:	ORIGINAL MEASUREMENTS:
(1) Dipotassium hydrogenphosphate; K_2HPO_4; [7758-11-4] (2) Water; H_2O; [7732-18-5]	Ravich, M.I. *Izv. AN SSSR, Ser. Khim.* <u>1938</u>, 141-6.
VARIABLES: Temperature and composition.	PREPARED BY: J. Eysseltová

EXPERIMENTAL VALUES:

Composition and crystallization temperatures of saturated solutions
in the K_2HPO_4-H_2O system.

	K_2HPO_4					K_2HPO_4			
t/°C.	mass%	mol%	mol/kg[a]	solid phase	t/°C.	mass%	mol%	mol/kg[a]	solid phase[b]
-4.2	16.78	2.04	1.16	ice	8.2	69.09	18.75	12.83	A[c]
-6.4	23.60	3.09	1.77	"	0	57.05	12.01	7.62	B[c]
-9	29.61	4.17	2.41	"	10.0	59.08	12.96	8.28	"
-11.7	34.10	5.07	2.97	"	15.0	60.16	13.49	8.66	B
-13.5	36.78	5.67	3.33	ice + A	20.0	61.52	14.16	9.17	"
0	46.11	8.12	4.91	A	25.0	62.74	14.83	9.66	"
4.95	50.12	9.40	5.76	"	30.0	64.13	15.60	10.26	"
9.7	54.43	10.99	6.85	"	35	65.68	16.51	10.98	"
13.15	57.89	12.44	7.89	"	39.5	67.54	17.68	11.94	"
14.3	-----	-----	----	A + B	44	69.83	19.29	13.28	"
14.6	60.82	13.82	8.91	A[c]	46	71.26	20.42	14.23	"
14.85	71.73	14.29	9.26	"	51	72.64	21.55	15.24	C
14.7	62.96	14.94	9.75	"	56	72.50	21.38	15.13	"
12.8	65.95	16.68	11.11	"	63	72.79	21.66	15.35	"

[a] The mol/kg H_2O values were calculated by the compiler.

[b] The solid phases are: A = $K_2HPO_4 \cdot 6H_2O$; B = $K_2HPO_4 \cdot 3H_2O$; C = KH_2PO_4.

[c] This is a metastable solution.

AUXILIARY INFORMATION

METHOD/APPARATUS/PROCEDURE:	SOURCE AND PURITY OF MATERIALS:
For the systems in which the trihydrate or the anhydrous salt was the solid phase, the solubility was determined by evaporating the solution to dryness. Where the hexahydrate was the solid phase, a visual polythermic method was used. Analyses were carried out gravimetrically: K_2O was determined as $KClO_4$; P_2O_5 was determined as $Mg_2P_2O_7$; and water was determined by weight loss during calcination.	The K_2HPO_4 was prepared from twice recrystallized KH_2PO_4 and KOH. Analysis: found calcd for $K_2HPO_4 \cdot 6H_2O$ K_2O 33.37 33.38 P_2O_5 25.20 25.26 H_2O 41.52 41.46
	ESTIMATED ERROR: No information is given.
	REFERENCES:

COMPONENTS:	ORIGINAL MEASUREMENTS:
(1) Dipotassium hydrogenphosphate; K_2HPO_4; [7758-11-4] (2) Diammonium hydrogenphosphate; $(NH_4)_2HPO_4$; [7783-28-0] (3) Water; H_2O; [7732-18-5]	Sokolov, S.J. *Kaliy* 1937, *2*, 28-32.

VARIABLES:	PREPARED BY:
Composition at 0°C.	J. Eysseltová

EXPERIMENTAL VALUES:

Solubility in the K_2HPO_4-$(NH_4)_2HPO_4$-H_2O system at 0°C.

K_2HPO_4		$(NH_4)_2HPO_4$		H_2O	solid[b]
mass%	mol/kg[a]	mass%	mol/kg[a]	mass%	phase
----	----	36.24	4.30	63.76	A
7.94	0.75	31.93	4.02	60.13	"
15.66	1.61	28.60	3.88	55.74	"
17.82	1.86	27.36	3.77	54.82	"
29.34	3.41	21.39	3.28	49.27	"
38.35	5.02	17.86	3.08	43.79	"
43.74	6.08	15.01	2.75	41.25	"
48.90	7.74	14.87	3.10	36.23	A + B
48.58	7.58	14.66	3.01	36.76	B
48.52	6.61	9.35	1.68	42.12	"
48.14	6.05	6.22	1.03	45.64	"
47.64	5.72	4.59	0.72	47.77	"
45.72	4.83	-----	----	54.28	"

[a] The mol/kg H_2O values were calculated by the compiler.

[b] The solid phases are: A = $(NH_4)_2HPO_4$; B = $K_2HPO_4 \cdot 6H_2O$.

AUXILIARY INFORMATION

METHOD/APPARATUS/PROCEDURE:	SOURCE AND PURITY OF MATERIALS:
The mixtures were equilibrated isothermally for several days. The P_2O_5 content was determined gravimetrically as $Mg_2P_2O_7$, potassium was determined gravimetrically as $KClO_4$, and nitrogen was determined by the Kjeldahl method.	Purified, commercial materials were used, but no details are given.
	ESTIMATED ERROR: No information is given.
	REFERENCES:

COMPONENTS:	ORIGINAL MEASUREMENTS:

COMPONENTS:

(1) Dipotassium dhydrogenphosphate; K_2HPO_4;
    [7758-11-4]

(2) Diammonium hydrogenphosphate;
    $(NH_4)_2HPO_4$; [7783-28-0]

(3) Water; H_2O; [7732-18-5]

ORIGINAL MEASUREMENTS:

1. Bergman, A.G.; Dzuev, A.D. *Uch. Zap. Kabardino-Balkar. Univ. Ser. Sel. Khoz. Khim.-Biol.* <u>1966</u>, *29*, 40-4.
2. Bergman, A.G.; Dzuev, A.D.; Opredelnikova, L.V. *Zh. Prikl. Khim.* <u>1967</u>, *40*, 1838-41.

VARIABLES:	PREPARED BY:

VARIABLES:

Temperature and composition.

PREPARED BY:

J. Eysseltová

EXPERIMENTAL VALUES:

Composition and crystallization temperature of invariant points
in the $K_2HPO_4-(NH_4)_2HPO_4-H_2O$ system.

| $t/°C.$ | K_2HPO_4 | | $(NH_4)_2HPO_4$ | | H_2O | solid[b] |
	mass%	mol/kg[a]	mass%	mol/kg[a]	mass%	phase
-18	30	3.02	13	1.73	57	A + D + F
-6	41	5.23	14	2.35	45	A + D + E
5	50	7.65	12.5	2.52	37.5	A + B + E
22	57	10.22	11	2.60	32	B + C + E

[a] The mol/kg H_2O values were calculated by the compiler.

[b] The solid phases are: A = $K_2HPO_4 \cdot 6H_2O$; B = $K_2HPO_4 \cdot 3H_2O$; C = K_2HPO_4;
    D = $(NH_4)_2HPO_4 \cdot 2H_2O$; E = $(NH_4)_2HPO_4$; F = ice.

AUXILIARY INFORMATION

METHOD/APPARATUS/PROCEDURE:	SOURCE AND PURITY OF MATERIALS:

METHOD/APPARATUS/PROCEDURE:

A visual polythermic method was used. The
disappearance of the last crystals was
observed.

SOURCE AND PURITY OF MATERIALS:

Reagent grade K_2HPO_4 was recrystallized
from water before use. Reagent grade
$(NH_4)_2HPO_4$ was recrystallized from
ammoniacal solutions before use.

ESTIMATED ERROR:

No details are given.

REFERENCES:

COMPONENTS:	ORIGINAL MEASUREMENTS:
(1) Dipotassium hydrogenphosphate; K_2HPO_4; [7758-11-4] (2) Potassium carbonate; K_2CO_3; [584-08-7] (3) Water; H_2O; [7732-18-5]	Bergman, A.G.; Velikanova, L.V. *Zh. Neorg. Khim.* <u>1968</u>, *13*, 557-61.

VARIABLES:	PREPARED BY:
Temperature and composition.	J. Eysseltová

EXPERIMENTAL VALUES:

Crystallization temperature and composition of invariant points
in the K_2HPO_4-K_2CO_3-H_2O system.

	K_2HPO_4		K_2CO_3		H_2O	
$t/°C.$	mass%	mol/kg^a	mass%	mol/kg^a	mass%	solid phases^b
-37	4	0.40	39	4.95	57	ice + A + D
-31.5	21.5	2.17	21.8	2.78	56.7	ice + A + B
72	44.2	8.02	24.2	5.54	31.6	B + C + F
53	32.5	5.15	31.3	6.25	36.2	B + E + F
-7	3.2	0.37	48.3	7.20	48.5	B + D + E

^{a}The mol/kg H_2O values were calculated by the compiler.

^{b}The solid phases are: A = $K_2HPO_4 \cdot 6H_2O$; B = $K_2HPO_4 \cdot 3H_2O$; C = K_2HPO_4;
D = $K_2CO_3 \cdot 6H_2O$; E = $2K_2CO_3 \cdot 3H_2O$; F = K_2CO_3.

Solubility isotherms in the temperature range -20 to +80°C are given in
graphical form only.

Relative areas of individual crystallization fields are: ice-16.06%;
$K_2CO_3 \cdot 6H_2O$-0.86%; $2K_2CO_3 \cdot 3H_2O$-3.26%; $K_2HPO_4 \cdot 6H_2O$-7.62%; $K_2HPO_4 \cdot 3H_2O$-11.38%;
K_2CO_3-40.82%; K_2HPO_4-20%.

AUXILIARY INFORMATION

METHOD/APPARATUS/PROCEDURE:	SOURCE AND PURITY OF MATERIALS:
A visual polythermic method was used in the temperature range of -37 to 80°C.	No information is given.
	ESTIMATED ERROR:
	The precision of the temperature was ± 0.2 to 0.4 K.
	REFERENCES:

COMPONENTS:	ORIGINAL MEASUREMENTS:
(1) Dipotassium hydrogenphosphate; K_2HPO_4; [7758-11-4] (2) Urea; CH_4N_2O; [57-13-6] (3) Water; H_2O; [7732-18-5]	Bergman, A.G.; Velikanova, L.V. *Zh. Neorg. Khim.* <u>1968</u>, *13*, 1158-62.

VARIABLES:	PREPARED BY:
Temperature and composition.	J. Eysseltová

EXPERIMENTAL VALUES:

Temperature and composition of invariant points in the
$K_2HPO_4-CO(NH_2)_2-H_2O$ system.

	K_2HPO_4		urea		H_2O	
$t/°C.$	mass%	mol/kg^a	mass%	mol/kg^a	mass%	solid phases[b]
-19.7	27.3	2.62	13	3.62	59.7	A + B + E
-9.8	33.4	3.49	11.7	3.54	54.9	B + E + F
63	59.3	13.09	14.7	9.41	26	D + F + G
77.3	40.2	11.09	39	31.21	20.8	D + G + H
12	56.5	8.31	4.5	1.92	39	B + C + F
38.3	66.2	13.76	6.2	3.74	27.6	C + D + F

[a] The mol/kg H_2O values were calculated by the compiler.

[b] The solid phases are: A = ice; B = $K_2HPO_4 \cdot 6H_2O$; C = $K_2HPO_4 \cdot 3H_2O$; D = K_2HPO_4;

E = α-urea; F = β-urea; G = γ-urea; H = δ-urea.

Solubility isotherms in the temperature range of -10 to +80°C are given in graphical form only.

Relative areas of the crystallization fields are: $K_2HPO_4 \cdot 6H_2O$-4.61%; ice-14.05%;

$K_2HPO_4 \cdot 3H_2O$-1.51%; K_2HPO_4-31.73%; α-urea-3.73%; β-urea-17.33%; γ-urea-13.27%;

δ-urea-13.77%.

AUXILIARY INFORMATION

METHOD/APPARATUS/PROCEDURE:	SOURCE AND PURITY OF MATERIALS:
Twelve parts of the system were studied by using a visual polythermic method. The cooling agent was solid carbon dioxide, either alone or with acetone.	No information is given.
	ESTIMATED ERROR: On the crystallization curves the temperature had a precision of ± 0.4°C. On the rest of the system it was ± 0.2°C.
	REFERENCES:

COMPONENTS:	ORIGINAL MEASUREMENTS:
(1) Dipotassium hydrogenphosphate; K_2HPO_4; [7758-11-4] (2) Urea; CH_4N_2O; [57-13-6] (3) Water; H_2O; [7732-18-5]	Bergman, A.G.; Dzuev, A.D. *Uch. Zap. Kabardino-Balkan. Univ., Ser. Sel.' -Khoz. Khim.-Biol.* 1969, 29, 40-4.

VARIABLES:	PREPARED BY:
Temperature and composition.	J. Eysseltová

EXPERIMENTAL VALUES:

The authors report that the crystallization surface in the K_2HPO_4-$CO(NH_2)_2$-H_2O system consists of ten fields: ice; four modifications of urea; three crystal forms of K_2HPO_4 (the hexahydrate, the trihydrate and the anhydrous form); and two ternary compounds (but these are not specified).

The eutectic temperature is -19°C. The composition of the eutectic point is: 28 mass% K_2HPO_4 (2.70 mol/kg H_2O -compiler); 12.5 mass% urea (3.50 mol/kg H_2O-compiler) and 59.5 mass% water.

The isotherms are given only in graphical form.

AUXILIARY INFORMATION

METHOD/APPARATUS/PROCEDURE:	SOURCE AND PURITY OF MATERIALS:
A visual polythermic method was used. The disappearance and the appearance of the first crystals was observed. Solid CO_2 was used as the cooling agent.	Reagent grade K_2HPO_4 was dried at 170°C before use. Reagent grade urea was recrystallized from ethanol and had a melting point of 132.5°C.
	ESTIMATED ERROR: Nothing is stated.
	REFERENCES:

AMO—T

COMPONENTS:	ORIGINAL MEASUREMENTS:
(1) Dipotassium hydrogenphosphate; K_2HPO_4; [7758-11-4] (2) Potassium nitrate; KNO_3; [7757-79-1] (3) Water; H_2O; [7732-18-5]	Endovitskaya, M.R.; Vereshchagina, V.I. *Zh. Neorg. Khim.* <u>1972</u>, *17*, 877-9.

VARIABLES:	PREPARED BY:
Temperature and composition.	J. Eysseltová

EXPERIMENTAL VALUES:

Composition and crystallization temperature of invariant points
in the K_2HPO_4-KNO_3-H_2O system.

$t/°C.$	K_2HPO_4		KNO_3		H_2O	solid[b] phase
	mass%	mol/kg[a]	mass%	mol/kg[a]	mass%	
31.5	31.5	3.28	13.5	2.42	55	B + D + E
-8	19.5	1.49	5.5	1.49	75	A + B + F
-11	38	3.69	3	0.50	59	B + C + D
-14	38	3.60	1.5	0.24	60.5	B + C + F

[a] The mol/kg H_2O values were calculated by the compiler.

[b] The solid phases are: A = α-KNO_3; B = β-KNO_3; C = $K_2HPO_4 \cdot 6H_2O$;

D = $K_2HPO_4 \cdot 3H_2O$; E = K_2HPO_4; F = ice.

Some of the data in the article are given only in graphical form.

AUXILIARY INFORMATION

METHOD/APPARATUS/PROCEDURE:	SOURCE AND PURITY OF MATERIALS:
A visual polythermic method was used. Solid carbon dioxide served as the cooling agent.	No information is given.
	ESTIMATED ERROR: The temperature had a precision of ± 0.1°C.
	REFERENCES:

COMPONENTS:	ORIGINAL MEASUREMENTS:
(1) Dipotassium hydrogenphosphate; K_2HPO_4; [7758-11-4] (2) Potassium chloride; KCl; [7747-40-7] (3) Water; H_2O; [7732-18-5]	Mráz, R.; Srb, V.; Tichý, S.; Vosolsobǎ, J. Chem. Prům. <u>1976</u>, 26, 511-4.

VARIABLES:	PREPARED BY:
Temperature and composition.	J. Eysseltová

EXPERIMENTAL VALUES:

Solubility isotherms in the K_2HPO_4-KCl-H_2O system.

K_2HPO_4		KCl		H_2O	solid[b]
mass%	mol/kg[a]	mass%	mol/kg[a]	mass%	phase
temp. = 25°C.					
0	0	26.4	4.81	73.6	A
0	0	26.6	4.86	73.4	"
1.3	0.10	26.0	4.79	72.7	"
2.0	0.15	24.8	4.54	73.2	"
2.1	0.16	24.2	4.40	73.7	"
15.7	1.36	18.1	3.66	66.2	"
20.2	1.83	16.6	3.52	63.2	"
17.3	1.49	16.3	3.29	66.4	"
17.5	1.49	15.5	3.10	67.0	"
26.4	2.49	12.9	2.85	60.7	"
31.5	3.15	11.2	2.62	57.3	"
42.2	5.05	9.9	2.77	48.0	"
38.3	4.17	9.0	2.29	52.7	"
42.2	4.83	7.7	2.06	50.1	A + B
44.4	5.34	7.9	2.22	47.7	B
46.8	5.79	6.8	1.96	46.4	"
50.9	6.81	6.2	1.93	42.9	"
59.0	8.77	2.4	0.83	38.6	B + C
61.8	9.93	2.5	0.93	35.7	C
66.4	12.10	2.1	0.89	31.5	"
62.6	10.15	2.0	0.75	35.4	"
66.5	12.11	2.0	0.85	32.5	"

(continued next page)

AUXILIARY INFORMATION

METHOD/APPARATUS/PROCEDURE:	SOURCE AND PURITY OF MATERIALS:
Saturated solutions were prepared at a temperature about 5 K higher than that of the isotherm to be studied. The samples were equilibrated for 4 hours with constant stirring. After being quiescent for 1 hour the phases were separated from each other and samples were taken for analysis. Silver content was determined by the Volhard method. HPO_4^{2-} ions were precipitated with excess $Bi(NO_3)_3$ and the Bi^{3+} ions were back titrated with Komplexon III.	No information is given.

ESTIMATED ERROR:

The temperature was controlled to within ± 0.2 K. The accuracy of the analysis for hydrogenphosphate ions was at least ± 3%.

REFERENCES:

COMPONENTS:	ORIGINAL MEASUREMENTS:
(1) Dipotassium hydrogenphosphate; K_2HPO_4; [7758-11-4]	Mráz, R.; Srb, V.; Tichý, S.; Vosolsobě, J. *Chem. Prům.* <u>1976</u>, *26*, 511-4.
(2) Potassium chloride; KCl; [7747-40-7]	
(3) Water; H_2O; [7732-18-5]	

EXPERIMENTAL VALUES cont'd:

Solubility isotherms in the K_2HPO_4-KCl-H_2O system.

K_2HPO_4		KCl		H_2O	solid
mass%	mol/kg[a]	mass%	mol/kg[a]	mass%	phase[b]
temp. = 25°C.					
59.3	8.64	1.3	0.44	39.4	C
66.7	11.70	0.6	0.24	32.7	"
67.9	12.18	0.1	0.04	32.0	"
60.2	8.68	0	0	39.8	"
temp. = 50°C.					
0	0	29.6	5.63	70.4	A
5.6	0.47	27.3	5.45	67.1	"
9.2	0.78	23.8	4.76	67.0	"
14.0	1.24	21.5	4.47	64.5	"
39.1	4.49	11.0	2.95	49.9	"
59.8	9.61	4.5	1.69	35.7	"
69.5	13.71	1.4	0.64	29.1	C
70.0	13.39	0	0	30.0	"
temp. = 75°C.					
0	0	33.2	6.66	66.8	A
5.9	0.52	29.6	6.15	64.5	"
10.0	0.91	27.1	5.77	62.9	"
13.7	1.28	25.0	5.46	61.3	"
20.2	2.02	22.4	5.23	55.4	"
47.3	6.38	10.2	3.21	42.5	"
60.1	9.85	4.9	1.87	35.0	"
69.4	13.93	2.0	0.93	28.6	"
73.9	16.25	0	0	26.1	C

[a] The mol/kg H_2O values were calculated by the compiler.

[b] The solid phases are: A = KCl; B = $2KCl \cdot K_2HPO_4 \cdot 5H_2O$; C = K_2HPO_4.

The authors state that, in fields where salts other than KCl exist as equilibrium solid phases, the precision of the results is poor because of the high viscosity of the saturated solutions. For the same reason, the authors could not determine the composition of eutonic solutions at 50°C and 75°C. They estimate that these solutions contain about 3% KCl at 50°C and less than 2% KCl at 75°C.

COMPONENTS:	ORIGINAL MEASUREMENTS:
(1) Dipotassium hydrogenphosphate; K_2HPO_4; [7758-11-4] (2) Potassium borate; KBO_2; [13709-94-9] (3) Water; H_2O; [7732-18-5]	Beremzhanov, B.A.; Voronina, L.V.; Savich, R.F. *Khim. Khim. Tekhnol.* (*Alma Ata*) <u>1978</u>, 29-36.

VARIABLES:	PREPARED BY:
Composition at 25° and 50°C.	J. Eysseltová

EXPERIMENTAL VALUES:

Solubility in the K_2HPO_4-KBO_2-H_2O system at 25°C.

K_2HPO_4			KBO_2			refr.		solid[b]
mass%	mol%	mol/kg[a]	mass%	mol%	mol/kg[a]	index	pH	phase
----	----	----	0.368[a]	0.081	0.045	1.441	14.0	A
86.03[a]	36.42	35.4	-----	-----	-----	1.380	9.45	B
67.13	17.25	11.74	0.056	0.032	0.020	1.421	10.12	C
56.35	11.68	7.43	0.117	0.051	0.032	1.423	10.81	"
49.49	9.27	5.64	0.152	0.059	0.036	1.424	11.15	"
45.57	7.95	4.82	0.176	0.064	0.039	1.426	11.76	"
44.59	7.55	4.64	0.211	0.075	0.046	1.428	12.85	"
44.10	7.50	4.55	0.257	0.076	0.056	1.430	13.48	A + C
27.93	3.86	2.23	0.281	0.075	0.047	1.431	13.49	"
15.68	1.89	1.07	0.298	0.073	0.042	1.432	13.50	"
14.21	1.66	0.95	0.295	0.074	0.042	1.432	13.52	"
12.25	1.42	0.80	0.304	0.075	0.042	1.433	13.54	"
11.27	1.21	0.73	0.316	0.076	0.043	1.433	13.58	"
10.78	0.99	0.70	0.323	0.077	0.044	1.434	13.64	"
6.86	0.77	0.42	0.328	0.077	0.043	1.434	13.81	"

(continued next page)

AUXILIARY INFORMATION

METHOD/APPARATUS/PROCEDURE:	SOURCE AND PURITY OF MATERIALS:
The isothermal method was used but no details are given.	No information is given.
	ESTIMATED ERROR: No details are given.
	REFERENCES:

COMPONENTS:	ORIGINAL MEASUREMENTS:
(1) Dipotassium hydrogenphosphate; K_2HPO_4; [7758-11-4]	Beremzhanov, B.A.; Voronina, L.V.; Savich, R.F.
(2) Potassium borate; KBO_2; [13709-94-9]	*Khim. Khim. Tekhnol.* (Alma Ata) <u>1978</u>, 29-36.
(3) Water; H_2O; [7732-18-5]	

EXPERIMENTAL VALUES cont'd:

Solubility isotherm in the K_2HPO_4-KBO_2-H_2O system at 50°C.

K_2HPO_4			KBO_2			refr.		solid[b]
mass%	mol%	mol/kg[a]	mass%	mol%	mol/kg[a]	index	pH	phase
----	----	----	0.369[a]	0.090	0.046	1.445	14.0	A
91.76[a]	53.50		-----	-----	-----	1.390	9.65	B
74.48	22.94	16.86	0.164	0.109	0.078	1.398	11.20	C
69.58	18.98	13.21	0.176	0.096	0.071	1.400	11.40	"
66.15	16.89	11.28	0.187	0.088	0.067	1.405	11.48	"
64.68	15.62	10.58	0.211	0.087	0.074	1.410	11.50	"
63.70	15.25	10.14	0.234	0.127	0.079	1.415	11.56	"
61.25	14.06	9.13	0.246	0.120	0.077	1.427	11.63	"
60.76	14.00	8.96	0.290	0.119	0.090	1.430	11.70	"
58.80	12.74	8.25	0.292	0.116	0.087	1.433	11.75	A + C
50.96	9.67	5.99	0.234	0.100	0.058	1.434	11.88	A
39.20	6.13	3.71	0.176	0.083	0.035	1.435	12.27	"
24.50	3.20	1.89	0.211	0.068	0.034	1.436	12.39	"
12.65	1.42	0.83	0.292	0.061	0.040	1.438	12.65	"
9.80	0.99	0.62	0.304	0.059	0.041	1.440	12.88	"
9.31	0.98	0.59	0.328	0.058	0.044	1.443	13.05	"
7.35	0.77	0.46	0.332	0.057	0.043	1.443	13.27	"
6.37	0.58	0.39	0.339	0.057	0.044	1.444	13.54	"
5.90	0.57	0.36	0.351	0.056	0.045	1.444	13.70	"
1.96	0.18	0.12	0.374	0.055	0.046	1.445	13.70	"

[a]These values were calculated by the compiler.

[b]The solid phases are: A = KBO_2; B = K_2HPO_4; C = $K_2HPO_4 \cdot 3H_2O$.

COMPONENTS:	ORIGINAL MEASUREMENTS:
(1) Dipotassium hydrogenphosphate; K_2HPO_4; [7758-11-4]	Velikanova, L.V.; Bergman, A.G.
(2) Potassium carbonate; K_2CO_3; [584-08-7]	*Izv. Vysch. Ucheb. Zaved., Khim. Khim.*
(3) Urea; CH_4N_2O; [57-13-6]	*Tekhnol.* 1974, *17*, 7-10 and 1513-6.
(4) Water; H_2O; [7732-18-5]	

VARIABLES:	PREPARED BY:
Temperature and composition.	J. Eysseltová

EXPERIMENTAL VALUES:

Monovariant points in the K_2HPO_4-K_2CO_3-$CO(NH_2)_2$-H_2O system.

$t/°C.$	K_2CO_3 mass%	K_2CO_3 mol/kg[a]	K_2HPO_4 mass%	K_2HPO_4 mol/kg[a]	$CO(NH_2)_2$ mass%	$CO(NH_2)_2$ mol/kg[a]	H_2O mass%	solid phase[b]
Section I: (25% K_2CO_3 + 75% K_2HPO_4)-$CO(NH_2)_2$-H_2O								
-22	7.25	0.89	21.75	2.13	12.4	3.52	58.6	ice + A + G
-10.5	8.75	1.17	26.25	2.80	11.3	3.50	53.7	A + G + H
3.7	12.05	1.95	36.15	4.65	7.2	2.68	44.6	A + B + H
34	13.65	2.19	30.95	3.94	10.4	3.84	35	B + H + I
50.6	15	3.93	45	9.36	12.4	7.48	27.6	B + C + I
54	14.7	3.92	43.9	9.29	14.3	8.78	27.1	C + I + J
-16.9	9.25	1.06	27.75	2.52	0	0	63	ice + A
5	13.58	2.14	40.72	5.11	0	0	45.7	A + B
57.5	16.82	3.72	50.48	8.86	0	0	32.7	B + C
Section II: (50% K_2CO_3 + 50% K_2HPO_4)-$CO(NH_2)_2$-H_2O								
-35	21	2.86	21	2.27	5	1.57	53	ice + B + G
-6[c]	22.2	3.29	22.2	2.61	7.2	2.46	48.5	B + G + H
37	50		50		13.7		33.8	B + D + H
39.6	26.3	5.74	26.3	4.56	14.3	7.19	33.1	D + H + I
61	25.4	7.23	25.4	5.74	23.8	15.60	25.4	D + I + J
-31.5	21.75	2.78	21.75	2.20	0	0	56.5	ice + B
51.5	31.5	6.15	31.5	4.88	0	0	37	B + D
Section III: (75% K_2CO_3 + 25% K_2HPO_4)-$CO(NH_2)_2$-H_2O								
-37.5	30.75	4.09	10.25	1.08	4.7	1.44	55	ice + B + G
-12.5	31.05	4.38	10.35	1.16	7.4	2.40	51.2	B + G + H
25.2	38.1	7.33	12.7	1.93	11.6	5.13	37.6	B + E + H

(continued next page)

AUXILIARY INFORMATION

METHOD/APPARATUS/PROCEDURE:	SOURCE AND PURITY OF MATERIALS:
A visual polythermic method was used. The nature of the solid phases was checked by microphotographical techniques.	No information is given.
	ESTIMATED ERROR:
	No information is given.
	REFERENCES:

COMPONENTS:	ORIGINAL MEASUREMENTS:
(1) Dipotassium hydrogenphosphate; K_2HPO_4; [7758-11-4] (2) Potassium carbonate; K_2CO_3; [584-08-7] (3) Urea; CH_4N_2O; [57-13-6] (4) Water; H_2O; [7732-18-5]	Velikanova, L.V.; Bergman, A.G. *Izv. Vysch. Ucheb. Zaved., Khim. Khim. Tekhnol. Russ.* <u>1974</u>, *17*, 7-10 and 1513-6.

EXPERIMENTAL VALUES cont'd:

Monovariant points in the K_2HPO_4-K_2CO_3-$CO(NH_2)_2$-H_2O system.

	K_2CO_3		K_2HPO_4		$CO(NH_2)_2$		H_2O	
$t/°C.$	mass%	mol/kg[a]	mass%	mol/kg[a]	mass%	mol/kg[a]	mass%	solid phase[b]
37.5	37.5	8.00	12.5	2.11	16.1	7.90	33.9	E + H + I
56.5	35.4	9.59	11.8	2.53	26.1	16.27	26.7	E + I + J
69.7	31.27	10.77	10.43	2.85	37.3	29.57	21	D + E + J
-35.5	32.55	4.16	10.85	1.10	0	0	56.6	ice + B
-29.2	42.6	7.13	14.2	1.88	0	0	43.2	B + E
86.5	48.75	10.07	16.25	2.66	0	0	35	D + E

Section IV: (45% K_2CO_3 + 55% K_2HPO_4)-$CO(NH_2)_2$-H_2O

-27	15.03	1.88	18.37	1.83	9	2.60	57.6	ice + A + G
-8.6	18.54	2.67	22.66	2.59	8.6	2.86	50.2	A + G + H
-7.5	19.17	2.76	23.43	2.68	7.3	2.42	50.2	A + B + H
41.8	24.3	5.46	29.7	5.29	13.8	7.13	32.2	B + D + H
45.6	24.07	5.56	29.43	5.39	15.2	8.08	31.3	D + H + I
63.7	22.59	6.68	27.61	6.49	25.4	17.33	24.4	D + I + J
-21.6	17.73	2.11	21.67	2.05	0	0	60.6	ice + A
-6.6	21.78	3.05	26.62	2.96	0	0	51.6	A + B
55.8	29.30	6.07	35.80	5.88	0	0	34.9	B + D

Section V: (85% K_2CO_3 + 15% K_2HPO_4)-$CO(NH_2)_2$-H_2O

-40	33.68	4.30	6.12	0.62	3.6	1.05	55.6	ice + B + G
-12.5	37.65	5.50	6.65	0.77	6.2	2.08	49.5	B + G + H
14.3	41.48	7.26	7.32	1.01	9.9	3.99	41.3	B + E + H
30	40.3	7.75	7.1	1.08	15	6.64	37.6	E + H + I
51.4	36.8	8.87	6.5	1.24	26.7	14.81	30	E + I + J
68.7	28.73	9.66	5.07	1.35	44.7	34.61	21.5	D + E + J
-36.5	36.98	4.73	6.52	0.66	0	0	56.5	ice + B
16.6	46.16	7.30	8.14	1.02	0	0	45.7	B + E
106.5	56.25	12.20	9.98	1.70	0	0	33.5	D + E

Section VI: (90% K_2CO_3 + 10% K_2HPO_4)-$CO(NH_2)_2$-H_2O

-40.7	36.54	4.77	4.06	0.42	4	1.20	55.4	ice + F + G
-22	40.77	5.89	4.53	0.52	4.7	1.56	50	B + F + G
7	45	7.75	5	0.68	8	3.17	42	B + E + G
15.5	44.37	7.98	4.93	0.70	10.5	4.34	40.2	E + G + H
40.2	41.13	8.65	4.57	0.76	19.9	9.63	34.4	E + H + I
58.7	34.83	9.36	3.87	0.82	34.4	21.29	26.9	E + I + J
-37	38.7	4.91	4.3	0.43	0	0	57	ice + B
-13.5	44.3	6.30	4.92	0.55	0	0	50.8	B + F
10.5	47.8	7.37	5.3	0.64	0	0	46.9	E + F
116.5	60.0	13.04	6.7	1.14	0	0	33.3	D + E

the quaternary eutectic point

-41.5	34.32	4.14	3.22	0.31	2.52	0.62	59.92	ice + A + F + G

The relative areas of crystallization of the individual phases are:

Section	ice	A	B	C	D	E	F	G	H	I	J
I	13.64	3.46	2.81	29.32	0	0	0	3.88	12.22	10.82	23.85
II	14.47	0	4	0	31.3	0	0	5.82	12.33	9.82	22.86
III	13.99	0	2.61	0	33.9	2.73	0	4.36	13	10.63	18.74
IV	13.81	1.83	3.79	0	32.35	0	0	5.01	11.69	12.33	19.19
V	14.13	0	1.43	0	36.21	0	0	5.89	10.56	10.43	16.8
VI	14.79	0	0.66	0	32.78	6.36	0.69	8.4	10.3	10.13	15.89

[a] The mol/kg H_2O values have been calculated by the compiler.

[b] The solid phases are: A = $K_2HPO_4 \cdot 6H_2O$; B = $K_2HPO_4 \cdot 3H_2O$; C = K_2HPO_4; D = K_2CO_3; E = $2K_2CO_3 \cdot 3H_2O$; F = $K_2CO_3 \cdot 6H_2O$; G = α-urea; H = β-urea; I = γ-urea; J = δ-urea.

[c] An obvious error - compiler.

COMPONENTS:	ORIGINAL MEASUREMENTS:

(1) Dipotassium hydrogenphosphate; K_2HPO_4;
    [7758-11-4]
(2) Ammonium dihydrogenphosphate; $NH_4H_2PO_4$;
    [7722-76-1]
(3) Diammonium hydrogenphosphate;
    $(NH_4)_2HPO_4$; [7783-28-0]
(4) Water; H_2O; [7732-18-5]
	Torochestnikov, N.S.; Rodionova, T.M.;
	  Kirsanova, L.D.
	
	VINITI <u>1979</u>, 2909, 17 p.

VARIABLES:	PREPARED BY:
Temperature and amount of K_2HPO_4 in solutions with a ratio of $NH_4H_2PO_4/(NH_4)_2$ HPO_4 = 2.34.	J. Eysseltová

EXPERIMENTAL VALUES:

Part 1. Solubility polytherm along the sections of the $NH_4H_2PO_4-(NH_4)_2HPO_4-K_2HPO_4-H_2O$
         system.

$t/°C.^a$	composition (g of each component)					mass% ammonium phosphates
	$NH_4H_2PO_4$	$(NH_4)_2HPO_4$	$K_2HPO_4 \cdot 2H_2O^b$	H_2O	A_i^c	
-7	----	----	12.18	25	25	-----
-6.9	1.4289	0.6100	12.18	25	25	5.2
-7.2	1.8607	0.7950	12.18	25	25	6.67
-7.7	2.3420	1.0013	12.18	25	25	8.25
-9	3.3088	1.4530	12.18	25	25	14.53
-10	(no data given)		12.18	25	25	18.1
-9	6.1565	2.6310	12.4626	25	25	19.0
-6.5	6.6565	2.8446	12.4626	25	25	20.23
-2.5	7.2679	3.1059	12.4626	25	25	21.68
-2	6.1219	2.6074	9.9422	20	25	22.57
4.5	7.1316	3.1039	9.9422	20	25	25.36
10.5	7.8846	3.3694	9.9422	20	25	27.32
9	8.0187	3.4293	9.9422	20	25	27.66
17	9.0374	3.8625	9.9422	20	25	30.11
2	----	----	22.4178	15	45	-----
-2	1	0.4274	22.4178	15	45	3.67
-5	2	0.8548	22.4178	15	45	7.09
-8	3	1.2822	22.4178	15	45	10.27
-11	4	1.7096	22.4178	15	45	13.24
-13	4.5	1.9233	22.4178	15	45	14.65
6	5	2.1370	22.4178	15	45	16.02
10	5.5	2.3507	22.4178	15	45	17.34

(continued next page)

AUXILIARY INFORMATION

METHOD/APPARATUS/PROCEDURE:	SOURCE AND PURITY OF MATERIALS:
A visual polythermic method was used in the temperature range of -20 to +20°C. The disappearance of the last crystal was observed. The mixtures were prepared by weight and heated, while being stirred, at a rate of 0.5 deg/min. The analyses have been described elsewhere (1).	Chemically pure salts were recrystallized, washed with ethanol, and dried below 60°C.

	ESTIMATED ERROR:
	The precision of the weighing was 0.005 g. No other information is given.

	REFERENCES:

	1. Vinnik, M.M.; Erbanova, L.N., et al.
	   Metody Analiza Fosfatnogo Syrja,
	   Moscow, <u>1975</u>, p. 215.

COMPONENTS:	ORIGINAL MEASUREMENTS:
(1) Dipotassium hydrogenphosphate; K_2HPO_4; [7758-11-4]	Torochestnikov, N.S.; Rodionova, T.M.; Korsanova, L.D.
(2) Ammonium dihydrogenphosphate; $NH_4H_2PO_4$; [7722-76-1]	*VINITI* <u>1979</u>, 2909, 17 p.
(3) Diammonium dydrogenphosphate; $(NH_4)_2HPO_4$; [7783-28-0]	
(4) Water; H_2O; [7732-18-5]	

EXPERIMENTAL VALUES cont'd:

Part 1. Solubility polytherm along the sections of the $NH_4H_2PO_4-(NH_4)_2HPO_4-K_2HPO_4-H_2O$ system.

	composition (g of each component)					mass% ammonium phosphates
$t/°C.^a$	$NH_4H_2PO_4$	$(NH_4)_2HPO_4$	$K_2HPO_4\cdot 2H_2O^b$	H_2O	A_i^c	
8	----	----	29.8211	15	50	-----
4	1	0.4274	19.8807	10	50	4.56
2	2	0.8548	29.8211	15	50	6.00
-0.5	3	1.2822	29.8211	15	50	8.72
-3	4	1.7096	29.8211	15	50	11.27
-3.5	4.5	1.9233	29.8211	15	50	12.50
-5	5	2.1370	29.8211	15	50	13.73
7	5.5	2.3507	29.8211	15	50	14.90
13.5	6	2.5644	29.8211	15	50	16.04
13	----	----	27.2952	10	55	-----
10.5	1	0.4274	27.2952	10	55	3.70
9	2	0.8548	27.2952	10	55	7.10
4.5	3	1.2822	27.2952	10	55	10.3
8	3.5	1.4970	27.2952	10	55	11.8
12	4	1.7096	27.2952	10	55	13.3
3	----	----	24.5162	15	46.4	-----
-8.5	4.0533	1.7322	24.5162	15	46.4	12.77
-10	4.6573	1.9903	24.5162	15	46.4	14.4
11.5	5.6232	2.4031	24.5162	15	46.4	16.85
12	----	----	24.3237	10	53	-----
1.7	2.8571	1.2210	24.3237	10	53	10.62
0	3.3453	1.4296	24.3237	10	53	12.21
10	2.4576	1.0503	34.5454	10	58	7.3

[a] The temperature at which the last crystal disappeared.

[b] This is probably a typographical error. The dihydrate is not mentioned anywhere in the text and on the basis of the compiler's recalculation the starting material appears to be the trihydrate.

[c] This is a constant, near to but not identical with the mass% of the binary solution of K_2HPO_4 lying on the section studied.

Part 2. The compiler has recalculated the data in Part 1 assuming that the starting dipotassium hydrogenphosphate is $K_2HPO_4\cdot 3H_2O$. The recalculated values are given below.

K_2HPO_4		$NH_4H_2PO_4$		$(NH_4)_2HPO_4$		H_2O	
mass%	mol/kg	mass%	mol/kg	mass%	mol/kg	mass%	$t/°C.$
25.0	1.9	----	----	----	----	75.0	-7
23.7	1.9	3.64	0.4	1.55	0.2	71.1	-6.9
23.3	1.9	4.67	0.6	2.00	0.2	70.0	-7.2
22.9	1.9	5.77	0.7	2.47	0.3	68.8	-7.7
22.2	1.9	7.88	1.0	3.46	0.4	66.5	-9
20.6	2.0	13.3	1.9	5.68	0.7	60.4	-9
20.2	2.0	14.2	2.1	6.05	0.8	59.5	-6.5
19.9	2.0	15.2	2.3	6.49	0.8	58.4	-2.5
19.6	1.9	15.8	2.4	6.74	0.9	57.8	-2
18.9	1.9	17.8	2.8	7.72	1.1	55.6	4.5
18.4	1.9	18.1	3.1	8.17	1.1	54.3	10.5
18.3	1.9	19.4	3.1	8.28	1.2	54.0	9
17.7	1.9	21.1	3.5	9.01	1.3	52.2	17

(continued next page)

COMPONENTS:	ORIGINAL MEASUREMENTS:
(1) Dipotassium hydrogenphosphate; K_2HPO_4; [7758-11-4] (2) Ammonium dihydrogenphosphate; $NH_4H_2PO_4$; [7722-76-1] (3) Diammonium hydrogenphosphate; $(NH_4)_2HPO_4$; [7783-28-0] (4) Water; H_2O; [7732-18-5]	Torochestnikov, N.S.; Rodionova, T.M.; Kirsanova, L.D. *VINITI* 1979, 2909, 17 p.

EXPERIMENTAL VALUES cont'd:

Part 2. The compiler has recalculated the data in Part 1 assuming that the starting dipotassium hydrogenphosphate is $K_2HPO_4 \cdot 3H_2O$. The recalculated values are given below.

K_2HPO_4		$NH_4H_2PO_4$		$(NH_4)_2HPO_4$		H_2O	
mass%	mol/kg	mass%	mol/kg	mass%	mol/kg	mass%	t/°C.
45.7	4.8	----	----	----	----	54.3	2
44.0	4.8	2.57	0.4	1.10	0.2	52.3	-2
42.5	4.8	4.96	0.8	2.12	0.3	50.4	-5
41.0	4.8	7.19	1.3	3.07	0.5	48.7	-8
39.7	4.8	9.27	1.7	3.96	0.6	47.1	-11
39.0	4.8	10.3	1.9	4.38	0.7	46.3	-13
38.4	4.8	11.2	2.1	4.79	0.8	45.6	6
37.8	4.8	12.1	2.4	5.19	0.9	44.9	10
50.8	5.9	----	----	----	----	49.2	8
48.5	5.9	3.19	0.6	1.36	0.2	47.0	4
47.7	5.9	4.19	0.8	1.79	0.3	46.3	2
46.3	5.9	6.10	1.2	2.61	0.4	44.9	-0.5
45.0	5.9	7.91	1.6	3.38	0.6	43.7	-3
44.4	5.9	8.78	1.8	3.75	0.7	43.1	-3.5
43.8	5.9	9.62	2.0	4.11	0.7	42.5	-5
43.2	5.9	10.4	2.2	4.46	0.8	41.9	7
42.6	5.9	11.2	2.4	4.80	0.9	41.3	13.5
55.8	7.3	----	----	----	----	44.2	13
53.8	7.3	2.58	0.5	1.10	0.2	42.5	10.5
51.9	7.3	4.98	1.1	2.12	0.4	41.0	9
50.1	7.3	7.21	1.6	3.08	0.6	39.6	4.5
49.2	7.3	8.27	1.8	3.53	0.7	38.9	8
48.4	7.3	9.30	2.1	3.97	0.8	38.3	12
47.3	5.2	----	----	----	----	52.7	3
41.3	5.2	8.94	1.7	3.82	0.6	45.9	-8.5
40.5	5.2	10.1	1.9	4.31	0.7	45.1	-10
39.3	5.2	11.8	2.3	5.05	0.9	43.8	11.5
54.1	6.8	----	----	----	----	45.9	12
48.3	6.8	7.44	1.6	3.17	0.6	41.0	1.7
47.5	6.8	8.55	1.8	3.65	0.7	40.3	0
54.9	8.3	5.11	1.2	2.18	0.4	37.8	10

Part 3. Graphically derived solubility isotherms in the $NH_4H_2PO_4$-$(NH_4)_2HPO_4$-K_2HPO_4-H_2O system.

authors' data[a]		K_2HPO_4		$NH_4H_2PO_4$		$(NH_4)_2HPO_4$		H_2O
K_2HPO_4	ammonium phosphates			compiler's recalculations				
mass%	mass%	mass%	mol/kg	mass%	mol/kg	mass%	mol/kg	mass%
				temp. = -10°C.				
0	21	0	0	14.7	1.6	6.28	0.6	79.0
45	12.25	39.5	4.7	8.58	1.5	3.66	0.6	48.3
46.4	14.4	39.7	5.0	10.1	1.9	4.31	0.7	45.9
45	14.85	38.3	4.7	10.4	1.9	4.44	0.7	46.9
25	18.3	20.4	1.9	12.8	1.8	5.47	0.7	62.3
36.5	0	36.5	3.3	0	0	0	0	63.5

(continued next page)

COMPONENTS:	ORIGINAL MEASUREMENTS:
(1) Dipotassium hydrogenphosphate: K_2HPO_4; [7758-11-4] (2) Ammonium dihydrogenphosphate; $NH_4H_2PO_4$; [7722-76-1] (3) Diammonium hydrogenphosphate; $(NH_4)_2HPO_4$; [7783-28-0] (4) Water; H_2O; [7732-18-5]	Torochestnikov, N.S.; Rodionova, T.M.; Kirsanova, L.D. *VINITI* <u>1979</u>, 2909, 17 p.

EXPERIMENTAL VALUES cont'd:

Part 3. Graphically derived solubility isotherms in the $NH_4H_2PO_4-(NH_4)_2HPO_4-K_2HPO_4-H_2O$ system.

authors' data[a]   compiler's recalculations

K_2HPO_4	ammonium phosphates	K_2HPO_4		$NH_4H_2PO_4$		$(NH_4)_2HPO_4$		H_2O
mass%	mass%	mass%	mol/kg	mass%	mol/kg	mass%	mol/kg	mass%

temp. = -5°C.

0	24.4	0	0	17.1	2.0	7.30	0.7	75.6
45	7.1	41.8	4.7	4.97	0.8	2.12	0.3	51.1
46.4	8.85	42.3	5.0	6.20	1.1	2.64	0.4	48.9
50.0	13.7	43.1	5.7	9.59	1.9	4.10	0.7	43.2
45.0	15.2	38.2	4.7	10.6	2.0	4.55	0.7	46.6
25.0	20.9	19.8	1.9	14.6	2.1	6.25	0.8	59.4
39.75	0	39.75	3.8	0	0	0	0	60.25

temp. = 0°C.

0	26.9	0	0	18.8	2.2	8.05	0.8	73.1
45.0	1.85	44.2	4.7	1.29	0.2	0.55	0.07	54.0
46.4	3.3	44.9	5.0	2.31	0.4	0.98	0.1	51.8
50.0	8.55	45.7	5.7	5.99	1.1	2.55	0.4	45.8
53.0	12.25	46.5	6.5	8.58	1.8	3.66	0.7	40.2
50.0	14.2	42.9	5.7	9.94	2.0	4.25	0.8	43.0
45.0	15.5	38.0	4.7	10.9	2.0	4.64	0.8	45.5
25.0	23.3	19.2	1.9	16.3	2.5	6.97	0.9	57.5
43.2	0	43.2	4.4	0	0	0	0	56.8

temp. = 10°C.

0	31.65	0	0	22.2	2.8	9.47	1.0	68.35
53.0	2.05	51.9	6.5	1.43	0.3	0.61	0.1	38.1
55.0	4.4	52.6	7.0	3.08	0.6	1.31	0.2	43.0
58.0	7.3	53.8	7.9	5.11	1.1	2.18	0.4	38.9
55.0	12.5	48.1	7.0	8.75	1.9	3.74	0.7	39.4
50.0	15.4	42.3	5.7	10.8	2.2	4.61	0.8	42.3
45.0	17.35	37.2	4.7	12.2	2.3	5.2	0.9	45.4
25.0	27.4	18.2	1.9	19.2	3.1	8.2	1.1	58.4
52.0	0	52.0	6.2	0	0	0	0	48.0

The authors state that the equilibrium solid phases are $NH_4H_2PO_4$, KH_2PO_4 and an unspecified double salt. There is no mention of the degree of hydration.

[a]Concerning the mass% of K_2HPO_4, see footnote *b* under Part 1.

Part 4. Solubility in the $NH_4H_2PO_4-(NH_4)_2HPO_4-K_2HPO_4-H_2O$ system at 0°C.

Nr	$NH_4H_2PO_4$		$(NH_4)_2HPO_4$		K_2HPO_4		H_2O
	mass%	mol/kg[a]	mass%	mol/kg[a]	mass%	mol/kg[a]	mass%
1	20.0	3.57	31.4	4.89	----	----	48.6
2	23.2	4.18	25.0	3.92	3.6	0.42	48.2
3	10.7	2.13	10.4	1.81	35.4	4.67	43.6
4	7.5	1.68	7.2	1.40	46.5	6.88	32.9
5	6.8	1.53	7.6	1.49	47.0	6.99	38.6
6	15.6	2.82	23.6	3.72	12.8	1.53	48.0
7	2.8	0.64	2.15	0.43	57.3	8.71	37.7

[a]The mol/kg H_2O values were calculated by the compiler.

COMPONENTS:	EVALUATOR:
(1) Tripotassium phosphate; K_3PO_4; [7778-53-2] (2) Water; H_2O; [7732-18-5]	J. Eysseltová Charles University Prague, Czechoslovakia May 1985

CRITICAL EVALUATION:

THE BINARY SYSTEM

The situation with this system is similar to that for the K_2HPO_4-H_2O system. There are insufficient data to use the solubility equation described in the section on NaH_2PO_4 (chap. 3). Solubility measurements were made by Ravich (1). However, there are only a few additional data: four experimental values in ref (2), two in ref (3), and one in each of two other papers (4,5). All these other values are 1-10% lower than those of Ravich (1). Therefore, no values can be recommended for the solubility of tripotassium phosphate in water.

There is also uncertainty with respect to the degree of hydration of the tripotassium phosphate. Ravich (1) reported the existence of a stable heptahydrate and trihydrate and a metastable enneahydrate. However, it is possible that there is some error in his assignment of stability and metastability to the eutonic solutions. Some authors (2,6) also report the existence of an octahydrate as the stable phase at room temperature, but neither Ravich (7,8) nor Berg (9-11) observed an octahydrate in their detailed studies of the K_2O-P_2O_5-H_2O system. Therefore, the evaluator concludes that the existence of the octahydrate has not been established.

MULTICOMPONENT SYSTEMS

Several ternary and one quaternary systems have been studied but there are insufficient solubility values to enable any to be recommended.

1. The K_3PO_4-NH_3-H_2O system. A miscibility gap was found in this system (2).

2. The K_3PO_4-KBO_2-H_2O system. Solubility measurements were made for this system at 298 K (6). The method of analysis for phosphate used in this study was incorrect, giving values that were in error by +30-80%.

3. The K_3PO_4-K_2SO_4-H_2O system. This system was studied at 343 K (4) and the existence of the compound $K_2SO_4 \cdot K_3PO_4 \cdot 9H_2O$ was reported.

4. The K_3PO_4-KNO_2-H_2O system. Solubility values were measured at 298 K (5). Neither new compounds, e.g., $K_3PO_4 \cdot KNO_2$, nor solid solutions are present in this system.

5. The K_3PO_4-K_2SO_4-KVO_3-H_2O system. A study was made of this system at 308 and 333 K (3). In addition to the components and their hydrates, the following were reported as equilibrium solid phases:

   (i) $4K_2O \cdot P_2O_5 \cdot V_2O_5 \cdot 30H_2O$;   (ii) $4K_2O \cdot P_2O_5 \cdot V_2O_5 \cdot 24H_2O$;

   (iii) $4K_2O \cdot P_2O_5 \cdot V_2O_5 \cdot 22H_2O$;   (iv) $4K_2O \cdot P_2O_5 \cdot V_2O_5 \cdot 18H_2O$;

   (v) $5K_2O \cdot P_2O_5 \cdot 2SO_3 \cdot 30H_2O$; and (vi) $5K_2O \cdot P_2O_5 \cdot 2SO_3 \cdot 22H_2O$. The ratio K:P:S for (v) and

   (vi) is the same as that reported by others (4).

References

1. Ravich, M.I. *Izv. Akad. Nauk SSSR, Ser. Khim.* <u>1938</u>, 141.
2. Janecke, E. *Z. Physik. Chem.* <u>1927</u>, *127*, 71.
3. Gasanova, Kh.D. ; Abduragimova, R.A. *Ukr. Khim. Zh.* <u>1978</u>, *44*, 158.
4. Rustamov, K.A.; Rza-Zade, P.F.; Abduragimova, R.A. *Issled. Obl. Neorg. Fiz. Khim.* <u>1971</u>, 167.
5. Protsenko, P.I.; Ivleva, T.I.; Rubleva, V.V.; Berdyukova, V.A.; Edush, T.V. *Zh. Prikl. Khim. (Leningrad)* <u>1975</u>, *48*, 1055.
6. Beremzhanov, B.A.; Voronina, L.V.; Savich, R.F. *Khim. Khim. Tekhnol. (Alma Ata)* <u>1978</u>, 29.
7. Ravich, M.I. *Kaliy* 1936, *10*, 33.
8. Ravich, M.I. *Izv. Akad. Nauk SSSR* <u>1938</u>, 167.
9. Berg, A.G. *Izv. Akad. Nauk SSSR* <u>1933</u>, 167.
10. Berg, A.G. *Izv. Akad. Nauk SSSR* <u>1938</u>, 147.
11. Berg, A.G. *Izv. Akad. Nauk SSSR* <u>1938</u>, 161.

COMPONENTS:	ORIGINAL MEASUREMENTS:
(1) Tripotassium phosphate; K_3PO_4; [7778-53-2]	Jänecke, E.
(2) Water; H_2O; [7732-18-5]	Z. Phys. Chem. 1927, 127, 71-92.

VARIABLES:	PREPARED BY:
Temperature and composition.	J. Eysseltová

EXPERIMENTAL VALUES:

Crystallization temperatures and composition of saturated solutions existing in equilibrium with crystalline $K_3PO_4 \cdot 8H_2O$.

$t/°C.$	H_2O conc[b]	mass%	K_3PO_4[a] mass%	mol/kg
45.1	68	40.3	59.7	6.98
43.2	75	43.0	57.0	6.24
23.3	104	51.0	49.0	4.52
7.5	125	55.8	44.2	3.73

[a] These values were calculated by the compiler.

[b] The concentration unit is: g/100 g K_3PO_4.

AUXILIARY INFORMATION

METHOD/APPARATUS/PROCEDURE:	SOURCE AND PURITY OF MATERIALS:
The salt was added to the water and the mixture was heated until total liquefaction occurred. A cooling curve of the mixture was then measured. The methods of analysis are not described.	Merck pure K_3PO_4 was used and was further purified by dissolving the salt in water and passing NH_3 through the solution for 2-3 hours. The octahydrate precipitated from the solution.

Analysis:

	found	calculated
H_2O	40.00%	40.58%
P_2O_5	19.10%	19.95%
K_2O	40.14%	39.47%

ESTIMATED ERROR:

No information is given.

REFERENCES:

COMPONENTS:	ORIGINAL MEASUREMENTS:
(1) Tripotassium phosphate; K_3PO_4; [7778-53-2] (2) Water; H_2O; [7732-18-5]	Ravich, M.I. *Izv. AN SSSR. ser. Khim.* <u>1938</u>, 141-6.

VARIABLES:	PREPARED BY:
Temperature and composition.	J. Eysseltová

EXPERIMENTAL VALUES:

Compositions and crystallization temperatures in the K_3PO_4-H_2O system.

$t/°C.$	mass%	K_3PO_4 mol%	mol/kg^a	solid phase^b	$t/°C.$	mass%	K_3PO_4 mol%	mol/kg^a	solid phase^b
-1.18	4.54	0.40	0.22	ice	42.6	59.46	11.06	6.90	B
-2.60	9.75	0.91	0.50	"	44.5	60.84	11.64	7.31	"
-4.6	15.43	1.52	0.85	"	45.4	61.94	12.13	7.66	"
-7.7	21.74	2.30	1.30	"	45.6	62.51	12.39	7.85	"
-12.0	27.34	3.09	1.77	"	45.6	63.12	12.68	8.06	"
-15.8	31.53	3.76	2.16	"	45.4				B + C^c
-20.0	35.12	4.39	2.54	"	25	63.17	12.70	8.07	C^c
-24.0	38.33	5.00	2.92	A + ice	30	63.19	12.71	8.08	"
-28.2	40.25	5.40	3.17	B^c+ ice	35	63.33	12.77	8.13	"
-8.8	42.92	6.00	3.54	B	40	63.41	12.81	8.16	"
0	44.26	6.31	3.74	"	45	63.56	12.89	8.21	"
10	46.83	6.95	4.14	"	50	63.80	13.00	8.30	C
20	49.62	7.71	4.63	"	60	64.08	13.14	8.40	"
25	51.42	8.23	4.98	"	-7.7	43.85	6.21	3.67	A^c
30	53.08	8.75	5.32	"	0	47.62	7.16	4.28	"
35	55.43	9.54	5.85	"	5.0	49.80	7.76	4.67	"
40	57.51	10.30	6.37	"	8.8	52.23	8.43	5.15	"
					12.3	57.72	10.00	6.43	"

^{a}The mol/kg H_2O values were calculated by the compiler.

^{b}The solid phases are: A = $K_3PO_4 \cdot 9H_2O$; B = $K_3PO_4 \cdot 7H_2O$; C = $K_3PO_4 \cdot 3H_2O$.

^{c}Metastable equilibrium.

AUXILIARY INFORMATION

METHOD/APPARATUS/PROCEDURE:	SOURCE AND PURITY OF MATERIALS:
The isothermal method was used. The solubility was determined by evaporating the saturated solutions and drying to constant weight. Cooling curves were determined for some of the mixtures.	The material used is reported as having been submitted by Berg. The compiler assumes the material is the same as that used in (1).
	ESTIMATED ERROR: No information is given.
	REFERENCES: 1. Berg. L.G. *Izv. AN SSSR. ser. Khim.* <u>1938</u>, 150.

COMPONENTS:	ORIGINAL MEASUREMENTS:
(1) Tripotassium phosphate; K_3PO_4; [7778-53-2]	Jänecke, E.
(2) Ammonia; NH_3: [7664-41-7]	Z. Phys. Chem. 1927, 127, 71-92.
(3) Water; H_2O; [7732-18-5]	

VARIABLES:	PREPARED BY:
Composition and temperature.	J. Eysseltová

EXPERIMENTAL VALUES:

A miscibility gap was found in the liquid phase of the K_3PO_4-NH_3-H_2O system.

Part 1. The isothermal binodal curve of the miscibility gap at 0°C.

H_2O conc.[a]	NH_3 conc.[a]	K_3PO_4[b] mass%	K_3PO_4[b] mol/kg	NH_3[b] mass%	NH_3[b] mol/kg	H_2O[b] mass%
238	13.2	25.68	0.51	3.90	0.96	70.41
258	15.9	23.49	0.43	4.44	1.01	72.07
257	20.9	22.16	0.41	5.85	1.34	71.99
276	27.6	19.26	0.33	7.34	1.56	73.40
294	36.6	16.09	0.26	9.29	1.86	74.62
307	41.6	14.35	0.22	10.22	1.96	75.43
320	47.0	12.62	0.18	11.19	2.06	76.19
329	53.9	10.74	0.15	12.56	2.25	76.69
349	74.2	5.75	0.08	16.52	2.78	77.73
285	90.6	2.44	0.04	23.53	4.86	74.02
235	91.6	2.51	0.05	27.34	6.84	70.15

[a] The concentration unit is: g/100 g of $(K_3PO_4 + NH_3)$.

[b] These values were calculated by the compiler.

AUXILIARY INFORMATION

METHOD/APPARATUS/PROCEDURE:	SOURCE AND PURITY OF MATERIALS:
No information is given.	No information is given.
	ESTIMATED ERROR:
	No information is given.
	REFERENCES:

COMPONENTS:	ORIGINAL MEASUREMENTS:
(1) Tripotassium phosphate; K_3PO_4; [7778-53-2]	Jänecke, E.
(2) Ammonia; NH_3; [7664-41-7]	Z. Phys. Chem. <u>1927</u>, *127*, 71-92.
(3) Water; H_2O; [7732-18-5]	

EXPERIMENTAL VALUES cont'd:

Part 2. Composition of solutions existing in equilibrium with $K_3PO_4 \cdot 8H_2O$.

		K_3PO_4			NH_3			H_2O	
t/°C.	layer	mass%	conc[a]	mol/kg[b]	mass%	conc[a]	mol/kg[b]	mass%	conc[a]
0	upper	3.2	13.85	0.04	19.9	86.15	3.52	76.9	332.6
	lower	39.1	95.7	1.27	1.8	4.3	0.71	59.1	144.7
15	upper	2.5	9.0	0.05	25.5	91.0	5.83	72.0	257.0
	lower	45.5	94.8	1.99	2.5	5.2	1.36	52.0	108.0
25	upper	4.2	14.6	0.09	24.5	85.4	6.52	71.3	232.0
	lower	50.0	97.3	2.47	1.4	2.7	0.86	48.6	95.0

[a] The concentration unit is: g/100 g of $(K_3PO_4 + NH_3)$.

[b] The mol/kg H_2O values were calculated by the compiler.

Part 3. Temperatures of the miscibility gap in some solutions of the K_3PO_4-NH_3-H_2O system.

	K_3PO_4			NH_3			H_2O		
gram	mass%[a]	mol/kg[a]	gram	mass%[a]	mol/kg[a]	gram	mass%[a]	t/°C.[b]
93.55	35.57	1.03	6.45	2.45	0.88	183	61.98	0
89.5	29.98	0.71	10.5	3.52	1.04	198.5	66.50	0
84.3	26.55	0.58	15.7	4.94	1.34	217.5	68.50	0.3
78.5	23.21	0.46	21.5	6.36	1.57	238.2	70.43	6.9
66.5	19.76	0.39	33.2	9.82	2.42	238.1	70.42	46.05
56.3	15.50	0.28	43.5	11.94	2.66	264.4	72.56	45.75
42.8	11.08	0.18	56.1	14.16	2.81	296.2	74.76	39.6
25.0	5.77	0.08	75.0	17.32	3.06	333	76.90	14.2
93.6	37.71	1.20	6.4	2.58	1.02	148.2	59.71	37.7
91.0	37.07	1.20	9.0	3.67	1.48	145.5	59.27	57.2
85.0	32.32	0.93	15.0	5.70	2.06	163	61.98	70.2
78.5	27.84	0.72	21.5	7.62	2.46	182	64.54	69.4
71.1	23.66	0.56	28.9	9.62	2.82	200.5	66.72	52.4
60.5	18.85	0.40	39.5	12.30	3.28	221	68.85	44.6
50.8	14.74	0.28	49.2	14.28	3.43	244.6	70.98	35.95
29.2	7.10	0.11	70.8	17.20	3.25	311.5	75.70	13

[a] These values were calculated by the compiler.

[b] When the temperature is raised, this is the temperature at which two layers are first observed.

COMPONENTS:	ORIGINAL MEASUREMENTS:
(1) Tripotassium phosphate; K_3PO_4; [7778-53-2] (2) Dipotassium sulfate; K_2SO_4; [10233-01-9] (3) Water; H_2O; [7732-18-5]	Rustamov, K.A.; Rza-Zade, P.F.; Abduragimova, R.A. *Issled. Obl. Neorg. Fiz. Khim.* 1971, 167-9. (Proceedings of the Institute of Inorganic and Physical Chemistry, Academy of Sciences of the Adzerbeidzhan SSR)
VARIABLES:	PREPARED BY:
Composition at 70°C.	J. Eysseltová

EXPERIMENTAL VALUES:

Solubility isotherm for the K_3PO_4-K_2SO_4-H_2O system at 70°C.

P_2O_5 mass%	K_3PO_4 mass%	K_3PO_4 mol/kg[a]	SO_3 mass%	K_2SO_4 mass%	K_2SO_4 mol/kg[a]	H_2O mass%	solid phase[b]
----	----	----	7.5906	16.5095	1.13	83.4905	A
0.6889	2.0599	0.11	6.9682	15.1558	1.05	82.7843	"
1.3413	4.0042	0.22	6.1820	13.4459	0.93	82.5499	A + B
1.4378	4.2932	0.24	5.4722	11.9020	0.81	83.8048	B
2.0000	5.9718	0.33	5.5432	9.8815	0.67	84.1467	"
3.5883	10.7143	0.61	3.2616	7.0941	0.49	82.1916	"
4.6607	13.9164	0.81	2.7395	5.9584	0.42	80.1252	"
6.8621	20.4895	1.27	1.6339	3.5537	0.26	75.9568	"
8.1070	24.2020	1.56	1.3371	2.9081	0.22	72.8899	"
8.8433	26.4033	1.71	0.5414	1.1835	0.09	72.4112	"
10.7585	32.1239	2.29	0.8320	1.8098	0.15	66.0663	B + C
11.9900	35.8011	2.65	0.2833	0.6163	0.05	63.5826	C
13.7434	41.0365	3.30	0.1879	0.4087	0.04	58.5548	"
17.3498	51.8049	5.08	0.0926	0.2014	0.02	47.9937	"
21.4349	64.0026	8.37	------	------	----	35.9974	"

[a] The mol/kg H_2O values were calculated by the compiler.

[b] The solid phases are: A = K_2SO_4; B = $K_2SO_4 \cdot K_3PO_4 \cdot 9H_2O$; C = $K_3PO_4 \cdot 7H_2O$.

AUXILIARY INFORMATION

METHOD/APPARATUS/PROCEDURE:	SOURCE AND PURITY OF MATERIALS:
The potassium phosphate was added to saturated solutions of potassium sulfate and the mixtures were equilibrated in vessels of Mo-glass placed in a water thermostat. Equilibrium was checked by repeated experiments. The P_2O_5 and SO_3 contents were determined photocolorimetrically. The solid phases were analyzed only occasionally.	"Pure" and "chemically pure" K_2SO_4 and K_3PO_4 were used.
	ESTIMATED ERROR: No details are given.
	REFERENCES:

COMPONENTS:	ORIGINAL MEASUREMENTS:
(1) Tripotassium phosphate; K_3PO_4; [7778-53-2] (2) Potassium nitrite; KNO_2; [7758-09-0] (3) Water; H_2O; [7732-18-5]	Protsenko, P.I.; Ivleva, T.I.; Rubleva, V.V.; Berdyukova, V.A.; Edush, T.V. *Zh. Prikl. Khim.* (*Leningrad*) <u>1975</u>, *48*, 1055-9.

VARIABLES:	PREPARED BY:
Composition at 25°C.	J. Eysseltová

EXPERIMENTAL VALUES: Solubility in the K_3PO_4-KNO_2-H_2O system at 25°C.

K_3PO_4			KNO_2			H_2O	
mass%	concn[a]	mol/kg[b]	mass%	concn[a]	mol/kg[b]	mass%	solid phase
50.71	87.24	4.85	0.00	0.00	0.00	49.29	$K_3PO_4 \cdot 7H_2O$
49.50	86.41	4.80	1.93	8.40	0.47	48.57	"
43.50	82.28	4.57	11.68	55.10	3.06	44.82	"
43.39	83.58	4.64	12.59	60.48	3.36	44.02	"
40.65	80.86	4.49	16.72	82.92	4.61	42.63	"
38.70	80.62	4.48	20.60	107.02	5.96	40.70	"
36.39	83.15	4.62	24.67	134.00	7.44	38.94	"
33.43	76.39	4.24	29.46	167.87	9.33	37.11	"
31.19	73.57	4.09	32.87	193.41	10.75	35.94	$K_3PO_4 \cdot 7H_2O$ + KNO_2
31.09	73.23	4.07	32.90	192.77	10.71	36.01	"
31.08	73.28	4.07	32.96	194.11	10.78	35.96	"
31.08	73.34	4.07	33.00	194.43	10.80	35.92	"
28.62	69.07	3.84	36.25	218.23	12.12	35.13	KNO_2
24.93	62.00	3.44	40.99	254.37	14.13	34.08	"
20.37	52.69	2.93	46.85	303.97	16.89	32.78	"
16.32	44.70	2.48	51.20	333.68	18.54	32.48	"
12.63	35.03	1.95	56.81	393.15	21.84	30.56	"
8.10	24.26	1.35	63.55	474.09	26.34	28.35	"
4.98	15.81	0.88	68.32	541.08	30.06	26.70	"
-----	-----	----	75.92	666.81	37.05	24.08	"

[a] The concentration unit is: mol/1000 mol water.

[b] The mol/kg H_2O values were calculated by the compiler.

AUXILIARY INFORMATION

METHOD/APPARATUS/PROCEDURE:	SOURCE AND PURITY OF MATERIALS:
The isothermal method was used. Ten to twelve hours were allowed for equilibration. The nitrite ion content was determined by the iodometric back titration of excess permanganate. The phosphate ion content was determined gravimetrically as $Mg_2P_2O_7$.	The $K_3PO_4 \cdot 7H_2O$ was recrystallized. It had a purity of 99.56%. The KNO_2 was synthesized by the reaction of $Ba(NO_2)_2$ with K_2SO_4. It had a purity of 99.65%.

	ESTIMATED ERROR:
	The temperature was controlled to within ± 0.1°C. The compiler estimates the reproducibility of the solubility values to be about ± 0.3%.

	REFERENCES:

COMPONENTS:	ORIGINAL MEASUREMENTS:
(1) Tripotassium phosphate; K_3PO_4; [7778-53-2]	Beremzhanov, B.A.; Voronina, L.V.; Savich, R.F.
(2) Potassium borate; KBO_2; [13709-94-9]	*Khim. Khim. Tekhnol.* (Alma Ata) 1978, 29-36.
(3) Water; H_2O; [7732-18-5]	

VARIABLES:	PREPARED BY:
Composition at 25°C.	J. Eysseltová

EXPERIMENTAL VALUES: Solubility in the KBO_2-K_3PO_4-H_2O system at 25°C.

K_3PO_4			KBO_2			refr.		solid[b]
mass%[a]	mol%[a]	mol/kg[a]	mass%[a]	mol%[a]	mol/kg[a]	index	pH	phase[b]
----	----	-----	0.368	0.081	0.045	1.441	13.95	A
97.4	77.5	180	-----	-----	-----	1.450	13.80	B
87.6	37.38	33.4	0.054	0.054	0.054	1.445	13.90	C
79.2	24.29	18.0	0.061	0.046	0.036	1.445	13.81	"
72.0	17.53	12.1	0.063	0.035	0.028	1.445	13.48	"
68.4	15.45	10.2	0.117	0.067	0.046	1.444	13.21	"
61.4	11.66	7.5	0.126	0.062	0.040	1.445	12.97	"
57.0	9.77	6.3	0.173	0.079	0.050	1.443	12.88	"
56.4	8.08	6.1	0.176	0.086	0.050	1.441	12.88	A + B
54.6	9.06	5.7	0.171	0.072	0.047	1.440	12.70	A
53.4	8.86	5.4	0.164	0.070	0.044	1.435	12.65	"
47.8	7.07	4.3	0.159	0.061	0.038	1.434	12.50	"
23.4	2.89	1.4	0.187	0.052	0.030	1.430	12.46	"
20.4	2.00	1.2	0.211	0.055	0.033	1.425	12.41	"
10.2	0.79	0.5	0.298	0.069	0.041	1.420	12.35	"
6.6	0.57	0.3	0.328	0.076	0.044	1.410	12.22	"

[a] These values were calculated by the compiler.

[b] The solid phases are: A = KBO_2; B = K_3PO_4; C = $K_3PO_4 \cdot 8H_2O$.

AUXILIARY INFORMATION

METHOD/APPARATUS/PROCEDURE:	SOURCE AND PURITY OF MATERIALS:
The isothermal method was used but no further details are given.	No information is given.
	ESTIMATED ERROR:
	No information is given.
	REFERENCES:

COMPONENTS:	ORIGINAL MEASUREMENTS:
(1) Tripotassium phosphate; K_3PO_4; [7778-53-2]	Gasanova, KH.G.; Abduragimova, R.A.
(2) Dipotassium sulfate; K_2SO_4; [10233-01-9]	*Ukr. Khim. Zh.* <u>1978</u>, *44*, 158-63.
(3) Potassium vanadate; KVO_3; [13769-43-2]	
(4) Water; H_2O; [7732-18-5]	

VARIABLES:	PREPARED BY:
Composition at 35° and 60°C.	J. Eysseltová

EXPERIMENTAL VALUES:

Invariant points in the K_3PO_4-K_2SO_4-KVO_3-H_2O system.

K_3PO_4		KVO_3		K_2SO_4		H_2O	
mass%	mol/kg^a	mass%	mol/kg^a	mass%	mol/kg^a	mass%	solid phase^b
				temp. = 35°C.			
0.00	0.00	12.83	1.06	0.00	0.00	87.17	A
0.00	0.00	0.00	0.00	12.23	0.79	87.77	B
52.90	5.29	0.00	0.00	0.00	0.00	47.10	C
0.00	0.00	1.88	0.15	10.28	0.67	87.84	A + C
21.97	1.43	6.01	0.60	0.00	0.00	72.02	C + D
2.49	0.13	10.58	0.88	0.00	0.00	86.93	A + D
4.89	0.26	2.28	0.18	5.90	0.38	86.93	A + D + E
1.85	0.09	0.00	0.00	10.51	0.68	87.54	B + F
2.79	0.15	0.46	0.03	10.48	0.69	86.27	B + E + F
4.97	0.25	1.03	0.08	2.58	0.16	91.42	D + E + F
17.54	1.07	2.98	0.28	2.88	0.21	76.60	C + D + F
24.21	1.58	0.00	0.00	3.84	0.30	71.95	C + F
2.29	0.12	0.72	0.06	10.14	0.66	86.85	A + B + E
				temp. = 60°C.			
0.00	0.00	22.46	2.09	0.00	0.00	77.54	A
0.00	0.00	0.00	0.00	15.38	1.04	84.62	B
61.55	7.54	0.00	0.00	0.00	0.00	38.45	G
0.00	0.00	3.37	0.29	13.84	0.95	82.79	A + G
4.32	0.26	18.11	1.69	0.00	0.00	77.57	A + H
24.93	1.78	9.12	1.00	0.00	0.00	65.95	G + H

(continued next page)

AUXILIARY INFORMATION

METHOD/APPARATUS/PROCEDURE:	SOURCE AND PURITY OF MATERIALS:
The method of invariant points was used. The third component was added to binary systems. No further details are given.	No information is given.
	ESTIMATED ERROR: Nothing is stated.
	REFERENCES:

COMPONENTS:	ORIGINAL MEASUREMENTS:
(1) Tripotassium phosphate; K_3PO_4; [7778-53-2] (2) Dipotassium sulfate; K_2SO_4; [10233-01-9] (3) Potassium vanadate; KVO_3; [13769-43-2] (4) Water; H_2O; [7732-18-5]	Gasanova, KH.G.; Abduragimova, R.A. *Ukr. Khim. Zh.* <u>1978</u>, *44*, 158-63.

EXPERIMENTAL VALUES cont'd:

Invariant points in the K_3PO_4-K_2SO_4-KVO_3-H_2O system.

K_3PO_4		KVO_3		K_2SO_4		H_2O	
mass%	mol/kg[a]	mass%	mol/kg[a]	mass%	mol/kg[a]	mass%	solid phase[b]
			temp. = 60°C.				
28.42	2.01	0.00	0.00	5.21	0.45	66.37	G + I
5.14	0.29	0.00	0.00	12.22	0.84	82.64	B + I
5.93	0.34	6.11	0.54	6.58	0.46	81.38	A + H + J
0.74	0.04	2.21	0.18	12.09	0.81	84.96	A + B + J
15.09	0.91	4.00	0.37	3.17	0.23	77.74	G + H + I
3.00	0.15	1.21	0.09	7.20	0.46	88.59	H + I + J
1.39	0.07	0.45	0.03	12.11	0.80	86.05	B + I + J

[a] The mol/kg H_2O values were calculated by the compiler.

[b] The solid phases are: A = $KVO_3 \cdot 3H_2O$; B = K_2SO_4; C = $K_3PO_4 \cdot 7H_2O$;

D = $4K_2O \cdot P_2O_5 \cdot V_2O_5 \cdot 30H_2O$; E = $4K_2O \cdot P_2O_5 \cdot V_2O_5 \cdot 24H_2O$;

F = $5K_2O \cdot P_2O_5 \cdot 2SO_3 \cdot 30H_2O$; G = $K_3PO_4 \cdot 3H_2O$; H = $4K_2O \cdot P_2O_5 \cdot V_2O_5 \cdot 22H_2O$;

I = $5K_2O \cdot P_2O_5 \cdot 2SO_3 \cdot 22H_2O$; J = $4K_2O \cdot P_2O_5 \cdot V_2O_5 \cdot 18H_2O$.

COMPONENTS:	EVALUATOR:
(1) Rubidium dihydrogen phosphate; RbH_2PO_4; [13774-16-8] (2) Water; H_2O; [7732-18-5]	J. Eysseltová Charles University Prague, Czechoslovakia December, 1983

CRITICAL EVALUATION:

Qualitative solubility studies were made of three rubidium orthophosphates (1): $Rb_3PO_4 \cdot 4H_2O$ [101056-52-4]; $Rb_2HPO_4 \cdot H_2O$ [79832-54-5]; and RbH_2PO_4 [13774-16-8]. It was estimated that all these compounds are highly soluble. There is also a reference to $Rb_3PO_4 \cdot 3H_2O$ [10156-51-3] but no solubility data are reported (2). When it was discovered that the crystals of rubidium dihydrogen phosphate had some desirable electrical characteristics, further solubility studies were made.

The binary system RbH_2PO_4-H_2O was studied by Bykova, et al. (3). All other solubility studies (4, 5) were for ternary systems. The solubility values were determined by a direct analytical method in all the studies, but some of the analytical procedures are questionable. One group (4) used a potentiometric titration with aqueous KOH and report an accuracy of ± 0.2 mass%. The others (3, 5) used gravimetric procedures. Bykova, et al. (3) weighed rubidium as the tetraphenylborate and discuss the problem of analyzing for phosphorus in the presence of rubidium. Literature data for the solubility of rubidium phosphomolybdate (6, 7) are cited (8.1×10^{-6} mol dm^{-3} in 0.1 mol dm^{-3} HNO_3 at 293 K) and the possible formation of $RbMgPO_4$ is discussed (8). Because of these facts the gravimetric determination of phosphorus in systems containing rubidium must be carried out under carefully defined and controlled conditions. Zvorykin, et al. (5) precipitated phosphorus as $(NH_4)_3PMO_{12}O_{40}$, reprecipitated it as NH_4MgPO_4 and then calcined the latter to form $Mg_2P_2O_7$. They made no comment about the consistency of their determinations of NH_3, Rb and P. The compiler found these values to be inconsistent with each other.

THE BINARY SYSTEM

The solubility of RbH_2PO_4 in water has been determined over the temperature range of 273 to 353 K (3). The temperature coefficient of solubility was also determined and the authors split the temperature interval in two parts: 273-313 and 323-353 K. The evaluator treated these data by the linear regression method. The results are summarized in Table I where the coefficients for equation [1] are given for concentrations expressed as mass% and as mol/kg. The results in Table I suggest that there is no

$$c_{RbH_2PO_4} = a(T-273) + b \qquad [1]$$

need to split the temperature interval.

Table I. Coefficients for equation [1]

temp. range	a	b	R
		c as mass%	
273 - 313 K	0.52 ± 0.01	30.4 ± 0.4	0.9988
273 - 323 K	0.50 ± 0.01	30.6 ± 0.5	0.9983
273 - 333 K	0.47 ± 0.02	31.2 ± 0.9	0.9939
273 - 353 K	0.40 ± 0.03	32.9 ± 1.7	0.9788
		c as mol/kg	
273 - 313 K	0.082 ± 0.002	2.33 ± 0.06	0.9990
273 - 323 K	0.087 ± 0.003	2.28 ± 0.10	0.9974
273 - 333 K	0.087 ± 0.002	2.28 ± 0.08	0.9983
273 - 353 K	0.084 ± 0.002	2.3 ± 0.1	0.9981

MULTICOMPONENT SYSTEMS

The solubility of RbH_2PO_4 has been measured in three ternary systems.

1. The RbH_2PO_4-RbCl-H_2O system. The solubility in this system was measured only at 298 K (3). The system is an eutonic one with the invariant solution having a composition of 4.34 mass% (0.47 mol/kg) RbH_2PO_4 and 45.12 mass% (7.41 mol/kg) RbCl.

(continued next page)

COMPONENTS:	EVALUATOR:
(1) Rubidium dihydrogen phosphate; RbH_2PO_4; [13774-16-8] (2) Water; H_2O; [7732-18-5]	J. Eysseltová Charles University Prague, Czechoslovakia December, 1983

CRITICAL EVALUATION: (continued)

2. The $RbH_2PO_4-NH_4H_2PO_4-H_2O$ system. The solubility in this system has been measured only at 298 K (5). There are legitimate questions about the analytical procedures used in this work and the results must be considered to be questionable. There is considerable scatter in the data, which are plotted on Figure 1. It appears that one series of solid solutions is formed. Figure 1 shows that they belong to Type I in the Roozeboom classification (9).

3. The $RbH_2PO_4-Rb_2O-H_3PO_4-H_2O$ system. Solubilities in this system have been measured at 298 and 323 K (4). The authors considered it as the ternary system $Rb_2O-P_2O_5-H_2O$. The compiler transformed the values to those for the quaternary system, Figure 2. In the $RbH_2PO_4-Rb_2O-H_2O$ part of the system the solubility of RbH_2PO_4 is only slightly affected by change in concentration of the solutions, especially at 323 K. However, in the $RbH_2PO_4-H_3PO_4-H_2O$ part of the system the solubility of the rubidium dihydrogenphosphate increases with increasing H_3PO_4 content until the invariant point is reached. Beyond this, the acid salt $RbH_5(PO_4)_2$ appears in the solid phase. Such acid phosphates are reported for most systems involving the alkaline metals (10-14).

   The solubility of RbH_2PO_4 in aqueous H_3PO_4 may be described by equation [2] where c is the concentration expressed as

$$c_{RbH_2PO_4} = a \cdot c_{H_3PO_4} + b \qquad [2]$$

mass% or as mol/kg. The value of the coefficients, calculated by linear regression, are given in Table II.

Table II. Values of coefficients for equation [2].

	c as mass%			c as mol/kg		
T/K	a	b	R	a	b	R
298	0.55 ± 0.02	44.1 ± 0.3	0.9950	0.93 ± 0.02	4.38 ± 0.07	0.9991
323	0.280 ± 0.007	54.8 ± 0.1	0.9979	0.983 ± 0.004	6.65 ± 0.03	1.0000

The authors (4) also linearized their data using equation [3].

$$w_{Rb_2O} = a + b \cdot w_{P_2O_5} \qquad [3]$$

However, they gave no details for the method they used. The compiler recalculated their values to give the following results:
for P/Rb > 1
T = 298 K a = 17.6 ± 0.7 mass%; b = 0.30 ± 0.02; R = 0.9825
T = 323 K a = 25.2 ± 0.5 mass%; b = 0.15 ± 0.02; R = 0.9647
for P/Rb < 1
T = 298 K a = -41 ± 5 mass%; b = 3.9 ± 0.3; R = 0.9832
T = 323 K a = -132 ± 35 mass%; b = 7.7 ± 1.7; R = 0.8688.

CONCLUSIONS

The results of two studies (3, 4) agree well with each other. Therefore, the tentative solubility values for RbH_2PO_4 in water in the temperature range 273-353 K can be described by equation [1]. There are insufficient data to use the method that was described in the Critical Evaluation for the solubility of NaH_2PO_4 (chap. 3).

More work is needed to describe the solubility of other rubidium phosphates.

References

1. von Berg, E. Ber. 1901, 34, 4182.
2. Lauffenburger, R. Thesis, Strasbourg 1932.
3. Bykova, I.N.; Kuznetsova, G.P.; Kolotilova, V.Ya.; Stepin, B.D. Zh. Neorg. Khim. 1968, 13, 540.
4. Rashkovich, L.N.; Momtaz, R.Sh. Zh. Neorg. Khim. 1978, 23, 1349.

(continued next page)

COMPONENTS:	EVALUATOR:
(1) Rubidium dihydrogen phosphate; RbH_2PO_4; [13774-16-8] (2) Water; H_2O; [7732-18-5]	J. Eysseltová Charles University Prague, Czechoslovakia December, 1983

CRITICAL EVALUATION:

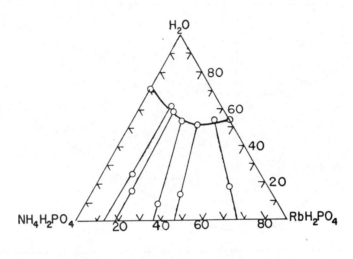

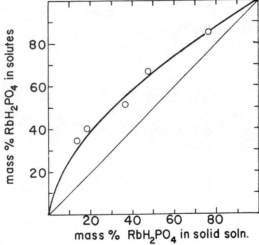

Figure 1. Phase diagram and distribution curve for the RbH_2PO_4-$NH_4H_2PO_4$-H_2O system at 298 K.

The data are from ref. (5). The distribution curve was constructed by the evaluator.

COMPONENTS:

(1) Rubidium dihydrogen phosphate; RbH_2PO_4;
    [13774-16-8]

(2) Water; H_2O; [7732-18-5]

EVALUATOR:

J. Eysseltová
Charles University
Prague, Czechoslovakia

December, 1983

CRITICAL EVALUATION:

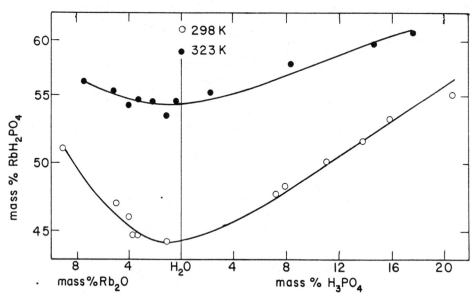

Figure 2. Solubility in the $Rb_2O-P_2O_5-H_2O$ system. The data have been
recalculated from ref. (4) by the compiler.

COMPONENTS:	EVALUATOR:
(1) Rubidium dihydrogen phosphate; RbH_2PO_4; [13774-16-8] (2) Water; H_2O; [7732-18-5]	J. Eysseltová Charles University Prague, Czechoslovakia December, 1983

CRITICAL EVALUATION: (continued)

5. Zvorykin, A.Ya.; Vetkina, L.S. *Zh. Neorg. Khim.* 1961, *6*, 2572.
6. Broadbank, R.W.C.; Dhabanandana, S.; Harding, R.D. *J. Inorg. Nucl. Chem.* 1961, *23*, 311.
7. Nikitina, E.A.; Sokolova, O.N. *Zh. Obshch. Chim.* 1954, *24*, 1123.
8. Erdmann, H.; Kothner, P. *Ann. Chem.* 1897, *294*, 72.
9. Roozeboom, B. *Z. Physik. Chem.* 1891, *8*, 521.
10. Muromtsev, B.A.; Nazarova, L.A. *Izv. AN SSSR (section of mathematics and natural sciences)* 1938, *1*, 177.
11. Flatt, R.; Brunisholz, G.; Chapuis-Goitreux, S. *Helv. Chim. Acta* 1951, *34*, 884.
12. Berg, L.G. *Izv. AN SSSR (section of mathematics and natural sciences)* 1938, *1*, 147.
13. Barkova, L.V.; Lopeshkov, I.N. *Zh. Neorg. Khim.* 1968, *13*, 1432.
14. Rashkovich, L.N.: Meteva, K.B.; Schevchik, J.E. *Zh. Neorg. Khim.* 1977, *22*, 1982.

COMPONENTS:	ORIGINAL MEASUREMENTS:
(1) Rubidium dihydrogenphosphate; RbH_2PO_4; [13774-16-8] (2) Water; H_2O; [7732-18-5]	Bykova, I.N.; Kuznetsova, G.P.; Kolotilova, V.Ya.; Stepin, B.D. *Zh. Neorg. Khim.* 1968, *13*, 540-4.

VARIABLES:	PREPARED BY:
Temperature.	J. Eysseltová

EXPERIMENTAL VALUES:

Solubility of RbH_2PO_4 in water.

$t/°C$	g/100 g H_2O	mass%[a]	mol/kg[a]
0	43.2	30.16	2.37
25	78.7	44.04	4.31
40	103.7	50.91	5.68
50	123.6	55.27	6.77
60	137.1	57.82	7.51
80	162.9	61.96	8.93

[a]These values were calculated by the compiler.

COMMENTS and ADDITIONAL DATA: The temperature coefficient of the solubility is reported to be constant in the temperature ranges 0 to 40°C and 50 to 80°C. The values are:

range/°C.	$dm_i/dT/mol\ kg^{-1}\ K^{-1}$
0 - 40	0.0803
50 - 80	0.070

AUXILIARY INFORMATION

METHOD/APPARATUS/PROCEDURE:	SOURCE AND PURITY OF MATERIALS:
The mixtures were equilibrated isothermally for 15 days. The apparatus has been described elsewhere (1). The rubidium content was determined gravimetrically as the tetraphenylborate. The temperature coefficient of the solubility was determined graphically.	RbH_2PO_4 was synthesized from reagent grade H_3PO_4 and Rb_2CO_3. The Rb_2CO_3 was obtained by calcining $Rb_2(COO)_2$. The maximum amount of impurity in the RbH_2PO_4 was 0.05 mass%.

ESTIMATED ERROR:

The temperature was controlled to within ± 0.1 K. No other information is given.

REFERENCES:

1. Kuznetsova, G.P.; Stepin, B.D. *Zh. Neorg. Khim.* 1965, *10*, 472.

COMPONENTS:	ORIGINAL MEASUREMENTS:
(1) Rubidium dihydrogenphosphate; RbH_2PO_4; [13774-16-8] (2) Ammonium dihydrogenphosphate; $NH_4H_2PO_4$; [7722-76-1] (3) Water; H_2O; [7732-18-5]	Zvorykin, A.Ya.; Vetkina, L.S. Zh. Neorg. Khim. 1961, 6, 2572-5.

VARIABLES:	PREPARED BY:
Composition at 25°C.	J. Eysseltová

EXPERIMENTAL VALUES:

Composition of saturated solutions in the
RbH_2PO_4-$NH_4H_2PO_4$-H_2O system at 25°C.

Rb mass%	NH_3 mass%	P mass%	RbH_2PO_4		$NH_4H_2PO_4$		solid phase
			mass%	mol/kg[a]	mass%	mol/kg[a]	
21.37	----	7.92	46.12	4.69	----	----	RbH_2PO_4
18.35	1.07	8.36	39.17	4.01	7.24	1.17	solid soln
15.94	2.43	7.53	32.05	4.41	16.43	2.77	"
11.26	3.48	9.40	24.03	2.51	23.54	3.90	"
8.32	2.89	6.52	17.72	1.62	22.3	3.23	"
8.05	3.14	6.84	17.19	1.61	24.42	3.64	"
7.3	2.97	8.06	15.58	1.40	23.27	3.31	"
6.78	3.47	7.61	14.48	1.28	23.47	3.29	"
-----	4.36	7.85	-----	----	29.31	3.60	$NH_4H_2PO_4$

[a]The mol/kg H_2O values were calculated by the compiler.

AUXILIARY INFORMATION

METHOD/APPARATUS/PROCEDURE:	SOURCE AND PURITY OF MATERIALS:
The components were mixed, dissolved in water at 65°C, cooled rapidly to 25°C and equilibrated by shaking for several days. P_2O_5 was determined gravimetrically as $Mg_2P_2O_7$, rubidium was weighed as $RbClO_4$, and ammonia was determined by the Kjeldahl method.	RbH_2PO_4 and $NH_4H_2PO_4$ were synthesized from chemically pure H_3PO_4 and Rb_2CO_3 or NH_3. The analyses were:

	found	calcd.
RbH_2PO_4	46.16% Rb	46.85% Rb
"	16.66% P	16.98% P
$NH_4H_2PO_4$	14.4% N	14.78% N
"	26.62% P	26.95% P

ESTIMATED ERROR:

No information is given.

REFERENCES:

COMPONENTS:	ORIGINAL MEASUREMENTS:
(1) Rubidium dihydrogenphosphate; RbH_2PO_4; [13774-16-8] (2) Rubidium chloride; RbCl; [7791-11-9] (3) Water; H_2O; [7732-18-5]	Bykova, I.N.; Kuznetsova, G.P.; Kolotilova, V.Ya.; Stepin, B.D. *Zh. Neorg. Khim.* <u>1968</u>, *13*, 540-4.

VARIABLES:	PREPARED BY:
Composition at 25°C.	J. Eysseltová

EXPERIMENTAL VALUES:

Solubility in the RbH_2PO_4–RbCl–H_2O system at 25°C.

RbH_2PO_4		RbCl		
mass%	mol/kg[a]	mass%	mol/kg[a]	solid phase
44.05	4.32	----	----	RbH_2PO_4
36.51	3.47	5.86	0.84	"
27.78	2.57	12.87	1.79	"
13.76	1.29	27.62	3.90	"
8.07	0.77	34.62	4.99	"
6.72	0.67	38.09	5.71	"
4.50	0.46	42.01	6.49	"
4.34	0.47	45.12	7.41	RbH_2PO_4 + RbCl
3.44	0.37	46.06	7.54	"
2.90	0.33	46.48	7.59	"
0.74	0.078	47.29	7.53	"
-----	-----	48.29	7.72	"

[a] The mol/kg H_2O values were calculated by the compiler.

AUXILIARY INFORMATION

METHOD/APPARATUS/PROCEDURE:	SOURCE AND PURITY OF MATERIALS:
The mixtures were equilibrated isothermally for 15 days in an apparatus described earlier (1). The rubidium and chloride contents were determined gravimetrically, Rb as the tetraphenylborate and Cl as AgCl. The composition of the solid phases was determined by the wet-residue method.	Chemically pure RbCl was heated to 400°C, recrystallized and dried at 120°C. RbH_2PO_4 was synthesized from H_3PO_4 and Rb_2CO_3. The latter was obtained by calcining $Rb_2(COO)_2$. The impurities in the RbH_2PO_4 were less than 0.05 mass%.

	ESTIMATED ERROR:
	The temperature was controlled to within ± 0.1 K. No other details are given.

	REFERENCES:
	1. Kuznetsova, G.P.; Stepin, B.D. *Zh. Neorg. Khim.* <u>1965</u>, *10*, 472.

COMPONENTS:	ORIGINAL MEASUREMENTS:
(1) Rubidium oxide; Rb_2O; [18008-11-4] (2) Phosphorus pentoxide; P_2O_5; [1314-56-3] (3) Water-d$_2$; D_2O; [7789-20-0] (4) Water; H_2O; [7732-18-5]	Rashkovich, L.N.; Momtaz, R.Sh. *Zh. Neorg. Khim.* 1978, *23*, 1349-55.

VARIABLES:	PREPARED BY:
Temperature and composition.	J. Eysseltová

EXPERIMENTAL VALUES:

Composition of saturated solutions in the $Rb_2O-P_2O_5-H_2O$ system.

$t/°C$	Rb_2O mass%	P_2O_5 mass%	solid phase	$t/°C$	Rb_2O mass%	P_2O_5 mass%	solid phase
25	43.2	22.2	$Rb(H,D)_2PO_4$	50	42.9	23.2	$Rb(H,D)_2PO_4$
25	35.5	21.2	"	50	36.5	22.2	"
25	32.3	20.3	"	50	34.5	22.4	"
25	31.0	20.1	"	50	33.5	22.3	"
25	28.8	19.9	"	50	32.0	22.3	"
25	26.4	20.8	"	50	31.6	22.2	"
25	28.2	29.2	"	50	29.9	22.5	"
25	28.9	32.6	"	50	29.6	23.2	"
25	28.8	32.9	"	50	30.5	30.2	"
25	29.4	35.9	"	50	30.8	33.1	"
25	26.1	38.2	$Rb(H,D)_5(PO_4)_2$	50	31.1	33.4	"
				50	31.5	36.5	"
				50	31.3	36.5	"
				50	31.1	37.8	$Rb(H,D)_5(PO_4)_2$

AUXILIARY INFORMATION

METHOD/APPARATUS/PROCEDURE:	SOURCE AND PURITY OF MATERIALS:
The mixtures were heated until nearly all the solid phase had disappeared. Then the temperature required was established and the mixtures were thermostated for at least 2 weeks. The composition of the liquid phase was determined by potentiometric titration with KOH after the samples had first been acidified with a known amount of H_3PO_4. The deuterium content was determined by a method described elsewhere (1).	Rb_2CO_3 and D_3PO_4 were used as received. The impurities in these substances was less than 0.05 mass%. The electrolytic conductivity of the heavy water was 6×10^{-6} S cm^{-1}. The extent of deuteration was 96% for D_3PO_4, 99% for D_2O, and 90-91% for the saturated solutions.

ESTIMATED ERROR:

The temperature was controlled to within ± 0.01 K. The analyses had a precision of ± 0.1 mass%.

REFERENCES:

1. Volkova, E.N.; Podshivalov, J.S.; Rashkovich, L.N.; Strukov, B.A. *Izv. AN SSSR, ser. fiz.* 1975, *39*, 288.

COMPONENTS:	ORIGINAL MEASUREMENTS:
(1) Rubidium dihydrogenphosphate; RbH_2PO_4; [13774-16-8]	Rashkovich, L.N.; Momtaz, R.Sh.
(2) Rubiduym oxide; Rb_2O; [18088-11-4]	*Zh. Neorg. Khim.* <u>1978</u>, *23*, 1349-55.
(3) Phosphoric acid; H_3PO_4; [7664-38-2]	
(4) Water; H_2O; [7732-18-5]	

VARIABLES:	PREPARED BY:
Temperature and composition.	J. Eysseltová

EXPERIMENTAL VALUES:

The authors express their data for the Rb_2O-P_2O_5-H_2O system. The compiler has recalculated their data to convert it to the RbH_2PO_4-H_3PO_4-H_2O system.

	authors' data		compiler's recalculated values						
	Rb_2O	P_2O_5	Rb_2O		H_3PO_4		RbH_2PO_4		solid
pH	mass%	mass%	mass%	mol%	mass%	mol%	mass%	mol%	phase
				temp. = 25°C.					
6.6	35.4	19.9	9.19	1.36	----	----	51.1	7.16	RbH_2PO_4
5.8	29.1	18.3	5.00	0.56	----	----	47.0	5.37	"
5.6	27.6	17.9	4.02	0.43	----	----	46.0	5.04	"
5.2	26.7	17.4	3.78	0.39	----	----	44.7	4.76	"
5.2	26.3	17.4	3.38	0.35	----	----	44.7	4.72	"
4.7	23.8	17.2	1.15	0.11	----	----	44.2	4.43	"
2.6	24.6	23.8	----	----	7.08	1.60	47.8	5.80	"
2.5	24.9	24.6	----	----	7.87	1.83	48.4	6.05	"
2.1	25.8	27.5	----	----	10.9	2.86	50.1	7.05	"
1.9	26.6	30.2	----	----	13.8	8.20	51.7	8.20	"
2.0	27.4	32.2	----	----	15.7	5.17	53.2	9.39	"
1.4	28.3	36.3	----	----	20.5	8.49	55.0	12.2	"

(continued next page)

AUXILIARY INFORMATION

METHOD/APPARATUS/PROCEDURE:	SOURCE AND PURITY OF MATERIALS:
The mixtures were heated until almost all the solid phase had dissolved and then they were thermostated at the specified temperature for at least 2 weeks. The composition of the liquid phase was determined by potentiometric titration with KOH after being acidified with known amounts of H_3PO_4. The solid phases were identified microscopically.	The Rb_2CO_3 and H_3PO_4 were used as received. The amount of impurities did not exceed 0.05 mass%. The electrolytic conductivity of the water was 5×10^{-5} S cm^{-1}.

	ESTIMATED ERROR:
	The temperature was controlled to within ± 0.1 K. The precision of the analyses was ± 0.1 mass%.

	REFERENCES:

COMPONENTS:	ORIGINAL MEASUREMENTS:
(1) Rubidium dihydrogenphosphate; RbH_2PO_4; [13774-16-8]	Rashkovich, L.N.; Momtaz, R.Sh.
(2) Rubidium oxide; Rb_2O; [18088-11-4]	$Zh.\ Neorg.\ Khim.$ <u>1978</u>, 23, 1349-55.
(3) Phosphoric acid; H_3PO_4; [7664-38-2]	
(4) Water; H_2O; [7732-18-5]	

EXPERIMENTAL VALUES cont'd:

authors' data		compiler's recalculated values						
Rb_2O	P_2O_5	Rb_2O		H_3PO_4		RbH_2PO_4		solid
mass%	mass%	mass%	mol%	mass%	mol%	mass%	mol%	phase
		temp. = 50°C.						
36.2	21.8	7.49	1.10	----	----	56.0	8.42	RbH_2PO_4
33.5	21.5	5.18	0.70	----	----	55.3	7.66	"
31.8	21.1	4.01	0.51	----	----	54.2	7.12	"
31.3	21.3	3.25	0.41	----	----	54.7	7.14	"
30.1	21.2	2.18	0.27	----	----	54.5	6.89	"
28.6	20.8	1.21	0.14	----	----	53.5	6.46	"
28.3	21.2	0.38	0.05	----	----	54.5	6.61	"
28.5	23.2	----	----	2.17	0.52	55.3	7.14	"
29.5	28.4	----	----	8.29	2.46	57.3	9.12	"
30.3	33.6	----	----	14.6	5.63	58.8	12.2	"
30.7	36.0	----	----	17.5	7.82	59.6	14.3	"
31.3	39.6	----	----	21.9	12.8	60.8	19.2	"
31.4	39.5	----	----	21.6	12.7	61.0	19.2	"
30.9	41.7	----	----	25.2	17.3	60.0	22.2	$RbH_5(PO_4)_2$

The authors linearized their data in the form:

$$w_{Rb_2O} = a + b\ w_{P_2O_5}$$

In the region where P/Rb >1 the constants have the following values:

for t = 25°C: a = 17.5 ± 0.4 mass% and b = 0.30 ± 0.01   ± σ = 0.1

for t = 50°C: a = 24.6 ± 0.1 mass% and b = 0.172 ± 0.004   ± σ = 0.06

In the region where Rb_2O is in excess

for t = 25°C: a = -40 ± 5 mass% and b = 3.8 ± 0.3   ± σ = 0.6

for t = 50°C: a = -124 ± 23 mass% and b = 7.3 ± 1.1   ± σ = 0.8

COMPONENTS:	EVALUATOR:
(1) Cesium dihydrogenphosphate; CsH_2PO_4; [18649-05-3] (2) Water; H_2O; [7732-18-5]	J. Eysseltová Charles University Prague, Czechoslovakia December, 1983

CRITICAL EVALUATION:

Three cesium orthophosphates are reported in the literature (1, 2): tricesium phosphate, pentahydrate, $Cs_3PO_4 \cdot 5H_2O$, [101056-43-3]; dicesium hydrogenphosphate, mono-hydrate, $Cs_2HPO_4 \cdot H_2O$, [50292-03-0]; and cesium dihydrogenphosphate, CsH_2PO_4, [18649-05-3]. No quantitative solubility data are available for the tricesium phosphate or the dicesium hydrogenphosphate. The solubility of the cesium dihydrogenphosphate was studied because its crystals had certain desirable electrical characteristics.

Bykova, et al. (3) report the solubility of cesium dihydrogenphosphate in the binary system while all the other reports of solubility data (3-6) are for ternary systems. A direct analytical solubility method was used in all the solubility studies. However, there is some disagreement about the validity of the experimental methods. In one study a potentiometric titration with aqueous KOH was used (4) and the results are reported to have an accuracy of ± 0.2 mass%. Gravimetric procedures were used in the other studies, phosphorus being weighed as $Mg_2P_2O_7$. In two of the studies (5, 6) phosphorus was pre-cipitated as $(NH_4)_3PMo_{12}O_{40}$ and then reprecipitated and transformed to $NH_4MgPO_4 \cdot 6H_2O$. Nothing is said about the accuracy or the precision of the results. In the other study (3), phosphorus was precipitated directly as $NH_4MgPO_4 \cdot 6H_2O$ under precisely defined and carefully controlled conditions. A thorough discussion of this analytical procedure is given. Literature data are presented for the co-precipitation of cesium phosphomolybdate (7, 8) and/or cesium magnesium phosphate (1, 9). The solubility of the cesium phospho-molybdate is reported to be 4.2×10^{-6} mol dm^{-3} in 0.1 mol dm^{-3} HNO_3 at 293 K. Because of this, Bykova, et al. (3) recommend that NH_4MgPO_4 be precipitated from solutions con-taining less than 0.3 g $H_2PO_4^-$ per 100 cm^3, and that the amount of $Mg_2P_2O_7$ produced be no greater than 0.06 g. They criticized the analytical procedures used by the other investigators (4-6). As a matter of fact, the results of Bykova, et al. (3) do agree with those obtained by potentiometric titration (4), but they are significantly lower than those reported by others (5, 6).

THE BINARY SYSTEM

Bykova, et al. (3) measured solubilities in the 273-353 K temperature range, and the temperature coefficient of molal solubility in the 273-333 K range was also determined. A recalculation of the results by the Evaluator, using linear regression, gives equation [1].

$$c_{CsH_2PO_4} = a \cdot (T-273) + b \qquad [1]$$

When c is expressed as mass%, the coefficients are:
a = 0.25 ± 0.01 and b = 52.2 ± 0.6 with R = 0.9916. When c is expressed as mol/kg the coefficients are: a = 0.0681 ± 0.0007 and b = 4.64 ± 0.03 with R = 0.9991.

MULTICOMPONENT SYSTEMS

Cesium dihydrogenphosphate was chosen as a component in four ternary systems. These systems are discussed individually.

1. The CsH_2PO_4-CsCl-H_2O system. Solubilities in this system have been measured at 298 K (3). The system is reported to be of the eutonic type and a saturated solution in equilibrium with both CsH_2PO_4 and CsCl has the composition 0.58 mol/kg CsH_2PO_4 and 10.5 mol/kg CsCl.

2. The CsH_2PO_4-MH_2PO_4-H_2O system where M is K^+ or NH_4^+. Solubility in the CsH_2PO_4-$NH_4H_2PO_4$-H_2O system was measured at 298 K (5) and solubility in the CsH_2PO_4-KH_2PO_4-H_2O system was also measured at 298 K (6). In both studies the analysis of phosphorus was made by precipitating it as $(NH_4)_3PMo_{12}O_{40}$ and this is the method that has been criticized (3). Therefore, the solubilities reported in these studies may be incorrect. This is especially true in the system containing $NH_4H_2PO_4$ where the solubility of pure CsH_2PO_4 was reported to be significantly larger than other reported results. Besides this, there is a discrepancy between tabular and graphical data in the article (5). The tabular data have been replotted, Figure 1, and there are intersections of some tie lines that make no physical sense. Therefore, these results (5) are not reliable and must be rejected.

(continued next page)

COMPONENTS:	EVALUATOR:
(1) Cesium dihydrogenphosphate; CsH_2PO_4; [18649-05-3] (2) Water; H_2O; [7732-18-5]	J. Eysseltová Charles University Prague, Czechoslovakia December, 1983

CRITICAL EVALUATION:

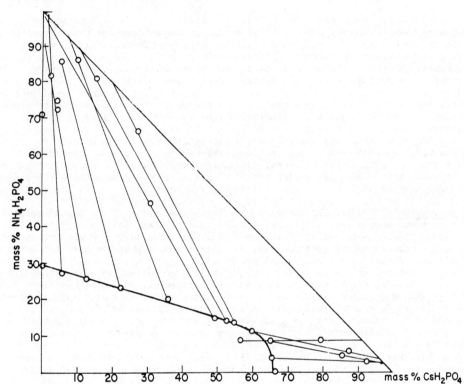

Figure 1.   Phase diagram for the CsH_2PO_4-$NH_4H_2PO_4$-H_2O system
at 298 K, ref. (5).

COMPONENTS:	EVALUATOR:
(1) Cesium dihydrogenphosphate; CsH_2PO_4; [18649-05-3]	J. Eysseltová Charles University Prague, Czechoslovakia
(2) Water; H_2O; [7732-18-5]	December, 1983

CRITICAL EVALUATION: (continued)

Although there is a question about the solubilities reported for these systems, the formation of solid solutions in these systems has been established. In the system containing KH_2PO_4 there are 3 crystallization fields. Solid KH_2PO_4 exists in equilibrium with solutions containing 1.32 to 1.55 mol/kg KH_2PO_4 and 0 to 0.71 mol/kg CsH_2PO_4. Solid CsH_2PO_4 exists in equilibrium with solutions containing 0 to 1.98 mol/kg KH_2PO_4 and 4.86 to 6.37 mol/kg CsH_2PO_4. Solid solutions are the equilibrium solid phases between these 2 crystallization fields. The data are given on Figure 2. It is impossible to place this system into one of Roozeboom's categories (10) because of the limited solid solubility.

For the system containing $NH_4H_2PO_4$ there is, in addition to the uncertainty discussed earlier, an uncertainty about the composition of the invariant solution. The data of Zvorykin, et al. (5) are plotted on Figure 3 and, again, because of the limited amount of data, this system cannot definitely be placed in one of Roozeboom's classifications (10).

3. The $Cs_2O-P_2O_5-H_2O$ system. Rashkovich, et al. (4) measured solubilities in solutions in equilibrium with solid CsH_2PO_4 and $CsH_5(PO_4)_2$ at 298 and 323 K along with a few solutions at 306.2, 312 and 317.7 K. They expressed their data in terms of Cs_2O and P_2O_5. The compiler converted their data to that of the quaternary system $CsH_2PO_4-Cs_2O-H_3PO_4-H_2O$. These values are given on Figure 4. In the $CsH_2PO_4-Cs_2O-H_2O$ part of the system, the solubility of cesium dihydrogenphosphate is affected only slightly by changes in solution composition. However, on the $CsH_2PO_4-H_3PO_4-H_2O$ side of the diagram, the solubility of cesium dihydrogenphosphate increases with increasing phosphoric acid content until the invariant point is reached. Beyond this, the solid phase may be $CsH_5(PO_4)_2$ but this has not been confirmed experimentally. In the 323 K isotherm the data points in the phosphoric acid region do not lie on a smooth line as do those on the 298 K isotherm. Therefore, it is possible that a new crystallization field comes into existence here. A similar situation exists with ammonium phosphates where the formation of $3NH_4H_2PO_4 \cdot H_3PO_4$ has been suggested (11).

The evaluator has used a linear regression method to treat the experimental data in ref. (4). The results for the 298 K isotherm, Figure 4, can be expressed by equation [2].

$$c_{CsH_2PO_4} = a \cdot c_{H_3PO_4} + b \qquad [2]$$

When c is expressed as mass%, a = 0.56 ± 0.02, b = 58.3 ± 0.2 and R = 0.9923. When c is expressed as mol/kg, a = 1.17 ± 0.02, b = 6.11 ± 0.06 and R = 0.9990. The authors (4) also linearized their results but they gave no details about the method that was used. Later (12) they linearized the results in the region where P/Cs < 1. However, the compiler's method of linear regression gave negative correlation coefficients for these results.

CONCLUSIONS

The reliability of the solubility data for cesium dihydrogenphosphate depends on the analytical method that was used. This matter was discussed by Bykova, et al. (3) and they were careful in their experimental work. Their data agree fairly well with those obtained by extrapolation of solubility isotherms measured by others (4). Therefore, the data of Bykova, et al, (3) as well as their equation for the temperature dependence of the solubility of cesium dihydrogenphosphate are recommended values. The solubility data published by others (5, 6) are rejected because of analytical uncertainties. The only information that can be considered to be proved is the occurrence of solid solutions in the $CsH_2PO_4-NH_4H_2PO_4-H_2O$ and $CsH_2PO_4-KH_2PO_4-H_2O$ systems at 298 K.

References

1. von Berg, E. Ber. 1901, 34, 4182.
2. Lauffenburger, R. Thesis, Strasbourg 1932.
3. Bykova, I.N.; Kuznetsova, G.P.; Kolotilova, V.Ya.; Stepin, B.D. Zh. Neorg. Khim. 1968, 13, 540.

(continued next page)

COMPONENTS:	EVALUATOR:
(1) Cesium dihydrogenphosphate; CsH_2PO_4; [18649-05-3] (2) Water; H_2O; [7732-18-5]	J. Eysseltová Charles University Prague, Czechoslovakia December, 1983

CRITICAL EVALUATION:

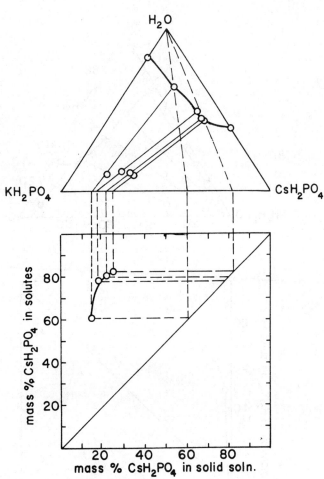

Figure 2. Phase diagram and distribution curve for
the CsH_2PO_4-KH_2PO_4-H_2O system at 298 K.
The data are from ref. (6).

COMPONENTS:

(1) Cesium dihydrogenphosphate; CsH_2PO_4;
    [18649-05-3]

(2) Water; H_2O; [7732-18-5

EVALUATOR:

J. Eysseltová
Charles University
Prague, Czechoslovakia

December, 1983

CRITICAL EVALUATION:

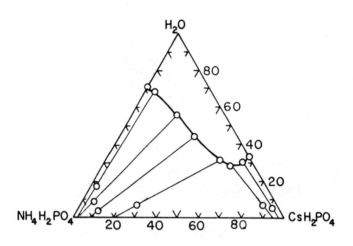

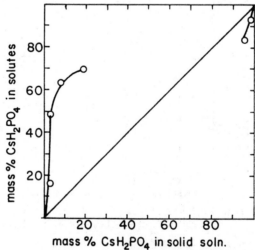

Figure 3. Phase diagram and distribution curve
for the CsH_2PO_4-$NH_4H_2PO_4$-H_2O system
at 298 K, ref. (5).

COMPONENTS:	EVALUATOR:
(1) Cesium dihydrogenphosphate; CsH_2PO_4; [18649-05-3]	J. Eysseltová Charles University Prague, Czechoslovakia
(2) Water; H_2O; [7732-18-5]	December, 1983

CRITICAL EVALUATION:

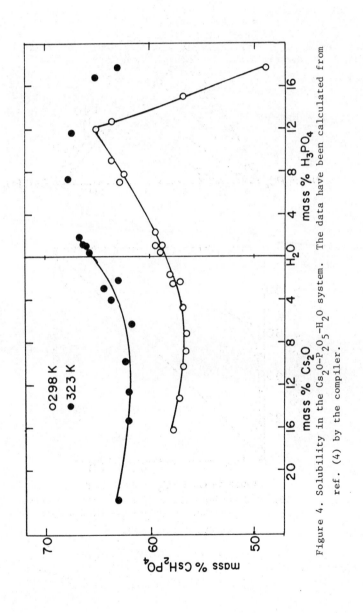

Figure 4. Solubility in the $Cs_2O-P_2O_5-H_2O$ system. The data have been calculated from ref. (4) by the compiler.

COMPONENTS:	EVALUATOR:
(1) Cesium dihydrogenphosphate; CsH_2PO_4; [18649-05-3] (2) Water; H_2O; [7732-18-5]	J. Eysseltová Charles University Prague, Czechoslovakia December, 1983

CRITICAL EVALUATION: (continued)

4. Rashkovich, L.N.; Meteva, K.B.; Shevchik, J.E. *Zh. Neorg. Khim.* <u>1977</u>, *22*, 1982.

5. Zvorykin, A.Ya.; Ratnikova, V.D. *Zh. Neorg. Khim.* <u>1963</u>, *8*, 1018.

6. Sayamyan, E.A.; Bashugyan, D.P.; Karapetyan, T.I.; Grigoryan, K.G.; Kohachikyan, A.V. *Zh. Neorg. Khim.* <u>1977</u>, *22*, 1119.

7. Broadbank, R.W.C.; Dhabanandana, S.; Harding, R.D. *J. Inorg. Nucl. Chem.* <u>1961</u>, *23*, 311.

8. Nikitina, E.A.; Sokolova, O.N. *Zh. Obshch. Chim.* <u>1954</u>, *24*, 1123.

9. Erdmann, H.; Kothner, P. *Ann Chem.* <u>1897</u>, *274*, 72.

10. Roozeboom, B. *Z. Physik. Chem.* <u>1891</u>, *8*, 521.

11. Flatt, R.; Brunisholz, G.; Chapuis-Goitreux, S. *Helv. Chim. Acta* <u>1951</u>, *34*, 884.

12. Rashkovich, L.N.; Momtaz, R.Sh. *Zh. Neorg. Khim.* <u>1978</u>, *23*, 1349.

COMPONENTS:	ORIGINAL MEASUREMENTS:
(1) Cesium dihydrogenphosphate; CsH_2PO_4; [18649-05-3] (2) Water; H_2O; [7732-18-5]	Bykova, I.N.; Kuznetsova, G.P.; Kolotilova, V.Ya.; Stepin, B.D. *Zh. Neorg. Khim.* <u>1968</u>, *13*, 540-4.

VARIABLES:	PREPARED BY:
Temperature.	J. Eysseltová

EXPERIMENTAL VALUES:

Solubility of CsH_2PO_4 in water.

$t/°C.$	g/100 g H_2O	mass%[a]	mol/kg[a]
0	106.0	51.43	4.61
25	146.97	59.5	6.39
40	169.4	62.88	7.37
50	185.3	64.96	8.06
60	199.7	66.63	8.69
80	258.0	72.07	11.2

[a] These values were calculated by the compiler.

The temperature coefficient of solubility is reported to be constant in the temperature range that was studied. The value is

$$dm_i/dT = 0.0683 \text{ mol kg}^{-1} \text{ K}^{-1}.$$

AUXILIARY INFORMATION

METHOD/APPARATUS/PROCEDURE:	SOURCE AND PURITY OF MATERIALS:
The mixtures were equilibrated isothermally for 15 days. The apparatus and procedure are described elsewhere (1). The solubility was determined by a gravimetric analysis for phosphorus. The phosphorus was weighed as $Mg_2P_2O_7$. The temperature coefficient of the solubility was determined graphically.	CsH_2PO_4 was synthesized by reacting H_3PO_4 with Cs_2CO_3. The latter was obtained by calcining $Cs_2(COO)_2$. The amount of impurities was no more than 0.05 mass%.

	ESTIMATED ERROR:
	The temperature was controlled to within ± 0.1 K. No other information is given.

	REFERENCES:
	1. Kuznetsova, G.P.; Stepin, B.D. *Zh. Neorg. Khim.* <u>1965</u>, *10*, 472.

COMPONENTS:	ORIGINAL MEASUREMENTS:
(1) Cesium dihydrogenphosphate; CsH_2PO_4; [18649-05-3]	Zvorykin, A.Ya.; Ratnikova, V.D.
(2) Ammonium dihydrogenphosphate; $NH_4H_2PO_4$; [7722-76-1]	Zh. Neorg. Khim. 1963, 8, 1018-9.
(3) Water; H_2O; [7732-18-5]	

VARIABLES:	PREPARED BY:
Composition at 25°C.	J. Eysseltová

EXPERIMENTAL VALUES:

Composition of saturated solutions in the CsH_2PO_4-$NH_4H_2PO_4$-H_2O system at 25°C.

CsH_2PO_4		$NH_4H_2PO_4$		
mass%	mol/kg[a]	mass%	mol/kg[a]	solid phase
66.26	8.55	----	----	CsH_2PO_4
65.29	9.15	3.65	7.86	solid soln A
59.73	8.91	11.09	25.3	"
64.85	10.7	8.72	22.0	"
56.76	7.18	8.85	17.1	solid soln A + solid soln B
49.12	5.92	14.77	27.2	solid soln B
52.79	6.93	14.07	28.3	"
54.83	7.61	13.80	29.3	"
35.91	3.56	20.16	30.6	"
22.50	1.80	23.14	28.3	"
12.88	0.91	25.57	27.7	"
5.42	0.35	27.20	26.9	"
-----	----	29.31	27.6	$NH_4H_2PO_4$

[a]The mol/kg H_2O values were calculated by the compiler.

AUXILIARY INFORMATION

METHOD/APPARATUS/PROCEDURE:	SOURCE AND PURITY OF MATERIALS:
The samples were dissolved at 65°C and equilibrated in a water bath by shaking for several days. P_2O_5 was determined gravimetrically as $Mg_2P_2O_7$. Ammonia was determined by the Kjeldahl method. The composition of the solid phases was determined by the wet-residue method.	Both phosphates were prepared by reacting H_3PO_4 with Cs_2CO_3 or with NH_3. Analysis: found CsH_2PO_4 13.40% P $NH_4H_2PO_4$ 26.62% P and 14.4% NH_3 calculated CsH_2PO_4 13.48% P $NH_4H_2PO_4$ 26.7% P and 15.5% NH_3.

	ESTIMATED ERROR:
	The temperature was controlled to within ± 0.1K. No other information is given.

	REFERENCES:

COMPONENTS:	ORIGINAL MEASUREMENTS:
(1) Cesium dihydrogenphosphate; CsH_2PO_4; [18649-05-3] (2) Cesium chloride; CsCl; [7647-17-8] (3) Water; H_2O; [7732-18-5]	Bykova, I.N.; Kuznetsova, G.P.; Kolotilova, V.Ya.; Stepin, B.D. Zh. Neorg. Khim. 1968, 13, 540-4.
VARIABLES: Composition at 25°C.	PREPARED BY: J. Eysseltová

EXPERIMENTAL VALUES:

Composition of saturated solutions in the CsH_2PO_4-CsCl-H_2O system at 25°C.

CsH_2PO_4		CsCl		
mass%	mol/kg[a]	mass%	mol/kg[a]	solid phase
59.51	6.39	----	----	CsH_2PO_4
52.31	5.63	7.30	1.07	"
45.16	4.72	13.22	1.88	"
38.28	3.86	18.62	2.57	"
33.80	3.40	22.91	3.14	"
20.93	2.12	36.09	4.99	"
7.19	0.84	55.42	8.80	"
4.60	0.58	60.99	10.5	CsH_2PO_4 + CsCl
3.42	0.43	61.73	10.5	CsCl
1.47	0.19	64.47	11.2	"
-----	----	65.77	11.4	"

[a]The mol/kg H_2O values were calculated by the compiler.

AUXILIARY INFORMATION

METHOD/APPARATUS/PROCEDURE:

The mixtures were equilibrated isothermally for 15 days. The apparatus is described elsewhere (1). The analyses were done gravimetrically: chloride was determined as AgCl; phosphorus was determined as $Mg_2P_2O_7$. The composition of the solid phases was determined by the wet-residue method.

SOURCE AND PURITY OF MATERIALS:

Chemically pure CsCl was heated to 400°C, recrystallized and dried at 120°C. CsH_2PO_4 was synthesized from H_3PO_4 and Cs_2CO_3. The latter was obtained by calcining $Cs_2(COO)_2$. The maximum amount of impurities in the CsH_2PO_4 was 0.05 mass%.

ESTIMATED ERROR:

The temperature was controlled to within ± 0.1 K. No other details are given.

REFERENCES:

1. Kuznetsova, G.P.; Stepin, B.D. Zh. Neorg. Khim. 1965, 10, 472.

COMPONENTS:	ORIGINAL MEASUREMENTS:
(1) Cesium dihydrogenphosphate; CsH_2PO_4; [18649-05-3] (2) Potassium dihydrogenphosphate; KH_2PO_4; [7778-77-0] (3) Water; H_2O; [7732-18-5]	Sayamyan, E.A.; Bashugyan, D.P.; Karapetyan, T.I.; Grigoryan, K.G.; Khachikyan, A.V. *Zh. Neorg. Khím.* <u>1977</u>, 22, 1119-23.

VARIABLES:	PREPARED BY:
Composition at 25°C.	J. Eysseltová

EXPERIMENTAL VALUES:

Composition of saturated solutions in the CsH_2PO_4-KH_2PO_4-H_2O system at 25°C.

CsH_2PO_4		KH_2PO_4		
mass%	mol/kg^a	mass%	mol/kg^a	solid phase
60.40	6.37	-----	-----	CsH_2PO_4
52.80	5.43	4.9	0.85	"
47.68	4.76	8.78	1.48	"
46.80	4.86	11.30	1.98	CsH_2PO_4 + solid soln
46.00	4.56	10.10	1.69	solid soln ($3.3KH_2PO_4 \cdot CsH_2PO_4$)
44.48	4.27	10.15	1.64	solid soln ($3.6KH_2PO_4 \cdot CsH_2PO_4$)
41.92	3.83	10.47	1.61	solid soln ($4.3KH_2PO_4 \cdot CsH_2PO_4$)
21.70	1.45	13.34	1.51	solid soln ($5.7KH_2PO_4 \cdot CsH_2PO_4$)
12.16	0.710	13.34	1.32	KH_2PO_4
8.32	0.476	15.69	1.52	"
-----	-----	17.4	1.55	"

^{a}The mol/kg H_2O values were calculated by the compiler.

The authors also present crystallographic and X-ray data for both phosphates and for the solid soln ($3.3KH_2PO_4 \cdot CsH_2PO_4$).

AUXILIARY INFORMATION

METHOD/APPARATUS/PROCEDURE:	SOURCE AND PURITY OF MATERIALS:
Saturated solutions of both phosphates were mixed at 50°C and the mixtures were thermostated at 25°C for a week. P_2O_5 was determined gravimetrically as $Mg_2P_2O_7$. Potassium content was determined by flame photometry. The composition of the solid phases was determined by the wet-residue method.	Both salts were of a "chemically pure" grade.
	ESTIMATED ERROR: No information is given.
	REFERENCES:

COMPONENTS:	ORIGINAL MEASUREMENTS:
(1) Cesium dihydrogenphosphate; CsH_2PO_4; [18649-05-3] (2) Cesium oxide; Cs_2O; [20281-00-9] (3) Phosphoric acid; H_3PO_4; [7664-38-2] (4) Water; H_2O; [7732-18-5]	Rashkovich, L.N.; Meteva, K.B.; Shevchik, J.E. *Zh. Neorg. Khim.* <u>1977</u>, *22*, 1982-8.

VARIABLES:	PREPARED BY:
Temperature and composition.	J. Eysseltová

EXPERIMENTAL VALUES:

Composition of saturated solutions in the $Cs_2O-P_2O_5-H_2O$ system.

		Cs_2O		P_2O_5		
Nr	pH	mass%	mol%	mass%	mol%	solid phase

temp. = 25°C.

Nr	pH	mass%	mol%	mass%	mol%	solid phase
1	7.7	51.6	9.12	17.8	6.24	CsH_2PO_4
2	7.2	48.2	7.80	17.6	5.65	"
3	6.65	45.0	6.76	17.5	5.20	"
4	6.4	43.4	6.29	17.4	5.00	"
5	6.05	41.8	5.85	17.4	4.83	"
6	5.6	39.5	5.28	17.5	4.64	"
7	5.1	37.9	4.95	17.8	4.61	"
8	5.0	37.3	4.80	17.6	4.49	"
9	5.0	37.2	4.81	17.9	4.61	"
10	4.1	36.1	4.61	18.5	4.69	"
11	3.9	36.0	4.61	18.9	4.81	"
12	3.6	36.4	4.80	20.0	5.23	"
13	2.85	38.5	5.77	24.5	7.31	"
14	2.7	38.2	5.73	24.8	7.39	"
15	2.6	38.9	6.09	26.2	8.14	"
16	2.2	39.9	6.82	28.8	9.77	"
17	2.1	38.9	6.48	28.8	9.52	$CsH_5(PO_4)_2$
18	1.7	34.8	5.23	28.4	8.47	"
19	1.3	29.9	3.98	27.9	7.34	"

(continued next page)

AUXILIARY INFORMATION

METHOD/APPARATUS/PROCEDURE:	SOURCE AND PURITY OF MATERIALS:
The mixtures were equilibrated isothermally by shaking twice daily for 10-15 days. They were thermostated for a month. Analysis was done by a potentiometric titration using KOH and H_3PO_4 solutions. The solid phases were identified crystallographically.	The Cs_2CO_3 and H_3PO_4 were used as received. The amount of impurities was less than 0.05 mass%.
	ESTIMATED ERROR:
	The temperature was controlled to within ± 0.05 K. The precision of the analyses for Cs_2O and P_2O_5 was ± 0.2 mass%
	REFERENCES:

COMPONENTS:	ORIGINAL MEASUREMENTS:
(1) Cesium dihydrogenphosphate; CsH_2PO_4; [18649-05-3]	Rashkovich, L.N.; Meteva, K.B. Shevchik, J.E.
(2) Cesium oxide; Cs_2O; [20281-00-9]	*Zh. Neorg. Khim.* 1977, 22, 1982-8.
(3) Phosphoric acid; H_3PO_4; [7664-38-2]	
(4) Water; H_2O; [7732-18-5]	

EXPERIMENTAL VALUES, cont'd:

Composition of saturated solutions in the Cs_2O-P_2O_5-H_2O system.

		Cs_2O		P_2O_5		
Nr	t/°C.	mass%	mol%	mass%	mol%	solid phase
20	33.2	42.2	6.00	17.8	5.03	CsH_2PO_4
21	33.2	37.2	4.89	19.0	4.96	"
22	33.2	31.6	4.52	29.5	8.38	$CsH_5(PO_4)_2$
23	39.0	42.8	6.22	18.3	5.28	CsH_2PO_4
24	39.0	38.2	5.16	19.4	5.21	"
25	39.0	32.4	4.82	30.6	9.04	$CsH_5(PO_4)_2$
26	44.7	43.5	6.47	18.7	5.53	CsH_2PO_4
27	44.7	39.0	5.41	19.9	5.48	"
28	44.7	33.4	5.20	31.7	9.80	$CsH_5(PO_4)_2$
29	50	61.1	151	19.4	95.2	CsH_2PO_4
30	50	53.3	102	19.1	72.3	"
31	50	50.6	89.8	19.1	67.2	"
32	50	47.9	79.8	19.2	63.5	"
33	50	44.0	66.6	19.0	57.1	"
34	50	43.0	64.5	19.6	58.3	"
35	50	42.3	62.8	19.8	58.5	"
36	50	40.7	57.9	19.4	54.8	"
37	50	40.2	57.8	20.5	58.5	"
38	50	40.4	59.0	21.1	61.2	"
39	50	40.6	59.8	21.3	62.3	"
40	50	40.8	61.0	21.9	65.0	"
41	50	41.5	69.3	26.2	86.9[a]	"
42	50	41.3	76.1	29.2	10.5[a]	"
43	50	39.9	74.0	32.3	11.9[a]	$CsH_5(PO_4)_2$
44	50	38.6	69.6	32.4	11.6[a]	"

[a] For the 50°C isotherm, these values are correct. All the other mol% values for Cs_2O and P_2O_5 are too large by a factor of ten.

The compiler has recalculated the above values to convert them to the following values for the CsH_2PO_4-Cs_2O-H_2O system.

	Cs_2O		H_3PO_4		CsH_2PO_4	
Nr	mass%	mol%	mass%	mol%	mass%	mol%
1	16.3	2.22	-----	----	57.7	9.63
2	13.3	1.59	-----	----	57.0	8.35
3	10.3	1.10	-----	----	56.7	7.47
4	8.88	0.91	-----	----	56.4	7.06
5	7.28	0.71	-----	----	56.4	6.75
6	4.78	0.44	-----	----	56.7	6.40
7	2.58	0.23	-----	----	57.7	6.31
8	2.38	0.21	-----	----	57.0	6.11
9	1.67	0.15	-----	----	58.0	6.26
10	----	----	0.40	0.19	58.9	6.29
11	----	----	1.02	0.26	58.7	6.35
12	----	----	2.27	0.60	59.4	6.74
13	----	----	7.01	2.37	62.8	9.06
14	----	----	7.64	2.60	62.3	9.03
15	----	----	9.08	3.38	63.5	10.1
16	----	----	12.0	5.33	65.1	12.3
17	----	----	12.7	5.42	63.5	11.6
18	----	----	15.0	5.41	56.8	8.74
19	----	----	17.7	5.39	48.8	4.87

(continued next page)

COMPONENTS:	ORIGINAL MEAUSREMENTS:
(1) Cesium dihydrogenphosphate; CsH_2PO_4; [18649-05-3]	Rashkovich, L.N.; Meteva, K.B.; Shevchik, J.E.
(2) Cesium oxide; Cs_2O; [20281-00-9]	*Zh. Neorg. Khim.* 1977, 22, 1982-8.
(3) Phosphoric acid; H_3PO_4; [7664-38-2]	
(4) Water; H_2O; [7732-18-5]	

EXPERIMENTAL VALUES cont'd:

	Cs_2O		H_3PO_4		CsH_2PO_4	
Nr	mass%	mol%	mass%	mol%	mass%	mol%
20	6.88	0.69	----	----	57.7	7.08
21	----	----	0.33	0.09	60.7	6.77
22	----	----	18.7	6.43	51.6	7.55
23	6.49	0.67	----	----	59.3	7.54
24	----	----	0.33	0.05	62.3	7.23
25	----	----	19.7	7.31	52.9	8.37
26	6.40	0.69	----	----	60.7	7.98
27	----	----	0.32	0.09	63.6	7.68
28	----	----	20.5	8.37	52.9	8.37
29	22.9	5.71	----	----	62.9	19.2
30	15.4	2.41	----	----	61.9	11.8
31	12.7	1.77	----	----	61.9	10.6
32	9.81	1.24	----	----	62.2	9.67
33	6.30	0.70	----	----	61.6	8.33
34	4.11	0.45	----	----	63.5	8.53
35	3.02	0.33	----	----	64.2	8.50
36	2.21	0.22	----	----	62.9	7.83
37	----	----	0.31	0.09	65.6	8.37
38	----	----	1.00	0.30	65.9	8.67
39	----	----	1.14	0.36	66.2	8.83
40	----	----	1.83	0.59	66.6	9.16
41	----	----	7.27	2.96	67.7	11.8
42	----	----	11.6	5.59	67.4	13.9
43	----	----	16.8	9.47	65.1	15.6
44	----	----	17.8	9.50	63.0	14.3

The authors linearized the data for the region containing H_3PO_4 and give the following equation

$$w_{Cs_2O} = a + b\ w_{P_2O_5}.$$

At 25°C: a = 28.8 ± 0.4 mass% and b = 0.39 ± 0.02   ±σ = 0.2.

At 50°C: a = 36.6 ± 0.4 mass% and b = 0.19 ± 0.02   ±σ = 0.1.

COMPONENTS:	ORIGINAL MEASUREMENTS:

COMPONENTS:

(1) Cesium dideuteriumphosphate; CsD_2PO_4;
   [28090-46-2]

(2) Cesium oxide; Cs_2O; [20281-00-9]

(3) Deuterium phosphate; D_3PO_4; [14335-33-2]

(4) Water-d_2; D_2O; [7789-20-0]

ORIGINAL MEASUREMENTS:

Rashkovich, L.N.; Meteva, K.B.;
   Shevchik, J.E.

Zh. Neorg. Khim. 1977, 22, 1982-8.

VARIABLES:

Temperature and composition.

PREPARED BY:

J. Eysseltová

EXPERIMENTAL VALUES:

The original experimental data are expressed for the $Cs_2O-P_2O_5-H_2O$ system.
These data are in the four left hand columns below. The compiler has converted
these data to values for the $CsD_2PO_4-D_3PO_4-D_2O$ system. These values are in
columns 5 to 10 below.

authors' data   compiler's calculations

Cs_2O		P_2O_5		CsD_2PO_4		D_3PO_4		Cs_2O		solid
mass%	mol%	mass%	mol%	mass%	mol%	mass%	mol%	mass%	mol%	phase
				temp. = 25°C.						
46.2	7.81	17.6	5.91	57.5	7.95	----	----	11.3	1.29	CsD_2PO_4
39.7	5.87	17.6	5.17	57.5	6.58	----	----	4.78	0.45	"
38.3	5.54	17.9	5.14	58.5	6.52	----	----	2.79	0.26	"
36.8	5.45	20.8	6.11	60.6	7.21	3.22	0.88	-----	----	"
37.6	5.86	22.7	7.02	61.9	8.15	5.35	1.62	-----	----	"
37.7	5.97	23.5	7.39	62.0	8.48	6.42	2.02	-----	----	"
37.9	6.05	23.7	7.52	62.4	8.66	6.56	2.10	-----	----	"
38.0	6.14	24.2	7.77	62.5	8.93	7.20	2.37	-----	----	"
37.9	6.13	24.3	7.80	62.4	8.90	7.42	2.43	-----	----	"
38.6	6.53	25.8	8.66	63.5	9.98	9.04	3.26	-----	----	"
36.8	6.29	28.3	9.61	60.6	10.2	13.9	5.40	-----	----	$CsD_5(PO_4)_2$

(continued next page)

AUXILIARY INFORMATION

METHOD/APPARATUS/PROCEDURE:

The mixtures were equilibrated isothermally
in sealed vessels. They were shaken twice
daily for 10-15 days and were thermostated
for a month. The solutions were analyzed
by means of a potentiometric titration
using solutions of KOH and of H_3PO_4. No
details are given. The content of deuterium
in the saturated solutions was determined by
a method described elsewhere (1).

SOURCE AND PURITY OF MATERIALS:

Cs_2CO_3 and D_3PO_4 were used as received.
The amount of impurities in each was less
than 0.05%. The extent of deuteration was
96% for D_3PO_4, 99.7% for D_2O, and 98 ±1%
for the saturated solutions.

ESTIMATED ERROR:

The temperature was controlled to within
± 0.05°C. The analyses for Cs_2O and P_2O_5
had a precision of ± 0.2 mass%.

REFERENCES:

1. Volkova, E.N.; Podshivalov, J.S.;
   Rashkovich, L.N.; Strukov, B.A.
   Izv. AN SSSR, ser. fiz. 1975, 39, 288.

COMPONENTS:	ORIGINAL MEASUREMENTS:
(1) Cesium dideuteriumphosphate; CsD_2PO_4; [28090-46-2]	Rashkovich, L.N.; Meteva, K.B.; Shevchik, J.E.
(2) Cesium oxide; Cs_2O; [20281-00-9]	Zh. Neorg. Khim. 1977, 22, 1982-8.
(3) Deuterium phosphate; D_3PO_4; [14335-33-2]	
(4) Water-d_2; D_2O; [7789-20-0]	

EXPERIMENTAL VALUES cont'd:

a u t h o r s ' d a t a   c o m p i l e r ' s c a l c u l a t i o n s

Cs_2O		P_2O_5		CsD_2PO_4		D_3PO_4		Cs_2O		solid
mass%	mol%	mass%	mol%	mass%	mol%	mass%	mol%	mass%	mol%	phase

temp. = 50°C.

42.2	6.75	19.1	6.06	62.4	8.08	----	----	4.32	0.46	CsD_2PO_4
40.0	6.17	19.6	6.00	64.0	7.91	----	----	1.11	0.11	"
39.5	6.27	21.8	6.87	65.0	8.67	2.71	0.84	----	----	"
39.6	6.33	21.9	6.95	65.2	8.78	2.73	0.86	----	----	"
39.5	6.39	22.6	7.26	65.0	9.00	3.05	1.23	----	----	"
40.3	6.89	24.5	8.32	66.3	10.3	5.98	1.15	----	----	"
40.5	7.03	25.0	8.61	66.7	10.7	6.55	2.43	----	----	"
40.4	7.01	25.1	8.65	66.5	10.7	6.76	2.51	----	----	"
40.8	7.35	26.4	9.44	67.1	11.8	8.33	3.37	----	----	"
38.0	7.20	31.7	11.9	62.5	13.8	17.9	9.04	----	----	$CsD_5(PO_4)_2$

The authors linearized the data in the region where P/Cs>1 as follows

$$w_{Cs_2O} = a + b \ w_{P_2O_5}.$$

At 25°C., a = 29.7 ± 0.6 and b = 0.34 ± 0.02 ±σ = 0.1
At 50°C., a = 32.9 ± 0.8 and b = 0.30 ± 0.03 ±σ = 0.1

SYSTEM INDEX

Page numbers preceded by E refer to evaluation text whereas those not
not preceded by E refer to compiled tables.

REGISTRY NUMBER INDEX

Page numbers preceded by E refer to evaluation texts whereas those not preceded by E refer to compiled tables.

AUTHOR INDEX

Page numbers precede by E refer to evaluation texts whereas those not
precede by E refer to compiled tables.

SOLUBILITY DATA SERIES

Selected Volumes in Preparation